Universitext

Springer

New York
Berlin
Heidelberg
Hong Kong
London
Milan
Paris
Tokyo

Universitext

Editors (North America): S. Axler, F.W. Gehring, and K.A. Ribet

Aguilar/Gitler/Prieto: Algebraic Topology from a Homotopical Viewpoint
Aksoy/Khamsi: Nonstandard Methods in Fixed Point Theory
Andersson: Topics in Complex Analysis
Aupetit: A Primer on Spectral Theory
Bachman/Narici/Beckenstein: Fourier and Wavelet Analysis
Badescu: Algebraic Surfaces
Balakrishnan/Ranganathan: A Textbook of Graph Theory
Balser: Formal Power Series and Linear Systems of Meromorphic Ordinary
Differential Equations
Bapat: Linear Algebra and Linear Models (2nd ed.)
Berberian: Fundamentals of Real Analysis
Boltyanskii/Efremovich: Intuitive Combinatorial Topology. (Shenitzer, trans.)
Booss/Bleecker: Topology and Analysis
Borkar: Probability Theory: An Advanced Course
Böttcher/Silbermann: Introduction to Large Truncated Toeplitz Matrices
Carleson/Gamelin: Complex Dynamics
Cecil: Lie Sphere Geometry: With Applications to Submanifolds
Chae: Lebesgue Integration (2nd ed.)
Charlap: Bieberbach Groups and Flat Manifolds
Chern: Complex Manifolds Without Potential Theory
Cohn: A Classical Invitation to Algebraic Numbers and Class Fields
Curtis: Abstract Linear Algebra
Curtis: Matrix Groups
Debarre: Higher-Dimensional Algebraic Geometry
Deitmar: A First Course in Harmonic Analysis
DiBenedetto: Degenerate Parabolic Equations
Dimca: Singularities and Topology of Hypersurfaces
Edwards: A Formal Background to Mathematics I a/b
Edwards: A Formal Background to Mathematics II a/b
Farenick: Algebras of Linear Transformations
Foulds: Graph Theory Applications
Friedman: Algebraic Surfaces and Holomorphic Vector Bundles
Fuhrmann: A Polynomial Approach to Linear Algebra
Gardiner: A First Course in Group Theory
Gårding/Tambour: Algebra for Computer Science
Goldblatt: Orthogonality and Spacetime Geometry
Gustafson/Rao: Numerical Range: The Field of Values of Linear Operators
and Matrices
Hahn: Quadratic Algebras, Clifford Algebras, and Arithmetic Witt Groups
Heinonen: Lectures on Analysis on Metric Spaces
Holmgren: A First Course in Discrete Dynamical Systems
Howe/Tan: Non-Abelian Harmonic Analysis: Applications of $SL(2, R)$
Howes: Modern Analysis and Topology
Hsieh/Sibuya: Basic Theory of Ordinary Differential Equations
Humi/Miller: Second Course in Ordinary Differential Equations
Hurwitz/Kritikos: Lectures on Number Theory
Jennings: Modern Geometry with Applications

(continued after index)

Hans Kurzweil   Bernd Stellmacher

The Theory of Finite Groups

An Introduction

Springer

Hans Kurzweil
Institute of Mathematics
University of Erlangen-Nuremburg
Bismarckstrasse 1½
Erlangen 91054
Germany
kurzweil@mi.uni-erlangen.de

Bernd Stellmacher
Mathematiches Seminar Kiel
Christian-Albrechts-Universität
Ludewig-Meyn Strasse 4
Kiel D-24098
Germany
stellmacher@math.uni-kiel.de

Mathematics Subject Classification (2000): 20-01, 20DXX

Library of Congress Cataloging-in-Publication Data
Kurzweil, Hans, 1942–
    The theory of finite groups: an introduction / Hans Kurzweil, Bernd Stellmacher.
        p.   cm. — (Universitext)
    Includes bibliographical references and index.

    1. Finite groups. I. Stellmacher, B. (Bernd) II. Title.
QA177.K87   2004
512´.2—dc21
                                                        2003054313

ISBN   978-1-4419-2340-0   e-ISBN   978-0-387-21768-0   Printed on acid-free paper.

Printed in the United States of America.

9 8 7 6 5 4 3 2 1

www.springer-ny.com

Springer-Verlag   New York Berlin Heidelberg
A member of BertelsmannSpringer Science+Business Media GmbH

Preface

Since its early beginnings in the nineteenth century the theory of finite groups has grown to be an extensive and diverse part of algebra. In the beginning of the 1980s, this development culminated in the classification of the finite simple groups, an impressive and convincing demonstration of the strength of its methods and results.

In our book we want to introduce the reader—as far as an introduction can do this—to some of the developments in this area that contributed to this success or may open new perspectives for the future.

The first eight chapters are intended to give a fast and direct approach to those methods and results that everybody should know who is interested in finite groups. Some parts, like nilpotent groups and solvable groups, are only treated as far as they are necessary to understand and investigate finite groups in general.

The notion of *action*, in all its facets, like action on sets and groups, coprime action, and quadratic action, is at the center of our exposition.

In the last chapters we focus on the correspondence between the local and global structure of finite groups. Our particular goal is to investigate non-solvable groups all of whose 2-local subgroups are solvable. The reader will realize that nearly all of the methods and results of this book are used in this investigation.

At least two things have been excluded from this book: the representation theory of finite groups and—with a few exceptions—the description of the finite simple groups. In both cases we felt unable to treat these two themes in an adequate way within the framework of this book.

For the more important results proved or mentioned in this book we tried to give the original papers as references, and in a few cases also some with alternative proofs. In the Appendix we state the classification theorem of the finite simple groups and also some of the fundamental theorems that are related to the subject of the last chapters.

The first eight chapters are accompanied by exercises. Usually they are not ordered by increasing difficulty and some of them demand serious thinking and persistence. They should allow the reader to get engaged with group theory and to find out about his or her own abilities.

The reader may want to postpone and revisit later some of the apparently more difficult exercises using the greater experience and insight gained from following chapters.

It should be pointed out here that—with the exception of the first chapter—all groups under consideration are meant to be finite.

Our special thanks go to our colleague H. Bender. Without him this book would not have been written, and without his encouraging support it would have taken a different shape.

We would like to thank J. Hall for reading the entire manuscript and A. Chermak for reading parts of it. We are also grateful to B. Baumann, D. Bundy, S. Heiss, and P. Flavell for their helpful comments and suggestions.

A German version of this book has been published in 1998 as a Springer-Lehrbuch.

Erlangen, Germany   Hans Kurzweil
Kiel, Germany   Bernd Stellmacher
February 2003

Contents

Contents   ix

List of Symbols

Chapter 1

Basic Concepts

In this first chapter we introduce some of the basic concepts of finite group theory. Most of these concepts apply to arbitrary groups, whether finite or infinite. For that reason we will make no assumption (as we will in the later chapters) that the objects under consideration are finite.

1.1 Groups and Subgroups

A nonempty set G is a **group**, if to every pair $(x, y) \in G \times G$ an element $xy \in G$ is assigned, the **product** of x and y, satisfying the following axioms:

Associativity: $x(yz) = (xy)z$ for all $x, y, z \in G$.

Existence of an identity: There exists an element $e \in G$ such that $ex = xe = x$ for all $x \in G$.[1]

Existence of inverses: For every $x \in G$ there exists an element $x^{-1} \in G$ such that

$$xx^{-1} = e = x^{-1}x.$$

A group G is **Abelian**[2] if, in addition:

Commutativity: $xy = yx$ for all $x, y \in G$.

---

[1]If also e' is such an identity, then $e' = ee' = e$. Thus, the identity of G is uniquely determined.

[2]Abelian groups are often written *additively*. In this case the element assigned to the pair (x, y) is denoted by $x + y$ and called the **sum** of x and y.

In the following, G is always a group. Associativity implies the **generalized associative law**: Every (reasonable) bracketing of the expression
$x_1 x_2 \cdots x_n$ of elements $x_i \in G$ gives the same element in G. This element
is then denoted simply by $x_1 x_2 \cdots x_n$.

The identity e of G may be denoted by 1 (or 1_G). If G is Abelian, written
additively, then e may be denoted by by 0 (or 0_G).

We write
$$G^{\#} := \{x \in G \mid x \neq e\}.$$

Let $x \in G$ and suppose that y_1 and y_2 are two inverses for x. Then
$$y_2 = (y_1 x) y_2 = y_1 (x y_2) = y_1.$$

Hence the inverse of x is uniquely determined. This shows that for $a, b \in G$
the equations
$$ya = b \quad \text{and} \quad ax = b$$
have unique solutions in G, namely
$$y = ba^{-1} \quad \text{and} \quad x = a^{-1} b.$$

Thus, the right and left *cancellation laws* hold in groups.

For $x, a \in G$ set
$$x^a := a^{-1} x a.$$

Such an element x^a is said to be a **conjugate** of x. More precisely, x^a is
the conjugate of x **by** a.

1.1.1 *For $a \in G$ the applications*
$$x \mapsto xa, \quad x \mapsto ax, \quad x \mapsto x^{-1}, \quad x \mapsto x^a$$
define bijective mappings from G to G.   □

For $x \in G$ we define the **powers** of x
$$x^0 := 1, \quad x^1 := x, \dots, x^{n+1} := (x^n) x \quad \text{for } n \in \mathbb{N}\ [3]$$
and
$$x^{-n} := (x^n)^{-1}.$$

---

[3] $nx = x + \cdots + x$ (n summands) for groups written additively.

Then

$$x^{-n} = \underbrace{x^{-1} \cdots x^{-1}}_{n\text{-times}},$$

and by induction on n one obtains the **laws of exponents**

$$x^{i+j} = x^i x^j \quad \text{and} \quad (x^i)^j = x^{i \cdot j}$$

for all $i, j \in \mathbb{Z}$.

A group G is **finite** if G contains only finitely many elements. In this case the number of elements is called the **order** of G, denoted by $|G|$. Every finite group $G = \{x_1, \ldots, x_n\}$ of order n can be described by its **group table** $T = (t_{ij})$; where $t_{ij} = x_i x_j \in G$. Thus, T is an $(n \times n)$-matrix with entries in G. For example,

	1	d	d^2	t	td	td^2
1	1	d	d^2	t	td	td^2
d	d	d^2	1	td^2	t	td
d^2	d^2	1	d	td	td^2	t
t	t	td	td^2	1	d	d^2
td	td	td^2	t	d^2	1	d
td^2	td^2	t	td	d	d^2	1

is the group table of a non-Abelian group of order 6.[4] We suggest the reader use this example as a test ground for the coming notation and definitions.

The group G is **cyclic** if every element of G is a power of a fixed element g. In this case we write

$$G = \langle g \rangle.$$

The multiplication in a cyclic group is determined by the laws of exponents; in particular, cyclic groups are Abelian.

For $i, j, k \in \mathbb{Z}$ we write $i | j$, if i is a divisor of j, and

$$i \equiv j \pmod{k} \quad \text{if} \quad k | (i - j).$$

Note that every integer is a divisor of 0.

---

[4] Putting $d := \begin{pmatrix} 1 & 2 & 3 \\ 2 & 3 & 1 \end{pmatrix}$ and $t := \begin{pmatrix} 1 & 2 & 3 \\ 2 & 1 & 3 \end{pmatrix}$ one can show that G is the group of permutations on $\{1, 2, 3\}$, the symmetric group S_3 (see **4.3**).

1.1.2   *Let $G = \langle g \rangle$ be a cyclic group of order n. Then*

$$G = \{1, g, \ldots, g^{n-1}\},$$

and the following hold:

(a) $n = \min\{ m \in \mathbb{N} \mid g^m = 1 \}$.

(b)   *For $z \in \mathbb{Z}$:* $g^z = 1 \iff n \mid z$.

(c)   *For $i, j, k \in \{0, 1, \ldots, n-1\}$:* $g^i g^j = g^k \iff i + j \equiv k \pmod{n}$.

Proof. Since $|\langle g \rangle| < \infty$ there exist $a, b \in \mathbb{N}$, $a < b$, such that $g^a = g^b$ and thus $g^{b-a} = 1$. Hence, there exists

$$l := \min\{ m \in \mathbb{N} \mid g^m = 1 \}.$$

If $g^i = g^j$ for $0 \leq i < j \leq l - 1$, then $g^{j-i} = 1$ which contradicts the minimality of l. Thus, all the elements $1, g, \ldots, g^{l-1}$ are distinct. Since every integer $z \in \mathbb{Z}$ can be written

$$z = lt + r \quad \text{with} \quad t \in \mathbb{Z}, \; r \in \{0, 1, \ldots, l-1\}$$

we obtain

$$g^z = g^{lt} g^r = (g^l)^t g^r = g^r.$$

Therefore $G = \{1, g, \ldots, g^{l-1}\}$ and $l = n$. Similarly we obtain (a) and (b) and thus also (c).   □

A nonempty subset U of G is a **subgroup** of G if U is a group with respect to the multiplication in G. Clearly, this is equivalent to saying that for all $x, y \in U$ also xy and x^{-1} are in U; and we then write $U \leq G$. If, in addition, $U \neq G$, then U is a **proper** subgroup of G, and we write $U < G$.

Every group possesses the **trivial** subgroup $U = \{1\}$. We then abuse notation and write simply $U = 1$.

Evidently, the intersection of any collection of subgroups of G is itself a subgroup.

A subgroup $U \neq 1$ is a **minimal** subgroup of G if no other nontrivial subgroup of G is contained in U, and a subgroup $U \neq G$ is a **maximal** subgroup if U is not contained in any other proper subgroup of G.

Evidently, every nontrivial finite group possesses minimal and maximal subgroups.

1.1.3 *A nonempty finite subset U of G is a subgroup if for all $x, y \in U$ also xy is in U.*

Proof. For $x \in U$ the mapping $\varphi\colon u \mapsto ux$ from U to U is injective and thus also surjective since U is finite. It follows that $1 = x^{\varphi^{-1}} \in U$ and $x^{-1} = 1^{\varphi^{-1}} \in U$. $\qquad\square$

For a nonempty subset X of G,

$$\langle X \rangle := \{ x_1^{z_1} \dots x_j^{z_j} \mid x_i \in X,\, z_i \in \mathbb{Z},\, j \in \mathbb{N} \}$$

is the subgroup **generated** by X. We set $\langle \varnothing \rangle := 1$. We also write

$$\langle X \rangle = \langle x_1, \dots, x_n \rangle$$

in the special case that X is a finite set $\{x_1, \dots, x_n\}$. If $\mathcal{X} = \{X_1, \dots, X_n\}$ is a finite set of subsets of G, we set

$$\langle \mathcal{X} \rangle := \langle X_1, \dots, X_n \rangle := \Big\langle \bigcup_{i=1}^{n} X_i \Big\rangle.$$

1.1.4 *Let X be a subset of G. Then $\langle X \rangle$ is a subgroup of G. More precisely, $\langle X \rangle$ is the smallest subgroup of G containing X.*

Proof. With $a, b \in \langle X \rangle$ also ab and a^{-1} are in $\langle X \rangle$. Thus $\langle X \rangle$ is a subgroup. Every subgroup of G containing X also contains $\langle X \rangle$. $\qquad\square$

Sometimes properties of the generating set X already determine the structure of $\langle X \rangle$. For example, if $xy = yx$ for all $x, y \in X \subseteq G$, then $\langle X \rangle$ is an Abelian group.

Let $g \in G$. The cyclic subgroup $\langle g \rangle$ is the smallest subgroup of G that contains g. If $\langle g \rangle$ is finite, then

$$o(g) := |\langle g \rangle|$$

is the **order** of g. According to 1.1.2 $o(g)$ is the smallest positive integer n such that $g^n = 1$.

For two nonempty subsets A, B of G let

$$AB := \{ab \mid a \in A, b \in B\} \quad \text{and} \quad A^{-1} := \{a^{-1} \mid a \in A\}.$$

AB is the **complex product** (or simply the **product**) of A and B. This product defines an associative multiplication on the set of nonempty subsets of G. In addition, we have

$$(AB)^{-1} = B^{-1}A^{-1}.$$

In the cases $A = \{a\}$ resp. $B = \{b\}$, we write aB resp. Ab instead of AB. Moreover, for $g \in G$ we set

$$B^g := g^{-1}Bg,$$

and say that B^g is a **conjugate** of B in G (more precisely, the conjugate of B **by** g). For any $A \subseteq G$ set

$$B^A := \{B^a \mid a \in A\}.$$

Note that for a nonempty subset U of G:

$$U \leq G \iff UU = U = U^{-1}.$$

1.1.5 *Let A and B be subgroups of G. Then AB is a subgroup of G if and only if $AB = BA$.*

Proof. From $AB \leq G$ we get

$$(AB) = (AB)^{-1} = B^{-1}A^{-1} = BA.$$

If $AB = BA$, then

$$(AB)(AB) = A(BA)B = A(AB)B = AABB = AB$$

and

$$(AB)^{-1} = B^{-1}A^{-1} = BA = AB.$$

Thus $AB \leq G$. $\square$

1.1.6 *Let A and B be finite subgroups of G. Then*

$$|AB| = \frac{|A|\,|B|}{|A \cap B|}.$$

Proof. We define an equivalence relation on the Cartesian product $A \times B$:

$$(a_1, b_1) \sim (a_2, b_2) \iff a_1 b_1 = a_2 b_2.$$

Then $|AB|$ is the number of equivalence classes in $A \times B$. Let $(a_1, b_1) \in A \times B$. The equivalence class

$$\{(a_2, b_2) \mid a_1 b_1 = a_2 b_2\}.$$

contains exactly $|A \cap B|$ elements since

$$a_2 b_2 = a_1 b_1 \iff a_1^{-1} a_2 = b_1 b_2^{-1} \; (\in A \cap B)$$
$$\iff a_2 = a_1 d \;\; \text{and} \;\; b_1 = d b_2 \;\; \text{for some } d \in A \cap B.$$

This gives the assertion.   □

Let U be a subgroup of G and $x \in G$. The complex product

$$Ux = \{ux \mid u \in U\} \quad \text{resp.} \quad xU = \{xu \mid u \in U\}$$

is a **right coset**, resp. a **left coset**, of U in G. The application

$$Ux \mapsto (Ux)^{-1} = x^{-1}U$$

defines a bijective mapping from the set of right cosets of U to the set of left cosets of U. If the set of right cosets of U in G is finite then the number of right cosets of U in G is the **index** of U in G, denoted by $|G : U|$.[5]

Since $u \mapsto ux$ is a bijective mapping from U to Ux (1.1.1) we get in addition

$$|U| = |Ux| = |xU|$$

for all $x \in G$. As

$$x = 1_G x \in Ux$$

the right cosets of U cover the set G. Moreover, for $y, x \in G$

$$Ux = Uy \iff yx^{-1} \in U \iff y \in Ux.$$

---

[5]The following statements hold for left cosets as well as for right cosets.

Hence any two right cosets of U are either equal or disjoint.[6]

This yields:

1.1.7   Lagrange's Theorem.[7] *Let U be a subgroup of the finite group G. Then*

$$|G| = |U|\,|G:U|.$$

In particular, the integers $|U|$ and $|G:U|$ are divisors of $|G|$. $\square$

Because $\langle g \rangle$ is a subgroup of G for all $g \in G$ we get from 1.1.7:

1.1.8   *For every finite group G and every $g \in G$, the order of g divides* $|G|$.

Let $U \le G$ and $S \subseteq G$. Then S is a **transversal** of U in G[8] if S contains exactly one element of every right coset Ux, $x \in G$; and S is a **left transversal** of U in G if S contains exactly one element of every left coset of U in G.

1.1.9   *Let $S \subseteq G$. Then S is a transversal of the subgroup U in G if and only if $G = US$ and $st^{-1} \notin U$ for all $s \ne t$ in S.*

If S is a transversal of U in G, then the mapping

$$U \times S \to G \quad \text{with} \quad (u,s) \mapsto us$$

is bijective.

Proof. $Us = Ut \iff st^{-1} \in U$. $\square$

An important special case is:

1.1.10   *Let U and S be subgroups of G such that $G = US$ and $U \cap S = 1$, then S is a transversal of U in G.* $\square$

---

[6] And the right cosets of U are the equivalence classes of the equivalence relation

$$y \sim x \iff yx^{-1} \in U.$$

[7] Compare with [75] and [42], p. 504.

[8] Or **set of right coset representatives** for U in G.

Such a subgroup S is called a **complement** of U in G.

The following observation is sometimes useful:

1.1.11 Dedekind Identity. *Let $G = UV$, where U and V are subgroups of G. Then every subgroup H satisfying $U \leq H \leq G$ admits the factorization $H = U(V \cap H)$.*

Proof. Every coset of U in G and thus every coset of U in H contains an element of V. □

According to Lagrange's Theorem the divisors of the order of a finite group are important invariants of G.

Let $\mathbb{P}$ be the set of all positive prime integers, and for $n \in \mathbb{N}$ set

$$\pi(n) := \{p \in \mathbb{P} \mid p \text{ divides } n\}.$$

For a finite group G set
$$\pi(G) := \pi(|G|).$$

An element $x \in G$ is a p-**element** $(p \in \mathbb{P})$ if $o(x)$ is a power of p, and G is a p-**group** if $\pi(G) = \{p\}$, i.e., $|G|$ is a power of p. Observe that the identity element (resp. the trivial group) is a p-element (resp. p-group) for every $p \in \mathbb{P}$. A p-**subgroup** is a subgroup which is a p-group.

It follows from 1.1.8 that in a p-group every element is a p-element. The converse is also true; this is a consequence of Cauchy's Theorem (3.2.1 on page 62).

Exercises

Let A, B, and C be subgroups of the finite group G.

1. If $B \leq A$, then $|A : B| \geq |C \cap A : C \cap B|$.

2. Let $B \leq A$. If $x_1, \ldots, x_n$ is a transversal of A in G and $y_1, \ldots, y_m$ a transversal of B in A, then $\{y_j x_i\}_{\substack{i=1,\ldots,n \\ j=1,\ldots,m}}$ is a transversal of B in G.

3. $|G : A \cap B| \leq |G : A| \, |G : B|$.

4. $A \cup B$ is a subgroup of G, if and only if $A \subseteq B$ or $B \subseteq A$.

5. Let $G = AA^g$ for some $g \in G$. Then $G = A$.

6.   Let $|G|$ be a prime. Then 1 and G are the only subgroups of G.

7. G has even order if and only if the number of involutions[9] in G is odd.

8.   If $y^2 = 1$ for all $y \in G$, then G is Abelian.

9.   Let $|G| = 4$. Then G is Abelian and contains a subgroup of order 2.

10.   If G contains exactly one maximal subgroup, then G is cyclic.

11.   Suppose that $A \neq 1$ and $A \cap A^g = 1$ for all $g \in G \setminus A$. Then

$$\left| \bigcup_{g \in G} A^g \right| \geq \frac{|G|}{2} + 1.$$

12.   If $A \neq G$, then $G \neq \bigcup_{g \in G} A^g$.

13.   Let $A^G = \{A_1, \ldots, A_n\}$. Then $\langle A_1, \ldots, A_n \rangle = A_1 \cdots A_n$.

1.2   Homomorphisms and Normal Subgroups

Let G and H be groups. A mapping

$$\varphi : G \to H,$$

(which may be written "exponentially," as $x \mapsto x^\varphi$) is a **homomorphism** from G to H, if

$$(xy)^\varphi = x^\varphi y^\varphi \quad \text{for all } x, y \in G.$$

1.2.1   *If the homomorphism $\varphi : G \to H$ is bijective, then also the inverse mapping φ^{-1} is a homomorphism.*

Proof. The equality

$$x^{\varphi^{-1}} y^{\varphi^{-1}} = (xy)^{\varphi^{-1}} \quad (x, y \in H)$$

follows from

$$\left(x^{\varphi^{-1}} y^{\varphi^{-1}} \right)^\varphi = \left(x^{\varphi^{-1}} \right)^\varphi \left(y^{\varphi^{-1}} \right)^\varphi = xy. \qquad \square$$

---

[9]Involutions are elements of order 2; see p. 34.

Let φ be a homomorphism from G to H, and let $X \subseteq G$ and $Y \subseteq H$. We set

$$X^\varphi := \{x^\varphi \mid x \in X\}, \quad Y^{\varphi^{-1}} := \{g \in G \mid g^\varphi \in Y\}, \quad \text{and}$$

$$\text{Ker } \varphi := \{x \in G \mid x^\varphi = 1_H\}, \quad \text{Im } \varphi := G^\varphi.$$

We refer to X^φ as the **image** of X and $Y^{\varphi^{-1}}$ as the **inverse image** of Y (with respect to φ). Further, $\text{Ker } \varphi \, (= 1_H^{\varphi^{-1}})$ is the **kernel** of φ and we write $\text{Im } \varphi$ for the image of φ.

The homomorphism φ is an **epimorphism** if $\text{Im } \varphi = H$, an **endomorphism** if $H = G$, a **monomorphism** if φ injective, an **isomorphism** if φ bijective, and an **automorphism** if φ is a bijective endomorphism.

If φ is an isomorphism, then G is said to be **isomorphic** to H; in which case we may write $G \cong H$.

The following points are immediate consequences of the group axioms:

- $(1_G)^\varphi = 1_H$.

- $(x^{-1})^\varphi = (x^\varphi)^{-1}$ for all $x \in G$.[10]

- If U is a subgroup of G, then U^φ is a subgroup of H.

- If V is a subgroup of H, then $V^{\varphi^{-1}}$ is a subgroup of G.

- $\langle X \rangle^\varphi = \langle X^\varphi \rangle$ for $X \subseteq G$.

1.2.2 *Let $N = \text{Ker } \varphi$. Then for all $x \in G$*

$$Nx = \{y \in G \mid y^\varphi = x^\varphi\} = xN.$$

Proof. $y^\varphi = x^\varphi \iff y^\varphi (x^\varphi)^{-1} = 1 \iff y^\varphi (x^{-1})^\varphi = 1$

$\qquad\qquad \iff (yx^{-1})^\varphi = 1 \iff yx^{-1} \in N$

$\qquad\qquad \iff y \in Nx,$

and similarly $y^\varphi = x^\varphi \iff (x^\varphi)^{-1} y^\varphi = 1 \iff \cdots \iff y \in xN.$ $\qquad$ □

A subgroup N of G that satisfies

$$Nx = xN \quad \text{for all } x \in G$$

---

[10]Instead of $(x^\varphi)^{-1}$ we often write $x^{-\varphi}$.

is a **normal subgroup** of G (or is **normal** in G). We write $N \trianglelefteq G$ if N is normal in G. If N is normal in G, then any right coset of N is also a left coset of N, and one may speak simply of the **cosets** of N in G.

Since

$$Nx = xN \iff N = x^{-1}Nx \ (= N^x)$$

one obtains:

1.2.3   *A subgroup N is normal in G if and only if $y^x \in N$ for all $y \in N$ and $x \in G$.* $\square$

The group G is itself a normal subgroup of G, and G always possesses the **trivial normal subgroup** 1. If $G \neq 1$, and 1 and G are the only normal subgroups of G, then G is **simple**. For example, a group of prime order is a simple group (1.1.7 on page 8).

The statements below, which follow directly from the definition of a normal subgroup, will be used frequently.

- For every homomorphism φ of G, the image (resp. inverse image) of any normal subgroup of G (resp. G^φ) is normal in G^φ (resp. G).

- The product and intersection of two normal subgroups of G is normal in G.

- If U is a subgroup of G and N is normal in G, then $U \cap N$ is normal in U.

- If U is a subgroup of G, then

$$U_G := \bigcap_{g \in G} U^g$$

  is the largest normal subgroup of G that is contained in U.

- If $X \subseteq G$, then $\langle X^G \rangle$ is the smallest normal subgroup of G that contains X.

Let N be a normal subgroup of G and G/N the set of all cosets of N in G. For $Nx, Ny \in G/N$

$$(Nx)(Ny) = N(xN)y = N(Nx)y = Nxy$$

and thus

$$(*) \qquad\qquad (Nx)(Ny) = Nxy \quad \text{for all} \quad x, y \in G.$$

Hence, this complex product defines an associative multiplication on the set G/N. Evidently, $N = N1_G$ is the identity of G/N (with respect to this multiplication), and Nx^{-1} is the inverse of Nx. Thus:

1.2.4 *Let N be a normal subgroup of G. Then G/N is a group with respect to the complex product. The mapping*

$$\psi : G \to G/N \quad \text{with} \quad x \mapsto Nx$$

is an epimorphism. $\qquad\qquad\qquad\qquad\qquad\qquad\qquad\qquad\qquad\square$

Here the second part of 1.2.4 follows from $(*)$.

The group G/N (one reads G modulo N) described in 1.2.4 is the **factor group** of N in G, and the corresponding φ is the **natural homomorphism** from G to G/N.

By 1.2.2 the normal subgroups of G are exactly the kernels of the homomorphisms of G. From 1.2.2 and 1.2.4 we derive the following:

1.2.5 **Homomorphism Theorem.** *Let φ be a homomorphism from G to H. Then*

$$G/\operatorname{Ker}\varphi \to H \quad \text{with} \quad (\operatorname{Ker}\varphi)x \mapsto x^{\varphi}$$

is a monomorphism. In particular

$$G/\operatorname{Ker}\varphi \cong \operatorname{Im}\varphi. \qquad\qquad\qquad\qquad\qquad\qquad\square$$

Let U be a subgroup and N a normal subgroup of G. Then by 1.1.5 UN is a subgroup of G, and thus N is a normal subgroup of UN.

Two direct consequences of 1.2.5 are the **Isomorphism Theorems**:

1.2.6 *Let U be a subgroup and N a normal subgroup of G. Then*

$$\varphi : U \to UN/N \quad \text{with} \quad u \mapsto uN$$

is an epimorphism with $\operatorname{Ker}\varphi = U \cap N$, and

$$U/U \cap N \cong UN/N. \qquad\qquad\qquad\qquad\qquad\qquad\square$$

1.2.7   _Let N and M be normal subgroups of G such that $N \leq M$. Then_

$$\varphi : G/N \to G/M \quad \text{with} \quad Nx \mapsto Mx$$

is an epimorphism with Ker $\varphi = M/N$, _and_

$$(G/N)/(M/N) \cong G/M. \qquad \square$$

It is important also to observe that the homomorphism theorem gives a bijection $(U \mapsto U^{\varphi})$ from the set of all subgroups $U \leq G$ containing Ker φ to the set of all subgroups of Im φ.

Often it is convenient to use the **bar convention** for subgroups and elements of G/N:

$$\overline{U} := UN/N \text{ for } U \leq G \quad \text{and} \quad \overline{x} := xN \text{ für } x \in G;$$

in particular $\overline{G} = G/N$.

In general, $A \trianglelefteq N \trianglelefteq G$ does not imply $A \trianglelefteq G$. A subgroup A is a **subnormal subgroup** of G (or is **subnormal** in G), if there exist subgroups $A_1, \ldots, A_d$ such that

$$\mathcal{S} \qquad A = A_1 \trianglelefteq A_2 \trianglelefteq \cdots \trianglelefteq A_{d-1} \trianglelefteq A_d = G.$$

We then write $A \trianglelefteq\trianglelefteq G$ and call $\mathcal{S}$ a **subnormal series** from A to G. Evidently, one gets

$$A \trianglelefteq\trianglelefteq B \trianglelefteq\trianglelefteq G \quad \Rightarrow \quad A \trianglelefteq\trianglelefteq G.$$

Because of this transitivity property the notion of subnormality plays an important role in the investigation of finite groups. We will use this notion later, from Chapter 5 on. Here we only give some elementary properties of subnormal subgroups which follow directly from the definition.

1.2.8   _Let A and B be subnormal subgroups of G._

(a) $U \cap A \trianglelefteq\trianglelefteq U$ _for $U \leq G$._

(b) $A \cap B \trianglelefteq\trianglelefteq G$.

(c)   _Let φ be a homomorphism of G. Then the image (resp. inverse image) of any subnormal subgroup of G (resp. G^{φ}) is subnormal in G^{φ} (resp. G)._

Proof. (a) Let $\mathcal{S}$ be a subnormal series from A to G. Then

$$U \cap A = U \cap A_1 \trianglelefteq \cdots \trianglelefteq U \cap A_{d-1} \trianglelefteq U \cap A_d = U$$

is a subnormal series from $U \cap A$ to U.

(b) From (a) it follows that $A \cap B \trianglelefteq\trianglelefteq B \trianglelefteq\trianglelefteq G$.

(c) This follows from the corresponding statements about normal subgroups. □

Let $B \trianglelefteq A \leq G$. Then A/B is a **section** of G.

Exercises

Let G be a group.

1. Every subgroup of index 2 is normal in G.

2. Show that there are exactly two nonisomorphic groups of order 4 and compute their group tables.

3. Let N be a normal subgroup of G and $|G : N| = 4$.

   (*a*) G contains a normal subgroup of index 2.

   (*b*) If G/N is not cyclic, then there exist three proper normal subgroups A, B, and C of G such that $G = A \cup B \cup C$.

4. Let G be simple, $|G| \neq 2$, and φ a homomorphism from G to H. If H contains a normal subgroup A of index 2, then $G^\varphi \leq A$.

5. Let $x \in G$, $D := \{x^g \mid g \in G\}$, and $U_i \leq G$ for $i = 1, 2$. Suppose that

$$\langle D \rangle = G \quad \text{and} \quad D \subseteq U_1 \cup U_2.$$

   Then $U_1 = G$ or $U_2 = G$.

6. Let $G \neq 1$ be a finite group. Suppose that every proper subgroup of G is Abelian. Then G contains a nontrivial Abelian normal subgroup.

1.3 Automorphisms

In the following, G is a group. The set $\operatorname{Aut} G$ of all automorphisms of G, with multiplication given by

$$\alpha\beta \colon x \mapsto (x^\alpha)^\beta \quad (x \in G, \ \alpha, \beta \in \operatorname{Aut} G)$$

is a group, the **automorphism group** of G. The identity mapping is the identity of $\operatorname{Aut} G$, and the inverse mapping α^{-1} the inverse of α (see 1.2.1).

Automorphisms map finite subgroups (resp. elements) to subgroups (resp. elements) of the same order. For $a \in G$, the mapping

$$\varphi_a \colon G \to G \quad \text{with} \quad x \mapsto x^a \quad (= a^{-1}xa)$$

is bijective by 1.1.1. Since

$$(xy)^a = a^{-1}xaa^{-1}ya = (a^{-1}xa)(a^{-1}ya) = x^a y^a,$$

φ_a is an automorphism of G, the **inner automorphism** induced by a. The mapping

$$\varphi \colon G \to \operatorname{Aut} G$$

given by $a \mapsto \varphi_a$ is a homomorphism from G to $\operatorname{Aut} G$, since

$$x^{ab} = b^{-1}a^{-1}xab = (x^a)^b.$$

Hence, the set of inner automorphisms of G,

$$\operatorname{Inn} G := \{\,\varphi_a \mid a \in G\,\},$$

is a subgroup of $\operatorname{Aut} G$. Moreover, the equality

$$\beta^{-1}\varphi_a\beta = \varphi_{a^\beta} \quad (\beta \in \operatorname{Aut} G, \quad a \in G)$$

shows that $\operatorname{Inn} G$ is a normal subgroup of $\operatorname{Aut} G$.
We set

$$\operatorname{Ker} \varphi = \{x \in G \mid x^a = x \text{ for all } a \in G\} =: Z(G).$$

The homomorphism theorem then yields

$$G/Z(G) \cong \operatorname{Inn} G.$$

The group $Z(G)$ is called the **center** of G.

For later use we note:

1.3.1   *Suppose that $G/Z(G)$ is cyclic. Then G is Abelian.*

Proof. There exists $g \in G$ such that $G/Z(G) = \langle gZ(G) \rangle$ and thus

$$G = Z(G)\langle g \rangle.$$

Since $\langle g \rangle$ is Abelian all pairs of elements of G commute. □

By definition a subgroup N of G is normal if and only if

$$N^a = N \text{ for all } a \in G.$$

Thus, a subgroup of G is normal if and only if it is mapped to itself by every inner automorphism of G.

A subgroup U of G is a **characteristic subgroup** of G (or is **characteristic in** G), if

$$U^\alpha = U \text{ for all } \alpha \in \text{Aut}\, G.$$

We write U char G in this case.

Evidently, characteristic subgroups are normal in G. Moreover, 1 and G are characteristic subgroups of G. Another example of a characteristic subgroup is $Z(G)$. Indeed, for $x \in Z(G)$, $g \in G$, $\alpha \in \text{Aut}\, G$,

$$x^\alpha g^\alpha = (xg)^\alpha = (gx)^\alpha = g^\alpha x^\alpha,$$

and since $G = \{g^\alpha \mid g \in G\}$ we have $x^\alpha \in Z(G)$.

We note two properties of characteristic subgroups, which we will use frequently.

1.3.2 *Let N be a normal subgroup of G and A be a characteristic subgroup of N.*

(a)   *A is normal in G.*

(b)   *If N is characteristic in G, then also A is characteristic in G.*

Proof. (a) Let $a \in G$ and φ_a be the inner automorphism of G induced by a. Then the restriction of φ_a to N is an automorphism of N since N is normal in G. Hence, A is invariant under φ_a for all $a \in G$, i.e., A is normal in G.

(b) Since N is now characteristic in G one can replace φ_a in the above argument by an arbitrary automorphism of G.   □

The above property (b) shows that being *characteristic* (as being *subnormal*, see page 14) is a *transitive* property.

We will now introduce a notion, which will prove to be greatly convenient. Let X be a group and

$$\varphi \colon X \to \operatorname{Aut} G$$

be a homomorphism from X to $\operatorname{Aut} G$. Then we say that X **acts on** G (with respect to φ). We set

$$g^x := g^{x^\varphi}$$

and get

$$(gh)^x = g^x h^x \quad \text{and} \quad (g^x)^y = g^{xy}$$

for all $g, h \in G$ and $x, y \in X$.

A subgroup U of G is X-**invariant** if for all $x \in X$:

$$U^x := \{ u^x \mid u \in U \} = U.$$

If U is an X-invariant subgroup of G, then X acts on U with respect to the homomorphism $X \to \operatorname{Aut}(U)$ induced by φ. If N is an X-invariant normal subgroup of G, then X acts on the factor group G/N by

$$(Ng)^x := Ng^x.$$

It is evident that every subgroup X of $\operatorname{Aut} G$ acts on G (with respect to $\varphi = \operatorname{id}$). In the cases $X = \operatorname{Aut} G$ resp. $X = \operatorname{Inn} G$ the X-invariant subgroups are the characteristic resp. normal subgroups of G.

Every subgroup X of G acts on G with respect to $\varphi|_X$, where φ is the homomorphism from G to $\operatorname{Inn} G$ defined on page 16 (*conjugation*). When we speak of X-invariant subgroups—where X is a subgroup of G—without mentioning φ, then we always mean the action by conjugation.

Let η be a homomorphism from G to the group H, and let X be a group that acts on both G and H. Then η is an X-**homomorphism**, if

$$(g^x)^\eta = (g^\eta)^x \quad \text{for all } g \in G, \quad x \in X.$$

(In the same way we define X-isomorphism and X-automorphism). Such an X-homomorphism maps X-invariant subgroups of G to X-invariant subgroups of H, and the inverse images of X-invariant subgroups of H are X-invariant subgroups of G. In particular, Ker η and Im η are X-invariant subgroups.

For example, $X := G$ acts by conjugation on G, but also on H by

$$h^x := h^{x^\eta} \quad (h \in H, \ x \in G).$$

This implies

$$(g^x)^\eta = (x^{-1}gx)^\eta = (g^\eta)^{x^\eta} = (g^\eta)^x,$$

and thus η is a G-homomorphism. If η is surjective, then the G-invariant subgroups of H are precisely the normal subgroups of H.

If $\eta \colon G \to H$ is an X-isomorphism, we write $G \cong_X H$.

The mappings introduced in the Homomorphism Theorem 1.2.5 and its two corollaries, 1.2.6, 1.2.7, yield the following results:

- Let η be an X-homomorphism of G. Then

$$G/\operatorname{Ker} \eta \cong_X \operatorname{Im} \eta.$$

- Let U and N be X-invariant subgroups of G. Then

$$U/U \cap N \cong_X UN/N.$$

- Let $N \leq M$ be X-invariant normal subgroups of G. Then

$$(G/N)/(M/N) \cong_X G/M.$$

Exercises

Let G be a group.

1. Let N be characteristic in G. The automorphisms α of G satisfying $\alpha|_N = 1$ form a normal subgroup of Aut G.

2. The automorphisms α of G satisfying $U^\alpha = U$ for all subgroups U of G form a normal subgroup of Aut G.

3. Let $\alpha \in \operatorname{Aut} G$ and $|\{x \in G \mid x^\alpha = x\}| > \frac{|G|}{2}$. Then $\alpha = 1$.

4. The group G is Abelian if and only if the mapping

$$G \to G \quad \text{with} \quad x \mapsto x^{-1} \quad (x \in G)$$

is an automorphism of G.

5. Let G be finite and $\alpha \in \operatorname{Aut} G$ such that $x^\alpha \neq x = x^{\alpha^2}$ for all $x \in G^\#$. Then the following hold:

(a) For every $x \in G$ there exists $y \in G$ such that $x = y^{-1}y^\alpha$.

(b) G is Abelian of odd order.

6. Let $N \trianglelefteq G$ and $U \leq G$ such that $G = NU$. Then there exists a bijection, preserving inclusion, from the set of subgroups X satisfying $U \leq X \leq G$ to the set of U-invariant subgroups Y satisfying $U \cap N \leq Y \leq N$.

7. Let G be finite with $Z(G) = 1$, and set $A := \operatorname{Aut} G$ and $I := \operatorname{Inn} G$.

(a) $C_A(I) = 1.$[11]

(b) Suppose that I is characteristic in A, i.e., $I = I^\alpha$ for all $\alpha \in \operatorname{Aut} A$. Then $\operatorname{Aut} A = \operatorname{Inn} A$.

(c) Suppose that G is simple. Then $\operatorname{Aut} A = \operatorname{Inn} A$.

8. Let $\operatorname{GL}_2(\mathbb{C})$ be the group of all invertible 2×2-matrices over the field of complex numbers $\mathbb{C}$, and let

$$G := \left\langle \begin{pmatrix} i & 0 \\ 0 & -i \end{pmatrix}, \begin{pmatrix} 0 & 1 \\ -1 & 0 \end{pmatrix} \right\rangle \leq \operatorname{GL}_2(\mathbb{C}).$$

The group G is called a **quaternion group** (of order 8).

(a) $|G| = 8$.

(b) $|Z(G)| = 2$.

(c) Every element of $G \setminus Z(G)$ has order 4.

(d) G contains exactly one element of order 2.

(e) Every subgroup of G is normal in G.

(f) G possesses an automorphism of order 3.

---

[11]$C_A(I) := \{\alpha \in \operatorname{Aut} G \mid \alpha\beta = \beta\alpha \text{ for all } \beta \in I\}$.

1.4 Cyclic Groups

Every finite cyclic group is already completely described by 1.1.2. But since there exists a "universal object" for the class of cyclic groups, namely the additive group $\mathbb{Z}$ of the integers, we may look at cyclic groups from a slightly more general point of view.

The group $\mathbb{Z}$ is an infinite cyclic group with identity $0 \in \mathbb{Z}$ and generating element $1 \in \mathbb{Z}$.

1.4.1 *Let U be a subgroup of $\mathbb{Z}$. Then*

$$U = \{\, nz \mid z \in \mathbb{Z} \,\} =: n\mathbb{Z}$$

for some $n \in \mathbb{N} \cup \{\, 0 \,\}$. Moreover

$$n\mathbb{Z} \leq m\mathbb{Z} \iff m \mid n.$$

Proof. If $U = 0$ then $U = 0\mathbb{Z}$; so we may assume that $U \neq 0$. Let $k \in U$. Then also $-k \in U$, and the minimum

$$n := \min\{\, i \in \mathbb{Z} \mid 0 < i \in U \,\}$$

exists. As the reader will know, there exist integers $z, r \in \mathbb{Z}$ such that

$$k = zn + r \quad \text{and} \quad r \in \{0, 1, \dots, n-1\}.$$

Then $r = k - zn \in U$ and thus $r = 0$ by the minimality of n, so $k = zn \in U$ and $U = n\mathbb{Z}$.

The additional statement is clear. □

Let $n \in \mathbb{N}$. The factor group

$$C_n := \mathbb{Z}/n\mathbb{Z} \;^{12}$$

is a cyclic group of order n consisting of the cosets modulo n

$$n\mathbb{Z}, \; 1 + n\mathbb{Z}, \dots, (n-1) + n\mathbb{Z}.$$

The integers $0, 1, \dots, n-1$ form a transversal of $n\mathbb{Z}$ in $\mathbb{Z}$.

---

[12]In future we will often use multiplicative notation for the group operation on C_n.

Let $G = \langle g \rangle$ be any cyclic group—now written multiplicatively. The laws of exponents show that

$$\varphi \colon \mathbb{Z} \to G \quad \text{with} \quad z \mapsto g^z$$

is an epimorphism. By 1.4.1 there exists $n \geq 0$ such that

$$\operatorname{Ker} \varphi = n\mathbb{Z}.$$

If $n = 0$, then G is isomorphic to $\mathbb{Z}$; if $n \geq 1$, then G is isomorphic to C_n (Homomorphism Theorem 1.2.5 on page 13). We obtain:

1.4.2   *A cyclic group of order n is isomorphic to C_n.*   □

Using φ and the second remark in 1.4.1 we also obtain:

1.4.3   Theorem. *Let $G = \langle g \rangle$ be a cyclic group of order n and $l_1, \ldots, l_k$ $\in \mathbb{N}$ the divisors of n, and set*

$$U_i := \langle g^{l_i} \rangle.$$

Then $U_1, \ldots, U_k$ are the only subgroups G. Moreover

(a)   *If $n = n_i l_i$, then U_i is a subgroup of order n_i $(i = 1, \ldots, k)$.*

(b)   *Let $0 \neq z \in \mathbb{Z}$. If $i \in \{1, \ldots, k\}$ such that $l_i = (z, n)$,[13] then $\langle g^z \rangle = U_i$.*

Proof. The subgroups of G correspond (with respect to φ) to the subgroups of $\mathbb{Z}$ that contain $n\mathbb{Z}$, and thus by 1.4.1 to the divisors of n. Hence $U_1 = (l_1 \mathbb{Z})^\varphi, \ldots, U_k = (l_k \mathbb{Z})^\varphi$ are the only subgroups of G.

(a) n_i is the smallest of the integers $m \in \mathbb{N}$ such that $(g^{l_i})^m = 1$. Hence (a) follows from 1.1.2 on page 4.

(b) Since $l_i | z$ we get $g^z \in U_i$, i.e., $\langle g^z \rangle \leq U_i$. Note that there exist integers $z_1, z_2 \in \mathbb{Z}$ such that $l_i = nz_1 + zz_2$. It follows that

$$g^{l_i} = g^{l_i}(g^{-n})^{z_1} = (g^z)^{z_2}$$

and thus also $U_i \leq \langle g^z \rangle$.   □

As a consequence, in every finite cyclic group G there exists *exactly* one subgroup of order m for every divisor m of $|G|$. Since automorphisms of G map subgroups to subgroups of the same order we have:

---

[13](z, n) is the **greatest common divisor** of z and n.

1.4.4   *Subgroups of cyclic groups are characteristic.*[14]   □

It is evident that in the situation of 1.4.3

$$U_i \leq U_j \iff l_j \,|\, l_i.$$

For cyclic p-groups this implies:

1.4.5   *Let $G = \langle g \rangle$ be a nontrivial cyclic group of order p^n, p a prime. Then*

$$1 < \langle g^{p^{n-1}} \rangle < \langle g^{p^{n-2}} \rangle < \cdots < \langle g^p \rangle < G$$

are the only subgroups of G. In particular, G contains exactly one minimal and one maximal subgroup.   □

Note that the converse of 1.4.5 is also true: A finite group that contains exactly one maximal subgroup is cyclic of prime power order.[15] In contrast to this, a finite group with exactly one minimal subgroup is not necessarily cyclic; compare with 2.1.7 on page 46 and 5.3.7 on page 114.

In an Abelian group G every subgroup is normal. If in addition G is simple, then G is cyclic of prime order.

1.4.6   *The cyclic groups of prime order are the only Abelian simple groups.*   □

Exercises

Let G be a group.

1.   Suppose that $U \leq N \trianglelefteq G$ and N is cyclic. Then $U \trianglelefteq G$.

2.   Let p, q be primes and G be cyclic of order pq. Then G contains more than three subgroups if and only if $p \neq q$.

3.   Let G be finite. Suppose that $|\{x \in G \mid x^n = 1\}| \leq n$ for all $n \in \mathbb{N}$. Then G is cyclic.

4.   Let G be finite. Suppose that all maximal subgroups of G are conjugate. Then G is cyclic.

---

[14]Indeed, this is also true for $\mathbb{Z}$ since here $z \mapsto -z$ is the only nontrivial automorphism.
[15]See Exercise 10 on p. 10.

1.5 Commutators

For any two elements x, y of the group G we define

$$[x, y] := x^{-1}y^{-1}xy \qquad (= y^{-x}y = x^{-1}x^y). \ ^{16}$$

Since

$$xy = yx\,[x, y]$$

the element $[x, y]$ is the **commutator** of x and y. One has

$$[x, y]^{-1} = [y, x].$$

The subgroup generated by all commutators

$$\langle\, [x, y] \mid x, y \in G \,\rangle =: G'$$

is the **commutator subgroup** of G.

1.5.1   *Let φ be a homomorphism of G. Then*

$$[x, y]^\varphi = [x^\varphi, y^\varphi]$$

for all $x, y \in G$, and so $(G')^\varphi = (G^\varphi)'$. In particular, G' is a characteristic subgroup of G.   □

Also the commutator subgroup G'' of G' [17] is characteristic in G (1.3.2).

1.5.2   *Let N be a normal subgroup of G. Then*

$$G/N \text{ is Abelian} \iff G' \le N.$$

Accordingly, G' is the smallest normal subgroup of G with Abelian factor group.

Proof. For $x, y \in G$

$$(xN)(yN) = (yN)(xN) \iff xyN = yxN \iff [x, y] \in N. \qquad \square$$

The group G is **perfect** if $G = G'$. In Section 6.5 we need:

----

[16] $y^{-x} := (y^{-1})^x$.
[17] That is, $(G')' =: G''$.

1.5.3 *Let N be an Abelian normal subgroup of G. If G/N is perfect, then also G' is perfect.*

Proof. From 1.5.1, applied to the natural epimorphism, we obtain

$$G/N = (G/N)' = G'N/N$$

and thus $G = G'N$. Since also $G'/N \cap G'$ ($\cong G/N$) is perfect, the same argument gives $G' = G''(N \cap G')$. It follows that $G = G''N$ and $G/G'' \cong N/N \cap G''$. Now 1.5.2 implies $G' = G''$ since N is Abelian. □

For $x, y, z \in G$ we define

$$[x, y, z] := [[x, y], z],$$

and for subsets $X, Y, Z \subseteq G$

$$[X, Y] := \langle [x, y] \mid x \in X,\ y \in Y \rangle,$$
$$[X, Y, Z] := [[X, Y], Z].$$

The following elementary properties are often expressed using commutators.

- For subsets X, Y of G:

$$[X, Y] = 1 \iff xy = yx \quad \text{for all } x \in X, y \in Y.$$

- For subgroups X, Y of G:

$$[X, Y] \leq Y \iff Y \text{ is } X\text{-invariant}.$$

Thus, for normal subgroups N and M of G we have:

- $[N, M] \leq N \cap M.$

We shall frequently use the following commutator relations, which can be verified easily.

1.5.4 *For $x, y, z \in G$:*

$$[x, yz] = [x, z]\,[x, y]^z \quad \text{and} \quad [xz, y] = [x, y]^z\,[z, y]. □$$

1.5.5   *For subgroups X and Y of G the subgroup $[X,Y]$ is normal in $\langle X, Y \rangle$.*

Proof. For $x, z \in X$ and $y \in Y$ 1.5.4 implies

$$[x,y]^z = [xz, y][z,y]^{-1} \in [X,Y];$$

and with a similar argument $[x,y]^z \in [X,Y]$ for $z \in Y$.   □

The next slightly more complicated relation can also be verified easily:

$$[x, y^{-1}, z]^y [y, z^{-1}, x]^z [z, x^{-1}, y]^x = 1 \quad (x, y, z \in G).^{18}$$

We will use this relation in the following form:

1.5.6   Three-Subgroups Lemma. *Let X, Y, Z be subgroups of G. Suppose that $[X, Y, Z] = [Y, Z, X] = 1$. Then also $[Z, X, Y] = 1$.*   □

Exercises

Let G be a group, $x \in G$, and set $C_G(x) := \{y \in G \mid yx = xy\}$. Obviously, $C_G(x)$ is a subgroup of G.

1. Let A be an Abelian normal subgroup of G and $x \in G$.

    (a)   The mapping $A \to A$ given by $a \mapsto [a, x]$ is a homomorphism.
    (b) $[A, \langle x \rangle] = \{[a, x] \mid a \in A\}$.

2. Let A and x be as in 1. Suppose that $G = AC_G(ax)$ for all $a \in A$. Then $[A, G] = [A, \langle x \rangle]$.

3. Let $|G| = p^n$, p a prime, and let $|G : C_G(x)| \leq p$ for all $x \in G$.

    (a) $C_G(x) \trianglelefteq G$ for all $x \in G$.
    (b) $G' \leq Z(G)$.
    (c)   (Knoche, [74]) $|G'| \leq p$.

4. Let $\alpha \in \operatorname{Aut} G$. Suppose that $x^{-1} x^\alpha \in Z(G)$ for all $x \in G$. Then $x^\alpha = x$ for all $x \in G'$.

---

[18]See [100] and [64].

5. (Ito, [70]) Let $G = AB$, where A and B are Abelian subgroups of G. Then G' is Abelian.

6. (Burnside, [4], p. 90) Let A be a normal subgroup of G. Suppose that every element in $G \setminus A$ has order 3. Then $[B, B^x] = 1$ for all Abelian subgroups $B \leq A$ and $x \in G \setminus A$.

1.6 Products of Groups

Products of groups are of twofold interest. On the one hand, they can be used to construct new groups from given ones (external products); on the other hand, they can be used to describe the structure of groups (internal products). One internal product we have already met: the complex product of two subgroups A and B. Indeed, AB is also a group if $AB = BA$ (1.1.5 on page 6).

Let $G_1, \ldots, G_n$ be groups. The Cartesian product of the sets G_i

$$\underset{i=1}{\overset{n}{\times}} G_i := G_1 \times \cdots \times G_n := \{ (g_1, \ldots, g_n) \mid g_i \in G_i \}$$

is a group with respect to componentwise multiplication

$$(g_1, \ldots, g_n)(h_1, \ldots, h_n) := (g_1 h_1, \ldots, g_n h_n).$$

This group is the (external) **direct product of the groups** $G_1, \ldots, G_n$. Obviously, for $j = 1, \ldots, n$ the embedding

$$\varepsilon_j \colon G_j \to \underset{i=1,\ldots,n}{\times} G_i \quad \text{with} \quad g \mapsto (1, \ldots, 1, \underset{j}{g}, 1, \ldots, 1)$$

is an isomorphism from G_j to

$$G_j^* := \{(g_1, \ldots, g_n) \mid g_i = 1 \text{ for } i \neq j\}.$$

For the subgroups $G_1^*, \ldots, G_n^*$ of $G := \underset{i=1,\ldots,n}{\times} G_i$ one has:

$\mathcal{D}_1 \quad G = G_1^* \cdots G_n^*,$

$\mathcal{D}_2 \quad G_i^* \trianglelefteq G, \; i = 1, \ldots, n,$

$\mathcal{D}_3 \quad G_i{}^* \cap \prod_{j \neq i} G_j{}^* = 1, \ i = 1, \dots, n.$

Conversely:

1.6.1   *Let G be a group with subgroups $G_1{}^*, \dots, G_n{}^*$ such that $\mathcal{D}_1$, $\mathcal{D}_2$ and $\mathcal{D}_3$ hold. Then the mapping*

$$\alpha \colon \ \underset{i=1}{\overset{n}{\times}} G_i{}^* \to G \quad \text{with} \quad (g_1, \dots, g_n) \mapsto g_1 \cdots g_n$$

is an isomorphism.

Proof. $\mathcal{D}_1$ shows that α is surjective. $\mathcal{D}_2$ gives

$$[G_i{}^*, G_k{}^*] \leq G_i{}^* \cap \prod_{j \neq i} G_j^* \text{ for } i \neq k,$$

and thus $[G_i{}^*, G_k{}^*] = 1$ because of $\mathcal{D}_3$. For $h_i, g_i \in G_i{}^*$, $i = 1, \dots, n$, this implies

$$(g_1 \cdots g_n)(h_1 \cdots h_n) = (g_1 h_1) \cdots (g_n h_n);$$

hence α is a homomorphism. Let $(g_1, \dots, g_n) \in \mathrm{Ker}\,\alpha$, i.e., $g_1 \cdots g_n = 1$. Then

$$g_i = \prod_{j \neq i} g_j^{-1} \in G_i{}^* \cap \prod_{j \neq i} G_j^* = 1,$$

again by $\mathcal{D}_3$, so $\mathrm{Ker}\,\alpha = 1$. Thus, α is a isomorphism. $\square$

If $\mathcal{D}_1$, $\mathcal{D}_2$ and $\mathcal{D}_3$ hold for the group G and subgroups $G_1^*, \dots, G_n^*$, then G is called the **(internal) direct product** of the subgroups $G_1{}^*, \dots, G_n{}^*$ (this notation is justified by 1.6.1); in this case we write as above

$$G = G_1{}^* \times \cdots \times G_n{}^* = \underset{i=1}{\overset{n}{\times}} G_i{}^*.$$

In particular, we have $[G_i{}^*, G_j{}^*] = 1$ for $i \neq j$, and every element $g \in G$ can be written in a unique way as a product

$$g = \prod_{i=1}^{n} g_i \quad \text{with} \quad g_i \in G_i{}^*.$$

We have met two versions of the direct product, the external and internal. The first is a product of (not necessarily distinct) groups, the other a product of (distinct) subgroups. Very often the factors $G_1, \ldots, G_n$ of the external direct product will be pairwise distinct. In these cases one usually identifies G_j with $G_j{}^*$ according to the embedding ε_j and no longer distinguishes between the external and internal direct products.

1.6.2   *Let $G = G_1 \times \cdots \times G_n$.*

(a) $Z(G) = Z(G_1) \times \cdots \times Z(G_n)$.

(b) $G' = G_1' \times \cdots \times G_n'$.

(c)   *Let N be a normal subgroup of G and $N_i = N \cap G_i$ $(i = 1, \ldots, n)$. Suppose that $N = N_1 \times \cdots \times N_n$. Then the mapping*

$$\alpha: \ G = G_1 \times \cdots \times G_n \to G_1/N_1 \times \cdots \times G_n/N_n$$

given by
$$g = (g_1, \ldots, g_n) \mapsto (g_1 N_1, \ldots, g_n N_n)$$

is an epimorphism, with $\operatorname{Ker} \alpha = N$. In particular

$$G/N \cong G_1/N_1 \times \cdots \times G_n/N_n.$$

(d)   *If the factors $G_1, \ldots, G_n$ are characteristic subgroups of G, then*

$$\operatorname{Aut} G \cong \operatorname{Aut} G_1 \times \cdots \times \operatorname{Aut} G_n.$$

Proof. (a) Componentwise multiplication in G gives (a).

(b) An easy induction using 1.5.4 gives (b). For example, for $n = 2$

$$G' = [G_1 G_2, G_1 G_2] = \prod_{i,j} [G_i, G_j] = G_1' \times G_2'.$$

(c) Apply 1.2.4 and 1.2.5.

(d) If α_i is an automorphism of G_i (for $i = 1, \ldots, n$), then

$$(g_1, \ldots, g_n)^\alpha := (g_1^{\alpha_1}, \ldots, g_n^{\alpha_n})$$

defines an automorphism of $G = G_1 \times \cdots \times G_n$, and

$$\varphi \colon \operatorname{Aut} G_1 \times \cdots \times \operatorname{Aut} G_n \to \operatorname{Aut} G \quad \text{with} \quad (\alpha_1, \ldots, \alpha_n) \mapsto \alpha$$

is a monomorphism. Moreover, φ is surjective if the factors $G_1, \ldots, G_n$ are characteristic subgroups of G. $\square$

1.6.3   *Let $G = G_1 \times \cdots \times G_n$ and N be a normal subgroup of G.*

(a)   *If N is perfect, then $N = (N \cap G_1) \times \cdots \times (N \cap G_n)$.*

(b)   *If $G_1, \ldots, G_n$ are non-Abelian simple groups, then there exists a subset $J \subseteq \{1, \ldots, n\}$ such that*

$$N = \underset{j \in J}{\times} G_j \quad \text{and} \quad G_k \cap N = 1 \text{ for } k \notin J.$$

Proof. (a) Since G_i and N are normal in G we get $[N, G_i] \leq N \cap G_i$, and then 1.5.4 yields

$$[N, G] = \prod_i [N, G_i] \leq \prod_i (N \cap G_i) =: N_0.$$

In particular $[N, N] \leq N_0$, and $N' = N$ gives $N = N_0$.

(b) The simplicity of the normal subgroups G_i yields $G_i \leq N$ or $G_i \cap N = 1$, so (b) follows from (a) if N is perfect. Thus, it suffices to prove by induction on $|G|$ that N is perfect.

According to 1.6.2 (b) we may assume that $G \neq N$. Hence, there exists $k \in \{1, \ldots, n\}$ such that $G_k \nleq N$, so $N \cap G_k = 1$ and thus $NG_k = N \times G_k$. Let $\overline{G} = G/G_k$. By 1.6.2 (c) $\overline{G} = \underset{i \neq k}{\times} \overline{G}_i$, and induction shows that $\overline{N} = \overline{N}'$. Now 1.5.1 on page 24 yields $N \times G_k = N' \times G_k$, so $|N| = |N'|$ and $N = N'$. $\square$

For Abelian simple groups $G_1, \ldots, G_n$ the statement in 1.6.3 (b) is wrong. For example, $C_2 \times C_2$ contains three minimal (normal) subgroups.

The following result, which is a consequence of the Homomorphism Theorem 1.2.5, shows that the external direct product can be used to get results about the internal structure of a group.

1.6.4   Let $N_1, \ldots, N_n$ be normal subgroups of G. Then the mapping

$$\alpha: \ G \to G/N_1 \times \cdots \times G/N_n$$

given by

$$g \mapsto (gN_1, \ldots, gN_n)$$

is a homomorphism with $\operatorname{Ker} \alpha = \bigcap_i N_i$. In particular, $G/\bigcap_i N_i$ is isomorphic to a subgroup of $G/N_1 \times \cdots \times G/N_n$. $\qquad\square$

Frequently one has the following situation:

1.6.5   Let G be a product of the normal subgroups $G_1, \ldots, G_n$. Suppose that

$$(\,|G_i|, |G_j|\,) = 1 \quad \text{for} \quad i \neq j \in \{1, 2, \ldots, n\}.$$

Then $G = G_1 \times \cdots \times G_n$.

Proof. We have to show that

$$D := \left(\prod_{j \neq i} G_j \right) \cap G_i = 1.$$

By Lagrange's theorem $|D|$ is a divisor of $|G_i|$ and of

$$k := \left| \prod_{j \neq i} G_j \right|.$$

Repeated application of 1.1.6 shows that k and thus also $|D|$ is a divisor of $\prod_{j \neq i} |G_j|$. Hence, k and $|G_i|$ are coprime, and $|D| = 1$. $\qquad\square$

This implies the fundamental observation:

1.6.6   Let a, b be elements of the finite group G such that $ab = ba$ and $(o(a), o(b)) = 1$. Then

$$\langle ab \rangle = \langle a \rangle \times \langle b \rangle$$

and $o(ab) = o(a)o(b)$.

Proof. Let $k := o(a)$ and $m := o(b)$. Note that $\langle a, b \rangle$ is an Abelian group, where the subgroups $\langle a \rangle$ and $\langle b \rangle$ are of coprime order. Hence

$$H := \langle a, b \rangle = \langle a \rangle \times \langle b \rangle$$

is a group of order mk. Let $g := ab$ $(\in H)$. The homomorphism

$$\varphi \colon \langle g \rangle \to H/\langle a \rangle \quad \text{with} \quad g^i \mapsto \langle a \rangle g^i = \langle a \rangle b^i$$

is surjective. Hence, $|\operatorname{Im} \varphi| = m$ is a divisor of $|\langle g \rangle|$ (Homomorphism Theorem). In the same way k is a divisor of $|\langle g \rangle|$. Now $(m, k) = 1$ implies $o(g) = mk = |H|$, i.e., $H = \langle g \rangle$. $\qquad\square$

Let G be a product of the subgroups $G_1, \ldots, G_n$, which satisfy

$\mathcal{Z}$ $\qquad\qquad\qquad [G_i, G_j] = 1 \quad \text{for } i \neq j \text{ in } \{1, \ldots, n\}.$

Then G is the **central product** of the subgroups $G_1, \ldots, G_n$. Because of $\mathcal{Z}$ the subgroups G_i are normal in G; moreover we have for $i = 1, \ldots, n$

$$G_i Z(G) \cap \prod_{j \neq i} G_j Z(G) = Z(G).$$

With the Homomorphism Theorem we obtain

1.6.7   *Let G be the central product of $G_1, \ldots, G_n$ and $\overline{G} := G/Z(G)$. Then $\overline{G}$ is a direct product of the groups $\overline{G}_1, \ldots, \overline{G}_n$, with*

$$\overline{G}_i \cong G_i/Z(G_i), \; i = 1, \ldots, n.$$

$\qquad\square$

We will now introduce the *semidirect product*. In contrast to our treatment of the direct product we first give the internal version.

Let G be group with subgroups X and H. Then G is called the **(internal) semidirect product** of X with H, if

$\mathcal{SD}_1$ $G = XH$,

$\mathcal{SD}_2$ $H \trianglelefteq G$,

$\mathcal{SD}_3$ $X \cap H = 1$.

Hence, in the semidirect product $G = XH$, the subgroup X is a complement of the normal subgroup H. If X is also normal in G, then G is the direct product $X \times H$.

1.6.8   *Let X and H be subgroups of G, which satisfy SD_1, SD_2, and SD_3.*

(a)   *Every $g \in G$ is in a unique way a product $g = xh$ with $x \in X$ and $h \in H$.*

(b)   *For $x_1, x_2 \in X$ and $h_1, h_2 \in H$*

$$(x_1 h_1)(x_2 h_2) = (x_1 x_2)(h_1^{x_2} h_2).$$

Proof. Part (a) follows from 1.1.9 since X is a transversal of H in G. Part (b) is obvious.   □

Now let X and H be groups, and let $\varphi \colon X \to \operatorname{Aut} H$ be a homomorphism. Then X acts on H (with respect to φ). As in Section 1.3 we set

$$h^x := h^{x^\varphi} \quad (x \in X, \ h \in H)$$

and thus

$$(h^x)^y = h^{xy} \quad (h \in H, \ x, y \in X).$$

The multiplication (compare with 1.6.8 (b))

$$(x_1, h_1)(x_2, h_2) := (x_1 x_2, h_1^{x_2} h_2) \quad (x_i \in X, \ h_i \in H),$$

turns the Cartesian product

$$G := \{(x, h) \mid x \in X, \ h \in H\}$$

into a group: The identity of G is $(1_X, 1_G)$, and the inverse of (x, h) is

$$(x^{-1}, (h^{-1})^{x^{-1}}).$$

Associativity is verified as follows:

$$
\begin{aligned}
((x_1, h_1)(x_2, h_2))\,(x_3, h_3) &= (x_1 x_2, h_1^{x_2} h_2)(x_3, h_3) = (x_1 x_2 x_3, (h_1^{x_2} h_2)^{x_3} h_3) \\
&= (x_1 x_2 x_3, h_1^{x_2 x_3} h_2^{x_3} h_3) = (x_1, h_1)(x_2 x_3, h_2^{x_3} h_3) \\
&= (x_1, h_1)\,((x_2, h_2)(x_3, h_3)).
\end{aligned}
$$

This group G is called the **(external) semidirect product** of X with H (with respect to φ); we write $G = X \ltimes_\varphi H$ or more simply $G = X \ltimes H$, or even $G = XH$, if there is no danger of confusion about which action is meant.

If φ is the trivial homomorphism; i.e., X acts trivially on H, then $X \ltimes H$ is the direct product $X \times H$.

As for the direct product, the embeddings

$$\varepsilon_X : X \to X \ltimes H \quad \text{with} \quad x \mapsto (x, 1)$$
$$\varepsilon_H : H \to X \ltimes H \quad \text{with} \quad h \mapsto (1, h)$$

are monomorphisms, and $X \ltimes H$ is the semidirect product of the subgroup X^{ε_X} with the subgroup H^{ε_H}. Usually one identifies X and X^{ε_X} (resp. H and H^{ε_H}) via ε_X (resp. ε_H). Then the action of X on H is conjugation in $X \ltimes H$.

Elements of order 2 are **involutions**, and a group generated by two involutions is a **dihedral group**. The following result shows that dihedral groups are semidirect products.

1.6.9   *Let G be a finite group of order $2n$. The following statements are equivalent:*

(i)   *G is a dihedral group.*

(ii)   *G is the semidirect product $X \ltimes H$ of two cyclic subgroups $X = \langle x \rangle$ and $H = \langle h \rangle$ such that*

(D) $o(x) = 2, \quad o(h) = n, \quad \text{and } h^x = h^{-1}.$

Proof. (i) $\Rightarrow$ (ii): Let x, y be involutions of G such that $G = \langle x, y \rangle$, and let

$$X := \langle x \rangle, \quad h := xy \text{ and } H := \langle h \rangle.$$

One has

$$h^x = xxyx = yx = h^{-1} = yxyy = h^y,$$

and so H is normal in G and $G = XH$. If $X \cap H \neq 1$, then x and thus also y is in H. Then $x = y$ since H is cyclic. Now $h = xy = 1$, and $H = 1$. But this contradicts $x \in H$.

(ii) $\Rightarrow$ (i): The element $y := xh$ is an involution since

$$y^2 = (xhx)h = h^{-1}h = 1.$$

Hence $G = \langle x, y \rangle$ is a dihedral group. $\square$

If G is as in 1.6.9 (ii), then the group table of G is uniquely determined by the relations in (D). Hence, there is only one dihedral group of order $2n$ (up to isomorphism). Such a group is denoted by D_{2n}. Clearly $D_2 \cong C_2$ and $D_4 \cong C_2 \times C_2$.

D_{2n} is the symmetry group of a regular n-gon. The reader is asked to verify this for $n = 3$ and $n = 4$.

It should be mentioned that in section 4.4 we will introduce a third type of product, the **wreath product**, which will be constructed by means of direct and semidirect products.

Exercises

Let A, B, and G be groups.

1. (a) Every normal subgroup of A is a normal subgroup of $A \times B$.

    (b) $U \leq A \times B$ does not imply $U = (A \cap U) \times (B \cap U)$.

    (c) If A and B are finite and $(|A|, |B|) = 1$, then A and B are characteristic subgroups of $A \times B$.

    (d) $\mathrm{Aut}(A \times B)$ contains a subgroup isomorphic to $\mathrm{Aut}\, A \times \mathrm{Aut}\, B$.

2. Let $G = A \times B$. Then $A \cong B$, if and only if there exists a subgroup D in G such that $G = AD = BD$ and $1 = A \cap D = B \cap D$.

3. Let G be finite. Suppose that every maximal subgroup of G is simple and normal in G. Then G is an Abelian group and $|G| \in \{1, p, p^2, pq\}$, where p, q are primes.

4. A group X is **semisimple** if X is a direct product of non-Abelian simple groups. Let G be a group and M, N normal subgroups of G. If G/M and G/N are semisimple, then also $G/(M \cap N)$ is semisimple.

5. Show that the non-Abelian group of order 6 on page 3 is the group D_6.

6. Let $n \geq 2$. Then

$$Z(D_{2n}) \neq 1 \iff n \equiv 0 \pmod{2}.$$

7. Let G_1 and G_2 be finite perfect groups such that $G_1/Z(G_1) \cong G_2/Z(G_2)$. Then there exists a finite perfect group G and subgroups $Z_1, Z_2 \leq Z(G)$ with
$$G/Z(G) \cong G_i/Z(G_i) \quad \text{and} \quad G/Z_i \cong G_i, \quad i = 1, 2.$$

In the following G is a dihedral group and $4 < |G| < \infty$.

8. Describe all subgroups of G.

9. (a) $|Z(G)| \leq 2$.

    (b) $\langle a \rangle Z(G) = \{g \in G \mid g^a = g\}$ for every involution $a \in G \setminus Z(G)$.

    (c) $|G : G'| = 2|Z(G)|$.

    (d) For every involution $a \in G \setminus Z(G)$ there exists an involution b such that $G = \langle a, b \rangle$.

10. Let $Z(G) \neq 1$ and a be an involution of $G \setminus Z(G)$. The elements in $aZ(G)$ are conjugate in G, if and only if 8 is a divisor of $|G|$.

11. The following statements are equivalent:

    (a) All involutions are conjugate in G.

    (b) $Z(G) = 1$.

    (c) There exists an involution $a \in G$ such that $|C_G(a)| = 2$.

    (d) $4 \nmid |G|$.

    (e) G contains a maximal subgroup of odd order.

1.7 Minimal Normal Subgroups

Let G be a group. A normal subgroup $N \neq 1$ of G is a **minimal normal subgroup** of G if 1 and N are the only normal subgroups of G that are contained in N. It is evident that every nontrivial finite group possesses minimal normal subgroups. Moreover, any nontrivial finite group is either simple or contains a proper minimal normal subgroup. In many proofs by induction on the group order, minimal normal subgroups play an important role.

In this section we collect some elementary properties of minimal normal subgroups. More information about the embedding of minimal normal subgroups can be found in Sections 6.5 and 6.6.

1.7.1 *Let N be a minimal normal subgroup of G.*

(a) For all normal subgroups M of G either $N \leq M$ or $N \cap M = 1$. In the second case $[N, M] = 1$.

(b) If N is Abelian, then $N \leq H$ or $N \cap H = 1$ for all subgroups H of G with $G = NH$.

(c) If φ is an epimorphism from G to a group H, then $N^\varphi = 1$ or N^φ is a minimal normal subgroup of H.

Proof. (a) follows directly from the minimality of N, noting that

$$[N, M] \leq M \cap N \trianglelefteq G.$$

(b) Here $M := H \cap N$ is normal in H but also normal in N since N is Abelian. Now since $G = HN$ it follows that $M \trianglelefteq G$ and thus $M \in \{1, N\}$.

(c) Let $A \neq 1$ be a normal subgroup of H that is contained in N^φ. Then $A^{\varphi^{-1}} \cap N$ is a normal subgroup of G, and $A^{\varphi^{-1}} \cap N \neq 1$ since $A \neq 1$. Hence $A^{\varphi^{-1}} \cap N = N$ and $N^\varphi = A$.   □

1.7.2   Let $\mathcal{M}$ be a finite set of minimal normal subgroups of G, and let $M = \prod\limits_{N \in \mathcal{M}} N$.

(a) Let U be a normal subgroup of G. Then there exist $N_1, \ldots, N_n \in \mathcal{M}$ such that

$$UM = U \times N_1 \times \cdots \times N_n.$$

(b) There exist $N_1, \ldots, N_n \in \mathcal{M}$ such that

$$M = N_1 \times \cdots \times N_n.$$

Proof. (a) By 1.7.1 (a) $U \cap N = 1$ for every $N \in \mathcal{M}$ with $N \not\leq U$ and thus $UN = U \times N$. Let $\{N_1, \ldots, N_n\}$ be a subset of $\mathcal{M}$ that is maximal with respect to the following property:

$$U\left(\prod_{i=1}^{n} N_i\right) = U \times N_1 \times \cdots \times N_n =: X.$$

Assume that $X \neq UM$. Then there exists $N \in \mathcal{M}$ such that $N \not\leq X$. By 1.7.1 (a)

$$XN = X \times N = U \times N_1 \times \cdots \times N_n \times N,$$

which contradicts the maximal choice of $\{N_1, \ldots, N_n\}$. Hence $X = UM$. Now (b) follows from (a), with $U = 1$. $\qquad\qquad\qquad\qquad\qquad\square$

1.7.3   *Let N be a minimal normal subgroup of G and E a minimal normal subgroup of N, and assume that the set $\mathcal{M} = \{E^g \mid g \in G\}$ is finite. Then E is simple, and there exist $E_1, \ldots, E_n$ in $\mathcal{M}$ such that*

$$N = E_1 \times \cdots \times E_n.$$

Proof. The subgroup $\prod_{g \in G} E^g$ is normal in G and thus equal to N. Hence $N = E_1 \times \cdots \times E_n$ follows from 1.7.2 (a). Every normal subgroup of E_1 also is a normal subgroup of N. This shows that E_1 is simple, and then E is simple as $E \cong E_1$. $\qquad\qquad\qquad\qquad\qquad\square$

If E in 1.7.3 is Abelian and thus isomorphic to C_p $(p \in \mathbb{P})$, one gets:

1.7.4   *Let N be an Abelian minimal normal subgroup of the finite group G. Then there exists $p \in \mathbb{P}$ such that N is a direct product of subgroups that are isomorphic to C_p.* $\qquad\qquad\qquad\qquad\qquad\square$

In the situation of 1.7.4 one knows the structure of the factors E_i. On the other hand, in general there are many different choices for these factors in N; compare with the remark after 1.6.3.

If the minimal normal subgroup N is not Abelian, then one has the opposite situation. Elementary methods do not yield any further properties of the structure of the factors E_i, but according to 1.6.3 (b) these factors are uniquely determined.

Together with 1.6.3 (b) we obtain:

1.7.5   *Let N be a non-Abelian minimal normal subgroup of the finite group G, and let $\mathcal{K}$ be the set of minimal normal subgroups of N.*

(a)   The elements of $\mathcal{K}$ are non-Abelian simple groups, which are conjugate in G.

(b)   For every $M \trianglelefteq N$ there exist $\mathcal{K}(M) \subseteq \mathcal{K}$ such that

$$M = \underset{E \in \mathcal{K}(M)}{\times} E \quad \text{and} \quad \mathcal{K}(M) = \{ E \in \mathcal{K} \mid E \leq M \}.$$

(c) $N = \underset{E \in \mathcal{K}}{\times} E.$ □

Exercises

Let G be a finite group and L a maximal subgroup of G.

1.   All minimal normal subgroups N of G that satisfy $N \cap L = 1$ are isomorphic.

2.   Let L be non-Abelian and simple. Then there exist at most two minimal normal subgroups in G.

3.   Let L and G be as in 2. Give an example where G possesses two minimal normal subgroups.

4.   Suppose that G contains two minimal normal subgroups, neither of which is contained in L. Then every minimal normal subgroup of L is contained in the product of all minimal normal subgroups of G.

5.   Let $(*)$ be the property:

   $(*)$ Every minimal normal subgroup is contained in the center.

   (a)   Let N and M be normal subgroups of G, which satisfy $(*)$. Then also NM satisfies $(*)$.

   (b)   If G satisfies $(*)$, then also every normal subgroup of G satisfies $(*)$.

1.8   Composition Series

In this section let G be a nontrivial *finite* group. By $(A_i)_{i=0,\dots,a}$ we denote a **subgroup series**

$$1 = A_0 < A_1 < \cdots < A_{i-1} < A_i < \cdots < A_{a-1} < A_a = G$$

of **length** a of G. A series $(A_i)_{i=0,1,\dots,a}$ is a **normal series**, if $A_i \trianglelefteq G$, and a **subnormal series**, if $A_{i-1} \trianglelefteq A_i$ for all $i = 1, \dots, a$.

A normal series $(A_i)_{i=0,\ldots,a}$ is a **chief series**, if each A_{i-1} is maximal among the normal subgroups of G that are properly contained in A_i.

A subnormal series $(A_i)_{i=0,\ldots,a}$ is a **composition series**, if each A_{i-1} is maximal among the proper normal subgroups of A_i.

The **composition factors** A_i/A_{i-1} of a composition series are simple groups. A composition series can be found by going downward—starting with G—and choosing A_{i-1} as a maximal normal subgroup of A_i. Similarly one can *refine* a normal series (resp. subnormal series) to obtain a chief series (resp. composition series).

Let $(A_i)_{i=0,\ldots,a}$ be a composition series for G. If all composition factors are Abelian[19] and thus cyclic of prime order (1.4.6 on page 23), then the structure of this composition series is determined by the order of G: The prime factor decomposition

$$|G| \;=\; p_1^{e_1} \cdots p_n^{e_n}$$

corresponds to

$$|G| \;=\; \prod_{i=1}^{a} |A_i/A_{i-1}|,$$

where $a = e_1 + \cdots + e_n$, and e_j is the number of factors A_i/A_{i-1} that are isomorphic to C_{p_j}.

The set of composition factors, for a given composition series of a finite group, forms an *invariant* of the group. This is the Jordan-Hölder Theorem. We will prove a version of this theorem that also gives nontrivial information in the above-mentioned special case, in particular for the case of Abelian groups. To do this, we use the notation introduced at the end of Section 1.3.

Let X be a group that acts on G, and let A and B be X-invariant subgroups of G such that $B \trianglelefteq A$. Then X also acts on A/B; we call A/B an X-**section** of G.

A subnormal series $(A_i)_{i=0,\ldots,a}$ is an X-**composition series** of G if all of the subgroups A_i are X-invariant and there are no X-invariant normal subgroups of A_i strictly between A_{i-1} and A_i. Then the factors A_i/A_{i-1} $(i = 1,\ldots,a)$ are X-**simple**.

An X-composition series of G is a composition series if $X = 1$, and a chief series if $X = G$.

---

[19]Such a group is said to be *solvable*; see 6.1 on p. 121.

1.8.1   Jordan-Hölder Theorem.[20] *Let X be a group that acts on G, and let $(A_i)_{i=0,...,a}$ and $(B_i)_{i=0,...,b}$ be two X-composition series of G. Then $a = b$ and there exists a permutation π on the set $\{A_i/A_{i-1} \mid i = 1, \ldots, a\}$ such that*

$$(A_i/A_{i-1})^\pi \cong_X B_i/B_{i-1}.$$

Proof. Let $N := B_{b-1}$. Then N is a maximal X-invariant normal subgroup of G, and G/N is X-simple. Hence, we may assume that $N \neq 1$ since in the other case the conclusion is obvious.

For $i \in \{1, \ldots, a\}$ and $A_i \not\leq N$, we get

$$N \trianglelefteq A_i N \trianglelefteq A_{i+1} N \trianglelefteq \cdots \trianglelefteq A_{a-1} N \trianglelefteq G$$

and thus $G = N A_i$ because of the maximality of N. Hence

(1) $\qquad\qquad\qquad A_i \leq N \quad \text{or} \quad G = N A_i.$

for $i = 0, \ldots, a$.

We set

$$A_i^* := A_i \cap N$$

and choose $j \in \{0, \ldots, a\}$ maximal such that $A_j \leq N$. Then

$$A_j \trianglelefteq A_{j+1}^* < A_{j+1}$$

and thus $A_j = A_{j+1}^*$ since $A_{j+1} \not\leq N$ and A_{j+1}/A_j is X-simple. Hence, we have

(2) $\qquad\qquad\qquad A_j = A_j^* = A_{j+1}^* \quad \text{and}$

(3) $\qquad\qquad\qquad A_{j+1}/A_j \cong_X G/N,$

the last statement because of

$$G/N \overset{(1)}{=} A_{j+1}N/N \overset{1.2.6}{\cong}_X A_{j+1}/A_{j+1}^*.$$

It follows for $k \geq j + 2$ that

$$A_k^* \cap A_{k-1} = A_k \cap N \cap A_{k-1} = A_{k-1}^*,$$

---

[20]Compare with [15], p. 42, and [68].

and by 1.2.6

$$A_k^*/A_{k-1}^* \cong_X A_k^* A_{k-1}/A_{k-1} \trianglelefteq A_k/A_{k-1}.$$

Now the X-simplicity of A_k/A_{k-1} implies either

(4) $$A_k^*/A_{k-1}^* \cong_X A_k/A_{k-1}, \quad k \geq j+2,$$

or

$$A_k^*/A_{k-1}^* = 1 \quad \text{and} \quad A_k^* = A_{k-1}^*.$$

In the second case $NA_k = G = NA_{k-1}$ yields

$$A_k/A_k^* \cong_X G/N \cong_X A_{k-1}/A_{k-1}^*$$

and the contradiction $A_k = A_{k-1}$.

Hence

$$1 = A_0^* < \cdots < A_j^* < A_{j+2}^* < \cdots < A_a^* = N$$

and

$$1 = B_0 < \cdots < B_{b-1} = N$$

are two X-composition series of N. By induction on $|G|$ we may assume that for these X-composition series there exists a permutation π with the desired property. In particular $a - 1 = b - 1$ and thus $a = b$.

We now extend π to a permutation on $\{A_i/A_{i-1} \mid i = 1, \ldots, a\}$ by setting

$$(A_{j+1} / A_j)^\pi := B_b/B_{b-1}.$$

Then (3) and (4) imply the assertion.   □

Chapter 2

Abelian Groups

In this chapter we determine the structure of the finite Abelian groups. As a starting point we use the structure of the cyclic groups described in Section 1.4. It will turn out that every finite Abelian group is the direct product of cyclic groups. In the second section of this chapter we will show that the automorphism groups of cyclic groups are examples of Abelian groups.

Compared with groups in general the structure of Abelian groups is much easier to investigate since commutativity implies many structural properties that almost never hold in non-Abelian groups. For example, in an Abelian group every subgroup is normal and the product of subgroups is again a subgroup (1.1.5 on page 6).

From this chapter on all groups considered are finite.

2.1 The Structure of Abelian Groups

If $G = \langle x \rangle$ is a cyclic group, then $|G| = o(x)$, and Lagrange's theorem implies

$$o(y) \text{ divides } o(x) \quad \text{for all } y \in G.$$

A more general property is true for Abelian groups, as one can show using 1.6.6 on page 31:

2.1.1 *Let G be an Abelian group and U a cyclic subgroup of maximal order in G. Then*

$$o(y) \text{ divides } |U| \quad \text{for all } y \in G.$$

43

Proof. Let $y \in G$. We show that every prime power p^r that divides $o(y)$ also divides $|U|$. Let $|U| = p^e m$ with $(p, m) = 1$. By 1.4.3 on page 22 there exist elements $a \in \langle y \rangle$ and $b \in U$ such that

$$o(a) = p^r \quad \text{and} \quad o(b) = m,$$

and 1.6.6 on page 31 implies $o(ab) = p^r m$. Now the maximality of $|U|$ gives $p^r \mid p^e m$. $\square$

2.1.2   *Let G and U be as in 2.1.1. Then there exists a complement V of U in G; in particular $G = U \times V$ and $|G| = |U|\,|V|$.*

Proof. If $G = U$, then $V = 1$ is the desired complement. Let $G \neq U$. Among all elements in $G \setminus U$ we choose y such that $o(y)$ is minimal. Then $y \neq 1$ and $\langle y^p \rangle < \langle y \rangle$ for every prime divisor p of $o(y)$ (1.4.3 on page 22), i.e., $\langle y^p \rangle \leq U$.

Let $U = \langle u \rangle$. By 2.1.1 and 1.4.3 on page 22 $o(y)$ is a divisor of $|U|$, and U contains exactly one subgroup for every such divisor. Hence, there exists a subgroup of order $\frac{o(y)}{p}$ in $\langle u^p \rangle$, namely $\langle y^p \rangle$. Let $i \in \mathbb{N}$ such that $u^{pi} = y^p$. Then $(yu^{-i})^p = 1$, but $yu^{-i} \notin U$ since $y \notin U$. The minimality of $o(y)$ gives

$$o(y) = p.$$

Thus, $N := \langle y \rangle$ is a nontrivial subgroup of G such that

$$U \cap N = 1.$$

Let $\overline{G} := G/N$.[1] For $\langle \overline{x} \rangle \leq \overline{G}$ we obtain

$$o(\overline{x}) = |\langle \overline{x} \rangle| = \min\{n \in \mathbb{N} \mid x^n \in N\} \leq |\langle x \rangle| = o(x),$$

and since

$$UN/N \cong U/U \cap N \cong U$$

we also have $|\overline{U}| = |U|$. Hence, $\overline{U}$ is a cyclic subgroup of maximal order in $\overline{G}$. By induction on $|G|$ we may assume that there exists a complement $\overline{V}$ of $\overline{U}$ in $\overline{G}$.

---

[1]For the "bar" convention, see p. 14.

Let $N \leq V \leq G$ such that $\overline{V} = V/N$. Then V is a complement of U in G since $U \cap V \leq U \cap N = 1$. □

The complement V in 2.1.2 can again by decomposed into a cyclic subgroup of maximal order and its complement. Hence, a repeated application of 2.1.2 gives:

2.1.3 Theorem. *Every Abelian group is the direct product of cyclic groups.* □

Thus, for every Abelian group G:

$$G \cong C_{n_1} \times \cdots \times C_{n_r} \quad \text{and} \quad |G| = n_1 \cdots n_r.^2$$

If m is a divisor of $|G|$, then there exist divisors m_i of n_i $(i = 1, \ldots, r)$ such that $m = m_1 \cdots m_r$. Hence $C_{m_1} \times \cdots \times C_{m_r}$ is isomorphic to a subgroup of order m of G. This implies:

2.1.4 *Let G be an Abelian group and m a divisor of $|G|$. Then G contains a subgroup of order m.* □

Let p be a prime. We set

$$G_p := \{x \in G \mid x \text{ is a } p\text{-element}\}.$$

2.1.5 *Let G be an Abelian group. Then G_p is a characteristic p-subgroup of order $|G|_p$.*[3]

Proof. For $x, y \in G_p$ also xy is a p-element; use $xy = yx$ and 1.1.2 on page 4. Thus G_p is a subgroup. Since automorphisms map p-elements to p-elements this subgroup is characteristic.

By 2.1.4 G contains a subgroup P of order $|G|_p$. Hence, P is a p-group, and thus every element of P is a p-element; in particular $P \leq G_p$.

If $P \neq G_p$, then

$$k := |G_p : P| \neq 1$$

and $(k, p) = 1$ (Lagrange's theorem). But now 2.1.4 gives a subgroup K of order k in G_p, which contradicts 1.1.8 on page 8 since every element of K is a p-element. □

---

[2] C_{n_i} is the cyclic group of order n_i; see **1.4**.
[3] For $n \in \mathbb{N}$ let n_p be the largest p-power dividing n.

2.1.6   Theorem. *Let G be an Abelian group. Then*

$$G = \underset{p\in\pi(G)}{\times} G_p.$$

Proof. By 1.6.5 on page 31 the product G_1 of the subgroups G_p, $p \in \pi(G)$, is a direct product; and 2.1.5 yields

$$|G_1| = \prod_{p\in\pi(G)} |G_p| = \prod_{p\in\pi(G)} |G|_p = |G|,$$

so $G_1 = G$.   □

In an Abelian group the product of two cyclic groups of coprime order is again cyclic (1.6.6 on page 31). Hence, the question whether an Abelian group is cyclic or not can already be decided in the subgroups G_p, $p \in \pi(G)$.

2.1.7   *For an Abelian group G the following statements are equivalent:*

(i)   *G is cyclic.*

(ii)   *For all $p \in \pi(G)$ there exists exactly one subgroup of order p in G.*

(iii)   *G_p is cyclic for all $p \in \pi(G)$.*

Proof. (i) $\Rightarrow$ (ii) follows from 1.4.3 on page 22 and (ii) $\Rightarrow$ (iii) from 2.1.3, both applied to G_p. Finally a repeated application of 1.6.6 on page 31 gives the implication (iii) $\Rightarrow$ (i).   □

Of course, in 2.1.3 more can be said about the factors of the decomposition. Because of the unique decomposition 2.1.6 it suffices to investigate Abelian p-groups.

An Abelian p-group is **elementary Abelian** if $x^p = 1$ for all $x \in G$.

2.1.8   *Let G be an elementary Abelian p-group of order $p^n > 1$.*

(a)   *G is the direct product of n cyclic groups of order p.*

(b)   If G is written additively, the scalar multiplication

$$\bar{k}x := \underbrace{x + \cdots + x}_{k\text{-times}}$$

for $\bar{k} := k + p\mathbb{Z} \in \mathbb{Z}/p\mathbb{Z}$ and $x \in G$, makes G into an n-dimensional vector space V over the prime field $\mathbb{Z}/p\mathbb{Z}$. The subgroups of G correspond to the subspaces of V and the automorphisms of G to the automorphisms of V.

Proof. (a) Since every nontrivial cyclic subgroup of G has order p, G is the direct product of such subgroups (2.1.3), and since $|G| = p^n$, n factors are required.

(b) There is nothing to prove. Clearly, the existence of a basis of V with n elements is equivalent to (a).   □

In a (not necessarily Abelian) p-group G, the group

$$\Omega_i(G) := \langle x \in G \mid x^{p^i} = 1 \rangle, \quad i = 0, 1, 2, \ldots$$

is a characteristic subgroup. Evidently

$$\Omega_{i-1}(G) \leq \Omega_i(G), \quad i = 1, 2, \ldots \quad .$$

We set

$$\Omega(G) := \Omega_1(G).$$

If G is Abelian, then

$$\Omega_i(G) = \{ x \in G \mid x^{p^i} = 1 \}$$

and

$$G \text{ elementary Abelian} \iff G = \Omega(G).$$

2.1.9   *Let G be an Abelian p-group such that*

(*) $G = A_1 \times \cdots \times A_n$

is the direct product of n cyclic groups $A_i \neq 1$. Then

$$|\Omega(G)| = p^n.$$

More precisely: If $n_i \in \mathbb{N}$ for $i = 1, 2, \ldots$ is defined by

$$|\Omega_i(G)/\Omega_{i-1}(G)| = p^{n_i},$$

then $n_i - n_{i+1}$ is the number of the factors of order p^i in (*).

Proof. From

$$\Omega_i(G) = \Omega_i(A_1) \times \cdots \times \Omega_i(A_n)$$

follows $|\Omega(G)| = p^n = p^{n_1}$. Since

$$\Omega_2(G)/\Omega(G) = \Omega(G/\Omega(G)) \overset{1.6.2(c)}{\cong} \Omega\Big(\underset{i}{\times}(A_i/\Omega(A_i))\Big) = \underset{i}{\times}\Omega(A_i/\Omega(A_i))$$
$$= \underset{i}{\times} \Omega_2(A_i)/\Omega(A_i),$$

n_2 is the number of factors in $(*)$ of order at least p^2. Thus, $n_1 - n_2$ is the number of factors of order p. In the same way one calculates $n_i - n_{i+1}$ for $i \geq 2$.   □

The minimal number of generators of a group G is the **rank** $r(G)$ of G. If G is an Abelian p-group, then $r(G) = n$, where n is as in 2.1.9.

The results 2.1.3, 2.1.6, and 2.1.9 allow a complete survey over all finite Abelian groups: Such a group is a direct product of cyclic groups of prime power order, and the isomorphism type is determined by the number and the order of these factors. For example, there are exactly 9 Abelian groups of order $1000 = 2^3 \cdot 5^3$, namely

$$C_2 \times C_2 \times C_2 \times C_5 \times C_5 \times C_5$$
$$C_2 \times C_2 \times C_2 \times C_5 \times C_{5^2}$$
$$C_2 \times C_2 \times C_2 \times C_{5^3}$$
$$C_2 \times C_{2^2} \times C_5 \times C_5 \times C_5$$
$$C_2 \times C_{2^2} \times C_5 \times C_{5^2}$$
$$C_2 \times C_{2^2} \times C_{5^3}$$
$$C_{2^3} \times C_5 \times C_5 \times C_5$$
$$C_{2^3} \times C_5 \times C_{5^2}$$
$$C_{2^3} \times C_{5^3}$$

Only the last of these groups is cyclic.

It should be mentioned that finitely generated Abelian groups have a structure similar to that of finite Abelian groups. They are direct products of finite Abelian groups and groups isomorphic to $\mathbb{Z}$ (e.g., see [19], p. 82).

Exercises

Let G be a finite Abelian group.

1. Let $e \in \mathbb{N}$ be minimal such that $a^e = 1$ for all $a \in G$ ($\exp G := e$ is the **exponent** of G). There exists an element $b \in G$ such that $o(b) = e$.

2. Let $\exp G = e$. Then G is cyclic, if and only if $|G| = e$.

3. Let p be a prime, $C = C_{p^3} \times C_{p^3}$, $B = C_p \times C_p \times C_p$, and $G = C \times B$. Then no subgroup of G has a complement isomorphic to C_{p^2} in G.

4. Every Abelian group of order 546 is cyclic.

5. Give an example of a non-Abelian group that satisfies the statement of 2.1.4.

6. Determine $\prod\limits_{g \in G} g$.

7. For every subgroup $U \leq G$ there exists an endomorphism φ of G such that $\operatorname{Im} \varphi = U$.

8. If $\operatorname{Aut} G$ is Abelian, then G is cyclic.

9. With the help of 6 show:

$$(p-1)! \equiv -1 \bmod p \quad (p \text{ prime}).^4$$

10. Let $a, p \in \mathbb{N}$, p a prime and $(a,p) = 1$. Then

$$a^{p-1} \equiv 1 \bmod p.^5$$

2.2 Automorphisms of Cyclic Groups

As examples of Abelian groups we determine in this section the automorphism groups of cyclic groups.

For an Abelian group G and every $k \in \mathbb{Z}$ the mapping

$$\alpha_k \colon \ G \to G \quad \text{such that} \quad x \mapsto x^k$$

is an endomorphism with

$$\operatorname{Ker} \alpha_k = \{\, x \in G \mid x^k = 1 \,\},$$

Thus, $\operatorname{Ker} \alpha_k$ contains all elements of G, whose orders divide k.

2.2.1 α_k *is an automorphism of the Abelian group G, if and only if* $(k, |G|) = 1$.

---

[4] Wilson's Theorem.
[5] Fermat's Little Theorem.

Proof. If $(k, |G|) = 1$, then Ker $\alpha_k = 1$ because of 1.1.8 on page 8. Conversely, if $(k, |G|) \neq 1$, then there exists a common prime divisor p of k and $|G|$. Now by 2.1.6 the p-subgroup G_p is nontrivial, and there exists a subgroup of order p in G. This subgroup is contained in Ker α_k.   □

Together with 1.4.3 on page 22 this gives for cyclic groups:

2.2.2   *The automorphisms of a cyclic group of order n are of the form α_k with $k \in \{1, \ldots, n-1\}$ and $(k, n) = 1$.*   □

From $\alpha_k \alpha_{k'} = \alpha_{k \cdot k'} = \alpha_{k' \cdot k} = \alpha_{k'} \alpha_k$ for $k, k' \in \mathbb{Z}$ we obtain:

2.2.3   *The automorphism group of a cyclic group is Abelian.*[6]   □

Because of the decomposition $G = \underset{p \in \pi(G)}{\times} G_p$ in 2.1.6 one has

$$\text{Aut } G \cong \underset{p \in \pi(G)}{\times} \text{Aut } G_p$$

(1.6.2 on page 29). Hence, it suffices to determine the automorphism group of cyclic p-groups.

If G is a cyclic p-group of order $p^e > 1$, then $|\text{Aut } G|$ is the number of integers k such that $1 \leq k < p^e$ and $(k, p) = 1$. Thus

$$|\text{Aut } G| = p^{e-1}(p-1).$$

In particular $|\text{Aut } G| = p - 1$ if $|G| = p$. In this case:

2.2.4   *The automorphism group of a group of order p is cyclic.*[7]

Proof.[8]   Let G be a (cyclic) group of prime order p. Then for $g \in G$ and $\alpha \in \text{Aut } G$

(1) $g^\alpha = g \iff g = 1$ or $\alpha = 1$.

---

[6]One can easily extend 2.2.2 and 2.2.3 to: The endomorphism ring of a cyclic group C_n is isomorphic to the ring $\mathbb{Z}/n\mathbb{Z}$.

[7]This also follows from the well-known result that the multiplicative group of a finite field is cyclic.

[8]The argument in this proof will be used again in 8.3.1 on p. 191 in a more general context.

We assume that $\operatorname{Aut} G$ is noncyclic and show that this leads to a contradiction. By 2.1.7 there exists $r \in \pi(\operatorname{Aut} G)$ and a subgroup $A \leq \operatorname{Aut} G$ such that

$$A \cong C_r \times C_r.$$

Let $\mathcal{B}$ be the set of all subgroups of order r of A. Then

(2) $\qquad |\mathcal{B}| = r + 1$ and $B_1 \cap B_2 = 1$ for $B_1 \neq B_2$ in $\mathcal{B}$.

For $1 \neq B \leq A$ and $g \in G^{\#}$ let

$$g_B := \prod_{\beta \in B} g^{\beta}.$$

Then

$$(g_B)^{\alpha} = \prod_{\beta \in B} g^{\beta \alpha} = g_B,$$

for $\alpha \in B^{\#}$ and thus $g_B = 1$ because of (1). Now (2) gives

$$1 = g_A = g^{-r} \prod_{B \in \mathcal{B}} g_B = g^{-r},$$

and $o(g) = r$. This implies $p = r$ (1.1.8). On the other hand by 2.2.2

$$r \text{ divides } |\operatorname{Aut} G| = p - 1,$$

a contradiction. $\qquad \square$

2.2.5 *Let G be a cyclic p-group of order $p^e > 1$ and $A := \operatorname{Aut} G$. Then*

$$A = S \times T,$$

where S is a group of order p^{e-1} and T is a cyclic group of order $p - 1$.

Proof. As we have already seen $|A| = p^{e-1}(p-1)$. Moreover, A is Abelian (2.2.3). The direct decomposition 2.1.6 gives

$$A = S \times T \text{ with } |S| = p^{e-1} \text{ and } |T| = p - 1.$$

Let H be the (characteristic) subgroup of order p in G (1.4.3 on page 22) and

$$\varphi \colon A \to \operatorname{Aut} H \text{ with } \alpha \mapsto \overline{\alpha} := \alpha|_H.$$

Then φ is an epimorphism since

$$\text{Aut}\,H = \{\overline{\alpha_k}\mid 1 \le k \le p-1\}.$$

Moreover, since $|\text{Aut}\,H| = p-1$ and $(|S|, p-1) = 1$ we get that $S \le \text{Ker}\,\varphi$ (1.1.8 on page 8). In fact $S = \text{Ker}\,\varphi$ since $|A| = |\text{Im}\,\varphi|\,|\text{Ker}\,\varphi|$. Now the Homomorphism Theorem gives

$$T \cong A/\text{Ker}\,\varphi \cong \text{Aut}\,H,$$

and T is cyclic by 2.2.4. □

2.2.6 *Let $G = \langle x \rangle$, $e \ge 2$, and A and S be as in 2.2.5.*

(a)   *The case $p \ne 2$ or $p = 2 = e$:*

$$S = \langle \alpha \rangle \quad \text{with} \quad x^\alpha = x^{1+p}.$$

In particular $\langle \alpha^{p^{e-2}} \rangle$ is the unique subgroup of order p in A, and for $\beta := \alpha^{p^{e-2}}$:

$$x^\beta = x^{1+p^{e-1}}.$$

(b)   *The case $p = 2 < e$:*

$$S = A = \langle \gamma \rangle \times \langle \delta \rangle \quad \text{with} \quad x^\gamma = x^{-1},\ x^\delta = x^5.$$

In particular γ, $\xi := \delta^{2^{e-3}}$, and $\eta := \gamma\xi$ are the only automorphisms of order 2, and

$$x^\xi = x^{1+2^{e-1}} \quad \text{and} \quad x^\eta = x^{2^{e-1}-1}.$$

Proof. (a) Since $(p, 1+p) = 1$ the mapping α is an automorphism of G (2.2.1). If $p = 2 = e$, then $x^\alpha = x^{1+p} = x^3 = x^{-1}$ is the only nontrivial automorphism of G. Hence, in the following we may assume that $p \ne 2$. The order of α is the smallest integer $m \in \mathbb{N}$ such that

$$(1+p)^m \equiv 1 \pmod{p^e}.$$

The binomial formula applied to $(1+p)^m$ shows that

$$(1+p)^{p^{e-1}} \equiv 1 \pmod{p^e}$$

and

$$(1 + p)^{p^{e-2}} \not\equiv 1 \pmod{p^e}$$

since $p \neq 2$. This gives $m = p^{e-1}$. Thus $\langle a \rangle$ has the same order as S, i.e., $S = \langle a \rangle$. The binomial formula also shows the statement for $\beta = a^{p^{e-2}}$.

(b) As in (a) the binomial formula applied to $(1 + 2^2)^{2^k}$, $k \in \mathbb{N}$, shows that

$$(1 + 2^2)^{2^{e-2}} \equiv 1 \pmod{2^e}$$

and

$$(1 + 2^2)^{2^{e-3}} \not\equiv 1 \pmod{2^e}.$$

This implies, much as in (a), that the automorphism δ defined by

$$x^\delta = x^5 = x^{1+2^2}$$

has order 2^{e-2}. From

$$(1 + 2)^k \not\equiv -1 \pmod{2^e} \quad (e \geq 3),$$

for all $k \in \mathbb{N}$, we finally conclude that no power of δ is equal to the automorphism γ defined by $x^\gamma = x^{-1}$. Hence $\langle \gamma \rangle$ and $\langle \delta \rangle$ generate a subgroup of order $2^{e-2} \cdot 2 = 2^{e-1}$ in $S(= A)$. This implies $A = \langle \gamma \rangle \times \langle \delta \rangle$. The equation $x^\xi = x^{1+2^{e-1}}$ follows from

$$(1 + 2^2)^{2^{e-3}} \equiv 1 + 2^{e-1} \pmod{2^e}.$$

Finally $x^\eta = x^{2^{e-1}-1}$ holds since

$$x^\eta = x^{\gamma\xi} = (x^{-1})^{1+2^{e-1}} = x^{-1-2^{e-1}} \quad \text{and} \quad x^{-2^{e-1}} = x^{2^{e-1}}. \qquad \square$$

It should be emphasized that in case 2.2.6 (b) the automorphism group A is not cyclic but contains a subgroup isomorphic to $Z_2 \times Z_2$.

Exercises

Let p be a prime and G a finite group.

1. Let $q \neq 1$ be a divisor of $p - 1$. Use a semidirect product to construct a non-Abelian group of order pq that contains a normal subgroup of order p. Also construct a non-Abelian group of order $p^{(e-1)e}$, $e \geq 2$, that contains a cyclic normal subgroup of order p^e.

2. Let p be the smallest prime divisor of $|G|$ and N be a normal subgroup of order p. Then $N \leq Z(G)$.

3. Let $p \neq 2$ and G a cyclic p-group. Then $\operatorname{Aut} G$ is cyclic.

4. With the idea used in the proof of 2.2.4 show: Let K be a field and U a finite subgroup of the multiplicative group of K. Then U is cyclic.

In the following let $G, \gamma, \eta, \varepsilon$ be as in 2.2.6 (b). Set

$$D := \langle \gamma \rangle \ltimes G, \quad H := \langle \eta \rangle \ltimes G,^9 \quad M := \langle \varepsilon \rangle \ltimes G.$$

5. D is a dihedral group.

6. All the involutions of M are contained in $\langle \epsilon, x^{2^{e-1}} \rangle$.

7. Let H_1 and H_2 be subgroups of H defined by

$$H_1 := \langle x^2, \eta \rangle \quad \text{and} \quad H_2 := \langle x^2, \eta x \rangle.$$

Then

(a) $H_1 \cap H_2 = \langle x^2 \rangle$ and $|H : H_i| = 2$, $i = 1, 2$.

(b) H_1 is a dihedral group and contains all of the involutions of H.

(c) H_2 contains exactly one involution.[10]

----

[9] H is a semidihedral group; see **5.3**.

[10] H_2 is called a (generalized) quaternion group; see **5.3**.

Chapter 3

Action and Conjugation

The notion of an *action* plays an important role in the theory of finite groups. The first section of this chapter introduces the basic ideas and results concerning group actions. In the other two sections the action on cosets is used to prove important theorems of Sylow, Schur-Zassenhaus and Gaschütz.

3.1 Action

Let $\Omega = \{\alpha, \beta, \ldots\}$ be a nonempty finite set. The set S_Ω of all permutations of Ω is a group with respect to the product

$$\alpha^{xy} := (\alpha^x)^y, \quad \alpha \in \Omega \quad \text{and} \quad x, y \in S_\Omega,$$

is the **symmetric group** on Ω. We denote by S_n the symmetric group on $\{1, \ldots, n\}$, which is the **symmetric group of degree** n. Evidently $S_n \cong S_\Omega$ if and only if $|\Omega| = n$.

A group G **acts** on Ω, if to every pair $(\alpha, g) \in \Omega \times G$ an element $\alpha^g \in \Omega$ is assigned[1] such that

$\mathcal{O}_1 \quad \alpha^1 = \alpha \quad$ for $1 = 1_G$ and all $\alpha \in \Omega$,

$\mathcal{O}_2 \quad (\alpha^x)^y = \alpha^{xy} \quad$ for all $x, y \in G$ and all $\alpha \in \Omega$.

---

[1] As in the definition of a group we are forming a *product*, but we write α^g instead of $\alpha\, g$.

The mapping

$$g^\pi : \Omega \to \Omega \quad \text{with} \quad \alpha \mapsto \alpha^g$$

describes the action of $g \in G$ on Ω. Because of

$$(\alpha^g)^{g^{-1}} \stackrel{\mathcal{O}_2}{=} \alpha^{gg^{-1}} = \alpha^1 \stackrel{\mathcal{O}_1}{=} \alpha,$$

$(g^{-1})^\pi$ is the inverse of g^π. In particular g^π is a bijection and thus a permutation on Ω. Now $\mathcal{O}_2$ implies that

$$\pi : G \to S_\Omega \quad \text{with} \quad g \mapsto g^\pi$$

is a homomorphism. The homomorphism theorem shows that $G/\operatorname{Ker} \pi$ is isomorphic to a subgroup of S_Ω and thus also to one of S_n, $n := |\Omega|$.

Conversely, every homomorphism $\pi : G \to S_\Omega$ gives rise to an action of G on Ω, if one defines $\alpha^g := \alpha^{g^\pi}$. A homomorphism $\pi : G \to S_\Omega$ is said to be an **action** of G on Ω.

If $\operatorname{Ker} \pi = 1$, then G acts **faithfully** on Ω; and if $\operatorname{Ker} \pi = G$, then G acts **trivially** on Ω.

Every action π of G on Ω gives rise to a faithful action of $G/\operatorname{Ker} \varphi$ on Ω, if we set

$$\alpha^{(\operatorname{Ker} \varphi)g} := \alpha^g.$$

Next we introduce some important actions, which we will frequently meet in the following chapters.

3.1.1 *The group G acts on*

(a)   *the set of all nonempty subsets A of G by conjugation:*

$$A \stackrel{x}{\mapsto} x^{-1} A x = A^x,$$

(b)   *the set of all elements g of G by conjugation:*

$$g \stackrel{x}{\mapsto} x^{-1} g x = g^x,$$

(c)   *the set of right cosets Ug of a fixed subgroup U of G by right multi-
      plication:*

$$Ug \stackrel{x}{\mapsto} Ugx.$$

Proof. In all cases $1 = 1_G$ acts trivially; this is $\mathcal{O}_1$. Associativity gives $\mathcal{O}_2$.

$\square$

In (a) and (b) the permutation x^π is the inner automorphism induced by x (see 1.3 on page 15).

Also *left* multiplication on the set Ω of all left cosets of a fixed subgroup U leads to an action $\pi\colon G \to S_\Omega$. But here one has to define

$$x^\pi \colon G \to S_\Omega \quad \text{with} \quad gU \mapsto x^{-1}gU$$

since $gU \mapsto xgU$ is not a homomorphism (but an *anti*-homomorphism).[2] Using (c) we obtain:

3.1.2 *Let U be a subgroup of index n of the group G. Then G/U_G is isomorphic to a subgroup of S_n.*[3]

Proof. As in 3.1.1 (c) let Ω be the set of all right cosets of U in G and $\pi\colon G \to S_\Omega$ the action by right multiplication. Then for $x, g \in G$

$$Ugx = Ug \iff gxg^{-1} \in U \iff x \in U^g,$$

and thus

$$x^\pi = 1_{S_\Omega} \iff x \in U_G,$$

i.e., $U_G = \operatorname{Ker} \pi$.

$\square$

In order to work with the actions given in 3.1.1 we first set some notation and collect some elementary properties of actions which follow more or less directly from the definition.

In the following, G is a group that acts on the set Ω. For $\alpha \in \Omega$

$$G_\alpha := \{x \in G \mid \alpha^x = \alpha\}.$$

The set G_α is the **stabilizer** of α in G; and $x \in G$ **stabilizes (fixes)** α if $x \in G_\alpha$.

Notice that G_α is a subgroup of G because of $\mathcal{O}_2$.

---

[2] We will use the action by left multiplication only in 3.3.

[3] $U_G = \bigcap_{g \in G} U^g$

3.1.3 $G_\alpha{}^g = G_{\alpha^g}$ *for* $g \in G$, $\alpha \in \Omega$.

Proof. $(\alpha^g)^x = \alpha^g \iff \alpha^{gxg^{-1}} = \alpha \iff gxg^{-1} \in G_\alpha \iff x \in (G_\alpha)^g$. $\square$

Two elements $\alpha, \beta \in \Omega$ are said to be **equivalent**, if there exists $x \in G$ such that $\alpha^x = \beta$. Then $\mathcal{O}_1$ and $\mathcal{O}_2$ show that this notion of equivalence does indeed define an equivalence relation on Ω. The corresponding equivalence classes are called the **orbits** of G (or G-orbits) on Ω. For $\alpha \in \Omega$

$$\alpha^G := \{\alpha^x \mid x \in G\}$$

is the orbit that contains α. G acts **transitively** on Ω, if Ω itself is an orbit of G, i.e., for all $\alpha, \beta \in \Omega$ there exists $x \in G$ such that $\alpha^x = \beta$.

3.1.4   Frattini Argument. *Suppose that G contains a normal subgroup, which acts transitively on Ω.*[4] *Then $G = G_\alpha N$ for every $\alpha \in \Omega$. In particular, G_α is a complement of N in G if $N_\alpha = 1$.*

Proof. Let $\alpha \in \Omega$ and $y \in G$. The transitivity of N on Ω gives an element $x \in N$ such that $\alpha^y = \alpha^x$. Hence $\alpha^{yx^{-1}} = \alpha$ and thus $yx^{-1} \in G_\alpha$. This shows that $y \in G_\alpha x \subseteq G_\alpha N$. $\square$

The following elementary result is similar to Lagrange's theorem:

3.1.5 $|\alpha^G| = |G : G_\alpha|$ *for* $\alpha \in \Omega$. *In particular, the* **length** $|\alpha^G|$ *of the orbit α^G is a divisor of $|G|$.*

Proof. For $y, x \in G$

$$\alpha^y = \alpha^x \iff \alpha^{yx^{-1}} = \alpha \iff yx^{-1} \in G_\alpha \iff y \in G_\alpha x. \qquad \square$$

Since Ω is the disjoint union of orbits of G we obtain:

3.1.6   *If n is an integer that divides $|G : G_\alpha|$ for all $\alpha \in \Omega$, then n also divides $|\Omega|$.* $\square$

---

[4]Of course, here we mean the action of N as a subgroup of G.

For $U \subseteq G$

$$C_\Omega(U) := \{\alpha \in \Omega \mid U \subseteq G_\alpha\}$$

is the set of **fixed points** of U in Ω. Obviously, $\Omega \setminus C_\Omega(G)$ is the union of all G-orbits of length > 1.

3.1.7 *Let G be a p-group. Then*

$$|\Omega| \equiv |C_\Omega(G)| \pmod p.$$

Proof. For $\alpha \in \Omega' := \Omega \setminus C_\Omega(G)$ the stabilizer G_α is a proper subgroup of G. Hence, p is a divisor of $|G : G_\alpha|$ (Lagrange's Theorem), and 3.1.6 implies

$$|\Omega'| \equiv 0 \pmod p.$$ □

We now apply 3.1.3 and 3.1.5 using the actions given in 3.1.1.

Let Ω be the set of all nonempty subsets of G and $H \leq G$. Then H acts by conjugation on Ω. For $A \in \Omega$ the set consisting of the subsets

$$A^x = x^{-1}Ax \quad (x \in H)$$

is an orbit of H. The stabilizer

$$N_H(A) := \{x \in H \mid A^x = A\}$$

of A in H is the **normalizer** of A in H.

By 3.1.5 $|H : N_H(A)|$ is the number of H-conjugates of A.

Let $B \in \Omega$. Then B **normalizes** A if $B \subseteq N_G(A)$.

By 3.1.1 (b) H acts by conjugation on the elements of G. For this action the stabilizer

$$C_H(g) := \{x \in H \mid g^x = g\}$$

of $g \in G$ is the **centralizer** of g in H. It is evident that this subgroup consists of those elements $x \in H$ that satisfy $xg = gx$.

Because of 3.1.5 $|H : C_H(g)|$ is the number of H-conjugates of g.

For a nonempty subset A of G

$$C_H(A) := \bigcap_{g \in A} C_H(g)$$

is the **centralizer** of A in H. Thus, $C_H(A)$ contains exactly those elements of H that commute with every element of A. For example, $C_G(A) = G$ if and only if A is a subset of $Z(G)$. A subset $B \subseteq G$ **centralizes** A, if $B \subseteq C_G(A)$ (or equivalently $[A, B] = 1$; see page 25).

3.1.3 implies for $x \in G$

$$N_G(A)^x = N_G(A^x) \quad \text{and} \quad C_G(A)^x = C_G(A^x);$$

and more generally

$$N_H(A)^x = N_{H^x}(A^x) \quad \text{and} \quad C_H(A)^x = C_{H^x}(A^x).$$

In the case $H = G$ the G-orbit g^G of the elements conjugate to g is the **conjugacy class** of g in G, and

$$|g^G| = |G : C_G(g)|.$$

The center $Z(G)$ contains exactly those elements of G whose conjugacy class has length 1, i.e., those elements that are only conjugate to themselves.

G is the disjoint union of its conjugacy classes since these classes are the G-orbits with respect to the action by conjugation. This gives:

3.1.8   Class Equation. *Let $K_1, \ldots, K_h$ be the conjugacy classes of G that have length larger than 1, and let $a_i \in K_i$ for $i = 1, \ldots, h$. Then*

$$|G| = |Z(G)| + \sum_{i=1}^{h} |G : C_G(a_i)|. \qquad \qquad \Box$$

We note:

3.1.9   *Let U be a subgroup of G. Then $N_G(U)$ is the largest subgroup of G in which U is normal. The mapping*

$$\varphi \colon N_G(U) \to \operatorname{Aut} U \quad \text{with} \quad x \mapsto (u \mapsto u^x)$$

is a homomorphism with $\operatorname{Ker} \varphi = C_G(U)$. In particular, $N_G(U)/C_G(U)$ is isomorphic[5] to a subgroup of $\operatorname{Aut} U$. $\qquad \Box$

We close this section with two fundamental properties of p-groups and p-subgroups, which follow from 3.1.7.

---

[5]Homomorphism Theorem 1.2.5 on p. 13.

3.1.10   *Let P be a p-subgroup of G and p be a divisor of $|G : P|$. Then $P < N_G(P)$.*

Proof. By 3.1.1 (c) P acts on the set Ω of right cosets Pg, $g \in G$, by right multiplication, and
$$|\Omega| = |G : P| \equiv 0 \pmod{p}.$$
From 3.1.7 we get (with P in place of G):
$$|C_\Omega(P)| \equiv |\Omega| \equiv 0 \pmod{p}.$$
Moreover $C_\Omega(P) \neq \varnothing$ since $P \in C_\Omega(P)$. Hence there exists $Pg \in C_\Omega(P)$ such that $P \neq Pg$. This implies $g \notin P$ and $PgP = Pg$. Thus $gPg^{-1} = P$ and $g \in N_G(P) \setminus P$. $\qquad\square$

3.1.11   *Let P be a p-group and $N \neq 1$ a normal subgroup of P. Then $Z(P) \cap N \neq 1$. In particular $Z(P) \neq 1$.*

Proof. P acts on $\Omega := N$ by conjugation, and
$$C_\Omega(P) = Z(P) \cap N.$$
Since N is a p-group we get from 3.1.7
$$|C_\Omega(P)| \equiv |\Omega| \equiv 0 \pmod{p}.$$
Now $1 \in C_\Omega(P)$ gives $|C_\Omega(P)| \geq p$. $\qquad\square$

Exercises

Let G be a group.

1.   Let G be the semidirect product of a subgroup K with the normal subgroup N, and let $\Omega := N$. Then
$$\omega^{kn} := \omega^k n \quad (\omega \in \Omega, \; k \in K, \; n \in N)$$
defines an action of G on Ω.

2. If G acts transitively on Ω, then $N_G(G_\alpha)$ acts transitively on $C_\Omega(G_\alpha)$ ($\alpha \in \Omega$).

3. Let p be the smallest prime divisor of $|G|$. Every subgroup of index p is normal in G.

4. Let $U \leq G$ and $1 \neq |G : U| \leq 4$. Then $|G| \leq 3$, or G is not simple.

5. Suppose that the class equation of G is

$$60 = 1 + 15 + 20 + 12 + 12.$$

Then G is simple.

6. Suppose that G acts faithfully on the set Ω. Let A be a subgroup of G that is transitive on Ω. Then $|C_G(A)|$ is a divisor of $|\Omega|$. If in addition A is Abelian, then $C_G(A) = A$.

7. Let $\varnothing \neq A \subseteq G$. Then $A \subseteq C_G(C_G(A))$ and $C_G(C_G(C_G(A))) \leq C_G(A)$.

8. Let A be a normal subgroup of G and $U \leq C_G(A)$. Then $[U, G] \leq C_G(A)$.

9. (a) $\quad A \trianglelefteq G,\ U \leq G \quad \Rightarrow \quad C_U(A) \trianglelefteq U$;

    (b) $\quad A\,\mathrm{char}\,G \quad \Rightarrow \quad C_G(A)\,\mathrm{char}\,G$ and $N_G(A)\,\mathrm{char}\,G$.

10. Let K be a field and V a vector space of dimension $|G|$ over K. Then G is isomorphic to a subgroup of $\mathrm{GL}(V)$.[6]

3.2   Sylow's Theorem

In this section Sylow's Theorem is proved. For every prime power divisor p^i of $|G|$ this theorem establishes the existence of a subgroup of order p^i in G.[7] This serves as the basis for a method, which turned out to be extremely successful in the theory of finite groups: The analysis of finite groups by means of the normalizers of nontrivial p-subgroups.

First a classical theorem from the first half of the nineteenth century.

3.2.1   Cauchy's Theorem.[8] *Let G be a group and p a prime dividing $|G|$. Then G contains an element of order p; in particular there exists a subgroup of order p in G.*

---

[6]$\mathrm{GL}(V)$ is the group of bijective linear mappings of V; see 8.6.

[7]If G is Abelian, then G possesses a subgroup of order n for *every* divisor n of $|G|$, and this is also true for *nilpotent* groups; see 2.1.4 on page 45 and 5.1 on page 99 .

[8]Compare with [37], p. 291.

Proof.[9]   Let

$$\Omega := \{\, \underline{x} := (x_1, \ldots, x_p) \mid x_1, \ldots, x_p \in G \quad \text{and} \quad x_1 x_2 \cdots x_p = 1 \,\}.$$

Since the components $x_1, \ldots, x_{p-1}$ of $\underline{x} \in \Omega$ can be chosen independently (after which x_p uniquely determined) we get

$$|\Omega| = |G|^{p-1} \equiv 0 \pmod{p}.$$

Notice that

$$x_1 x_2 \cdots x_p = 1 \quad \Leftrightarrow \quad x_2 \cdots x_p = x_1^{-1} \quad \Leftrightarrow \quad x_2 \cdots x_p x_1 = 1,$$

so the cyclic group $C_p = \langle a \rangle$ acts on Ω by

$$(x_1, x_2, \ldots, x_p) \overset{a}{\mapsto} (x_2, \ldots, x_p, x_1).$$

Hence 3.1.7 implies

$$|C_\Omega(\langle a \rangle)| \equiv |\Omega| \equiv 0 \pmod{p}.$$

Since $\underline{1} = (1, \ldots, 1) \in C_\Omega(\langle a \rangle)$ there exists $\underline{x} = (x_1, \ldots, x_p) \neq \underline{1} \in C_\Omega(\langle a \rangle)$. This shows that $x_1 = \ldots = x_p \neq 1$ and $x_1{}^p = 1$. $\square$

In the following let p be a prime and G a group. A p-subgroup P is called a **Sylow p-subgroup** of G if no p-subgroup of G contains P properly. Thus, the Sylow p-subgroups of G are the maximal elements of the set of p-subgroups of G (ordered by inclusion). We denote the set of Sylow p-subgroups of G by $\mathrm{Syl}_p G$.

For example, $\mathrm{Syl}_p G = \{1\}$ if p is not a divisor of G (Lagrange's Theorem); and $\mathrm{Syl}_p G = \{G\}$ if G is a p-group. Since automorphisms of G map Sylow p-subgroups to Sylow p-subgroups the subgroup

$$O_p(G) := \bigcap_{P \in \mathrm{Syl}_p G} P$$

is a characteristic p-subgroup of G. More precisely:

3.2.2   *Let N be a normal p-subgroup of G.[10] Then $N \leq O_p(G)$.*

---

[9]Following J.H. McKay.
[10]That is, a normal subgroup that is a p-group.

Proof. Let $P \in \mathrm{Syl}_p G$. Then NP is a p-subgroup (1.1.6). The maximality of P and $P \le PN$ gives $N \le P$.   □

In particular, if a Sylow p-subgroup P of G is normal in G, then $\mathrm{Syl}_p G = \{P\}$ and P is the set of all p-elements of G. Such a group is said to be p-**closed**. Moreover, if G is p-closed and xP, $x \in G$, is a p-element of G/P, then $\langle x \rangle P$ is a p-group and thus $x \in P$. Now Cauchy's Theorem shows that p does not divide $|G/P|$.

If P is a p-subgroup of G and p does not divide $|G : P|$, then Lagrange's Theorem shows that $|P|$ the largest p-power dividing $|G|$, and $P \in \mathrm{Syl}_p G$. In general:

3.2.3   Sylow's Theorem [89]. *Let p^e be the largest p-power dividing the order of G.*

(a)   *The Sylow p-subgroups of G are exactly the subgroups of order p^e.*

(b)   *The Sylow p-subgroups of G are conjugate in G. In particular*

$$|\mathrm{Syl}_p G| = |G : N_G(P)| \text{ for } P \in \mathrm{Syl}_p G.$$

(c) $|\mathrm{Syl}_p G| \equiv 1 \pmod p$.

Proof.[11] Let P be a Sylow p-subgroup of G. Then P is also a Sylow p-subgroup of

$$U := N_G(P).$$

Hence as mentioned above

(1) U is p-closed   and $|U : P| \not\equiv 0 \pmod p$,

We claim

(2) $|G : U| \equiv 1 \pmod p$.

To prove this we investigate the action of P on the set Ω of all cosets Ug, $g \in G$. Then $|\Omega| = |G : U|$ and by 3.1.7

$$|C_\Omega(P)| \equiv |\Omega| \pmod p.$$

---

[11]Other proofs can be found in [97], [40], and [41], for example.

U is in $C_\Omega(P)$ since $P \leq U$. Let $Ug \in C_\Omega(P)$. Then $UgP = Ug$, and this implies $gPg^{-1} \leq U$. As U is p-closed, we get $gPg^{-1} = P$ and thus $g \in N_G(P) = U$. It follows that $C_\Omega(P) = \{U\}$. This shows (2).

Let S be another Sylow p-subgroup. Then also S acts on Ω by right multiplication. Thus (2) and 3.1.7 give a coset Ug such that $UgS = Ug$. It follows that $gSg^{-1} \leq U$ and thus $gSg^{-1} = P$ by (1). This implies (b). Now (c) follows from (2) and

$$|\operatorname{Syl}_p G| = |G : N_G(P)| = |G : U|.$$

In addition (a) holds since

$$|G| = |P||U : P||G : U|,$$

where the second and third factors are not divisible by p ((1) and (2)). $\square$

We note some consequences.

3.2.4 *Let p^i be a divisor of the order of G. Then G possesses a subgroup of order p^i.*

Proof. We proceed by induction on $|G|$. Because of Sylow's Theorem we may assume that G is a nontrivial p-group. By 3.1.11 $Z(G) \neq 1$. Let N be a subgroup of order p in $Z(G)$. Then induction, applied to G/N, gives a subgroup U/N, $N \leq U \leq G$, such that $|U/N| = p^{i-1}$. Hence $|U| = p^i$. $\square$

3.2.5 *Let N be a normal subgroup of G and $P \in \operatorname{Syl}_p G$. Then*

$$PN/N \in \operatorname{Syl}_p G/N \quad \text{and} \quad P \cap N \in \operatorname{Syl}_p N.$$

Proof. Both claims follow from 3.2.3 (a): In the chain of subgroups

$$1 \leq P \cap N \leq N \leq PN \leq G$$

we get $|P \cap N||PN/N| = |P|$ since $PN/N \cong P/P \cap N$ (1.2.6 on page 13). Hence $|N : P \cap N|$ and $|G : PN|$ are not divisible by p. $\square$

3.2.6   *Let U be a p-subgroup but not a Sylow p-subgroup of G. Then $U < R$ for every Sylow p-subgroup R of $N_G(U)$.*

Proof. If $P \in \mathrm{Syl}_p\, G$ with $U < P$, then $U < N_P(U)$ by 3.1.10. Hence U is not a maximal p-subgroup of $N_G(U)$. On the other hand, U is contained in every Sylow p-subgroup of $N_G(U)$ since $U \trianglelefteq N_G(U)$ (3.2.2).   □

A variant of 3.1.4 yields an important factorization:

3.2.7   **Frattini Argument.** *Let N be a normal subgroup of G and $P \in \mathrm{Syl}_p\, N$. Then $G = N_G(P)N$.*

Proof. G acts on the set $\Omega = \mathrm{Syl}_p\, N$ by conjugation, and the stabilizer of P is $N_G(P)$. Moreover, by Sylow's Theorem N is transitive on Ω. Hence, the claim follows from 3.1.4.   □

The following result is an application of the Frattini argument:

3.2.8   *Let N be a normal subgroup of G with factor group[12] $\overline{G} := G/N$, and let P be a p-subgroup of G. Assume that $(|N|, p) = 1$. Then*

$$N_{\overline{G}}(\overline{P}) = \overline{N_G(P)} \quad \text{and} \quad C_{\overline{G}}(\overline{P}) = \overline{C_G(P)}.$$

Proof. The definition of a factor group gives

$$N_{\overline{G}}(\overline{P}) = \overline{N_G(NP)}.$$

P is a Sylow p-subgroup of NP since $(|N|, p) = 1$, and NP is a normal subgroup of $N_G(NP)$. Hence, the Frattini argument gives

$$N_G(NP) = NP\, N_{N_G(NP)}(P) = NP\, N_G(P) = N\, N_G(P)$$

and thus the claim $N_{\overline{G}}(\overline{P}) = \overline{N_G(P)}$.

It is evident that $\overline{C_G(P)} \leq C_{\overline{G}}(\overline{P})$. Let $\overline{c} \in C_{\overline{G}}(\overline{P})$. Since $C_{\overline{G}}(\overline{P}) \leq N_{\overline{G}}(\overline{P})$ there exists $n \in N$ and $y \in N_G(P)$ such that $c = ny$. Hence $\overline{c} = \overline{y}$ and

$$Nx = (Nx)^y = Nx^y \quad \text{for all } x \in P.$$

---

[12]Bar convention, see p. 14.

The commutator $y^{-1}xyx^{-1}$ is in $N \cap P = 1$. It follows that $y \in C_G(P)$ and thus $C_{\overline{G}}(\overline{P}) = \overline{C_G(P)}$. □

This yields a result that we will need later in Chapter 11:

3.2.9 Let $G = NH$ be a factorization of G with $H \leq G$, $N \trianglelefteq G$ and $(p, |N|) = 1$. Then for every p-subgroup P of H

$$N_G(P) = (N \cap N_G(P))(H \cap N_G(P)).$$

Proof. Let $\overline{G} := G/N$ and $N_1 := N \cap H$. By 3.2.8

$$N_{H/N_1}(PN_1/N_1) = N_H(P)N_1/N_1.$$

The isomorphism (1.2.6 on page 13)

$$H/N_1 \cong HN/N \quad (= \overline{G})$$

shows that

$$\overline{N_G(P)} = N_{\overline{G}}(\overline{P}) = \overline{N_H(P)}.$$

This implies

$$N_H(P) \leq N_G(P) \leq N N_H(P),$$

and the claim follows from the Dedekind identity 1.1.11. □

The alternating group A_5 is a simple group of order

$$60 = 2^2 \cdot 3 \cdot 5$$

(for the definition, order and simplicity see Section 4.3). On the other hand, using Sylow's Theorem one can show that there are no non-Abelian simple groups of order less than < 60.

We will now use Sylow's Theorem—in particular 3.2.3 (c)—to determine the structure of a group of order 60, which is not 5-closed. First two remarks:

3.2.10 Let G be not 3-closed and $|G| = 12$. Then G is 2-closed.

Proof. Let $S \in \mathrm{Syl}_3 G$. Then

$$n := |\mathrm{Syl}_3 G| \overset{3.2.3(\mathrm{b})}{=} |G : N_G(S)|$$

is a divisor of $\frac{|G|}{|S|} = 4$. Now 3.2.3 (c) implies $n = 4$. Since different Sylow 3-subgroups ($\cong C_3$) of G intersect trivially there are exactly

$$4 \cdot (3 - 1) = 8$$

elements of order 3 in G. Hence, the number of 2-elements in G is at most 4, and G contains a unique Sylow 2-subgroup. $\square$

3.2.11   *Let $|G| \in \{5, 10, 15, 20, 30\}$. Then G is 5-closed.*

Proof. We show $|\mathrm{Syl}_5 G| = 1$. For $|G| \neq 30$ this follows directly from 3.2.3 (b), (c). In the case $|G| = 30$ we show that the assumption $n := |\mathrm{Syl}_5 G| > 1$ leads to a contradiction: Again 3.2.3 (c) gives $n = |\mathrm{Syl}_5 G| = 6$. As different subgroups of order 5 intersect trivially there are $6 \cdot 4 = 24$ elements of order 5 in G. Let t be an involution of G (3.2.1) and $S \in \mathrm{Syl}_5 G$. Then $|t^S| = 5$ since $N_G(S) = S$. Hence there is no element of order 3 in G, which contradicts 3.2.1. $\square$

3.2.12   *Let G be a group of order 60, which is not 5-closed.*

(a)   *G is simple.*

(b)   *Let $\mathcal{M}$ be the set of maximal subgroups of of G. Then*

$$\mathcal{M} = \{N_G(G_p) \mid G_p \in \mathrm{Syl}_p G, \quad p \in \{2, 3, 5\}\},$$

and

$$|N_G(G_p)| = \begin{cases} 12, & p = 2 \\ 6, & \text{if} \quad p = 3 \\ 10, & p = 5 \end{cases}.$$

Proof. In the following G_p, $p \in \pi(G)$, always denotes a Sylow p-subgroup of G. By our hypothesis $|G : N_G(G_5)| \neq 1$. Hence 3.2.3 implies

$$(1) \qquad\qquad |N_G(G_5)| = 10.$$

(a) Assume that G is not simple. Then G contains a a proper nontrivial normal subgroup N. If $5 \in \pi(N)$, then N contains a Sylow 5-subgroup G_5 of G, which by 3.2.11 is normal in N. But then G_5 is characteristic in N and thus normal in G (1.3.2). This contradicts the hypothesis. Hence $5 \notin \pi(N)$ and thus $5 \in \pi(G/N)$. By 3.2.11

$$1 \neq G_5 N/N \trianglelefteq G/N,$$

and $NG_5 \trianglelefteq G$. As seen above, $NG_5 = G$ since $5 \in \pi(NG_5)$. Thus every proper nontrivial normal subgroup N of G has order 12. But then 3.2.10 shows that N contains a normal and thus characteristic Sylow subgroup, which has to be normal in G but is not of order 12. This final contradiction shows that G is simple.

(b) By (a) $|G : N_G(G_p)| \neq 1$ for $p \in \{2, 3\}$, and 3.2.3 and (1) imply

$$(2) \qquad\qquad |N_G(G_3)| = 6$$

and $|N_G(G_2)| \in \{4, 12\}$. Together with (2) and 3.2.10 we obtain:

(3)   Every subgroup of order 12 in G is 2-closed.

Next we show:

$$(4) \qquad\qquad |N_G(G_2)| = 12.$$

More precisely, we show that $|N_G(G_2)| = 4$ leads to a contradiction: Then $|\mathrm{Syl}_2 G| = 15$, and G_2 is Abelian since it has order 4. Let $S_1, S_2 \in \mathrm{Syl}_2 G$ such that

$$1 \neq S_1 \cap S_2 < S_1.$$

Then

$$\langle S_1, S_2 \rangle \leq N_G(S_1 \cap S_2) =: L$$

and $L \neq G$ by (a). If $5 \in \pi(L)$, then L is 5-closed (3.2.11) and $L \leq N_G(G_5)$, which contradicts (1). It follows that $|L| = 12$. Since $|\mathrm{Syl}_2 L| \geq 2$ this contradicts (3).

Hence $S_1 \cap S_2 = 1$ for any two different $S_1, S_2 \in \mathrm{Syl}_2\, G$. It follows that there are $3 \cdot 15 = 45$ 2-elements in G. On the other hand, by (1) there are also $4 \cdot 6 = 24$ elements of order 5 in G. This contradiction proves (4).

Let M be a maximal subgroup of G. By (1), (2), and (4) M is not a Sylow subgroup of G. If $5 \in \pi(M)$, then M is 5-closed by 3.2.10. Hence $M \leq N_G(G_5)$ for $G_5 \leq M$ and thus $M = N_G(G_5)$.

We now may assume that $5 \notin \pi(M)$ and $|M| \in \{6, 12\}$. If $|M| = 6$, then M is 3-closed and $M = N_G(G_3)$ for $G_3 \leq M$. If $|M| = 12$, then by (3) M is 2-closed and $M = N_G(G_2)$ for $G_2 \leq M$. □

Let G be as in 3.2.12 and $U := N_G(G_2)$. Then $|G : U| = 5$, and 3.1.2 yields a monomorphism from G into the symmetric group S_5. Hence G is isomorphic to a subgroup A of index 2 in S_5. Let A_5 be the alternating group of degree 5 (see Section 4.3). Then also $|S_5 : A_5| = 2$, so A and A_5 or both normal subgroups of order 60 in S_5. In particular by the Homomorphism Theorem $|A : A \cap A_5| \leq 2$, and the simplicity of A yields $A = A_5$. This shows that every group of order 60 that is not 5-closed is isomorphic to A_5.

Exercises

Let G be a group, p a prime and $S \in \mathrm{Syl}_p\, G$.

1. Let $N_G(S) \leq U \leq G$. Then $|G : U| \equiv 1 \bmod p$.

2. $|\{g \in G \mid g^p = 1\}| \equiv 0 \bmod p$ for all prime divisors p of $|G|$.

3. Let $|G| = 168$. How many elements of order 7 are in G?

4. Every group of order 15 is cyclic.

5. Let p, q, r be different primes.

    (a) A group of order pq contains a normal Sylow p-subgroup, if $p > q$.

    (b) A group of order pqr contains at least one nontrivial normal Sylow subgroup.

6. Let $H \leq G$ and $|G : H| = p^n$. Then the following hold:

    (a) $O_p(H) \leq O_p(G)$.

    (b) If $H \cap H^x = 1$ for all $x \in G \setminus H$, then G is p-closed.

7. Every non-Abelian group of order less than 60 is not simple.

8. Every simple group of order 168 contains a subgroup of index 7.

9. Let $S \cap S^g = 1$ for all $g \in G \setminus N_G(S)$. Then $|\operatorname{Syl}_p G| \equiv 1 \bmod |S|$.

10. Let $S \neq 1$ and $|G : S| = p + 1$. Then $O_p(G) \neq 1$, or $p + 1 = q^r$, $q \in \mathbb{P}$, and there exists an elementary Abelian normal subgroup of order $p + 1$ in G.

11. (Brodkey, [33]) Let S be Abelian and $O_p(G) = 1$. Then there exists $g \in G$ such that $S \cap S^g = 1$.

12. Suppose that $G \neq 1$ and $|G : M| \in \mathbb{P}$ for every maximal subgroup M of G. Then G contains a normal maximal subgroup or $G = 1$.

3.3 Complements of Normal Subgroups

To find complements of normal subgroups is one of the basic problems of group theory. In general such complements do not exist. For example, the center of a quaternion group Q does not have a complement since there is only one involution in Q (see Exercise 8 on page 20); also, proper subgroups of cyclic groups do not have complements.

This leaves the problem of finding suitable conditions that allow us to establish the existence of such complements. For example, let G be a group and K a normal subgroup of G such that G/K is a p-group and p does not divide $|K|$. Then Sylow's Theorem shows that the Sylow p-subgroups of G are the complements of K; in particular, all complements of K are conjugate in G.

In this section we use a method of Wielandt to prove similar results in a more general situation.[13]

In the following let K be an *Abelian* subgroup of the group G and $\mathcal{S}$ the set of all transversals of K in G (1.1.9 on page 8).

For $R, S \in \mathcal{S}$ define

$$R|S := \prod_{\substack{(r,s)\in R\times S \\ Kr=Ks}} (rs^{-1}) \quad (\in K);$$

note that

$$Kr = Ks \iff rs^{-1} \in K.$$

The ordering of the factors in this product is unimportant since K is Abelian.

---

[13]The idea for the proof of Gaschütz's Theorem was communicated to us by G. Glauberman.

For $R, S, T \in \mathcal{S}$ we get

(1) $$(R|S)^{-1} = S|R,$$

(2) $$(R|S)\,(S|T) = R|T.$$

We now assume in addition that K is a *normal subgroup* of G. Then every $S \in \mathcal{S}$ is also a left transversal of K in G, i.e.,

$$G = \overset{\cdot}{\underset{s \in S}{\bigcup}}\, sK.$$

G acts by left multiplication on $\mathcal{S}$ (see page 57):

$$(S, x) \mapsto xS \quad (x \in G,\ S \in \mathcal{S}).$$

In particular

(3) $$kR|S = k^{|G:K|}\,(R|S) \quad \text{for} \quad k \in K.$$

Now

$$xR \,|\, xS = \prod_{\substack{Kxr = Kxs \\ (r,s) \in R \times S}} x(rs^{-1})x^{-1} = x(R|S)x^{-1}$$

implies

(4) $$R|S = 1 \quad \Rightarrow \quad xR \,|\, xS = 1.$$

We now assume in addition that $|K|$ and $|G/K|$ are coprime. Then the mapping

$$\alpha \colon K \to K \quad \text{with} \quad k \mapsto k^{|G/K|}$$

is an automorphism of K since K is Abelian (see 2.2.1 on page 49). Thus (3) implies

(5) $$kR|S = 1 \quad \text{for} \quad k := (R|S)^{-\alpha^{-1}}$$

(i.e. $k^{|G/K|} = (R|S)^{-1}$), and

(6) $$R|S = 1 = kR|S \quad \Rightarrow \quad k = 1.$$

The statements (1)–(6) are the crucial steps in the proof of our main result:

3.3.1 Theorem of Schur-Zassenhaus.[14] *Let K be an Abelian normal subgroup of G such that $(|K|, |G : K|) = 1$. Then K has a complement in G, and all the complements of K are conjugate in G.*

Proof. Because of (1) and (2) the relation

$$R \sim S \iff R|S = 1$$

is an equivalence relation on $\mathcal{S}$. Let $\tilde{R}$ be the equivalence class, which contains R. By (4)

$$\tilde{S}^x := \widetilde{x^{-1}S}, \quad (x \in G)$$

defines an action of G on $\mathcal{S}/\sim$. If $R, S \in \mathcal{S}$ and k is as in (5), then $\tilde{R}^k = \tilde{S}$; i.e., K acts transitively on $\mathcal{S}/\sim$. On the other hand, by (6) the stabilizer of $\tilde{R}$ in K is trivial. Hence, the Frattini argument shows that the stabilizer

$$G_{\tilde{R}} = \{\, x \in G \mid xR \,|\, R = 1 \,\}$$

is a complement of K in G. Conversely, if X is a complement of K in G, then $xX = X$ and $xX|X = 1$ for all $x \in X$. Thus $X = G_{\tilde{R}}$ for $X = R$, and all complements of K are conjugate by 3.1.3 since K acts transitively on $\mathcal{S}/\sim$. □

In chapter 6 we will generalize this Theorem of Schur-Zassenhaus allowing K to be non-Abelian (see 6.2 on page 125).

We now investigate a more general situation. Let

$$K \leq U \leq G \quad \text{and} \quad K \trianglelefteq G.$$

If H is a complement of K in G, then $H \cap U$ is a complement of K in U (Dedekind identity). The opposite implication is treated in Gaschütz's Theorem. For $K = U$ this theorem coincides with the Theorem of Schur-Zassenhaus, but in contrast to that result, Gaschütz's Theorem does not generalize to non-Abelian K.

3.3.2 Gaschütz's Theorem [48]. *let K be an Abelian normal subgroup of G and U a subgroup of G such that*

$$K \leq U \quad \text{and} \quad (|K|, |G : U|) = 1.$$

---

[14]Compare with [82] and [19], p. 126.

(a)   Suppose that K has a complement in U. Then K has a complement in G.

(b)   Suppose that H_0 and H_1 are two complements of K in G such that

$$H_0 \cap U = H_1 \cap U.$$

Then H_0 and H_1 are conjugate in G.

Proof. It should be mentioned that the following proof coincides with that of 3.3.1 if $U = K$. Let A be a complement of K in U, i.e.,

(i) $U = KA, \quad K \cap A = 1.$

Let $\mathcal{L}$ be the set of left transversals of U in G, and let S_0 be a fixed element of $\mathcal{L}$. Then for every left transversal $L \in \mathcal{L}$ and $\ell \in L$:

(ii) $\ell = s_\ell k_\ell a_\ell$ with $s_\ell \in S_0$, $k_\ell \in K$, $a_\ell \in A$ and $s_\ell U = \ell U$.

Moreover, because of (i) the factorization of ℓ in (ii) is unique. In particular, for every $\ell \in L$ there exists exactly one $\ell_0 \in S_0 K$ such that $\ell U = \ell_0 U$, namely $\ell_0 := s_\ell k_\ell$.

Hence, every $L \in \mathcal{L}$ is associated with an element $L_0 := \{\ell_0 \mid \ell \in L\}$ in

$$\mathcal{S} := \{L \in \mathcal{L} \mid L \subseteq S_0 K\}$$

such that $LA = L_0 A$. The uniqueness of the factorization in (ii) also gives:

(iii) L_0 is the unique element of $\mathcal{S}$ such that $LA = L_0 A$.

For $x \in G$ and the left transversal $xL \in \mathcal{L}$ one gets

$$(xL)_0 A = xLA = xL_0 A = (xL_0)_0 A,$$

and thus by (iii)

(iv) $(xL)_0 = (xL_0)_0$ for all $L \in \mathcal{L}$.

We now define

(v) $S^x := (x^{-1}S)_0$ for $S \in \mathcal{S}$ and $x \in G$.

Since

$$(S^x)^y = (y^{-1}(x^{-1}S)_0)_0 \overset{\text{(iv)}}{=} (y^{-1}(x^{-1}S))_0 = ((xy)^{-1}S)_0 = S^{(xy)}$$

(v) defines an action of G on $\mathcal{S}$. In the following we always write $(xS)_0$ instead of $S^{x^{-1}}$ since we want to use the notion

$$R|S := \prod_{\substack{(r,s)\in R\times S \\ Kr=Ks}} (rs^{-1}) \qquad (R, S \in \mathcal{S})$$

which is slightly more general than that introduced at the beginning of this section. First we discuss the statements (1)–(6) given there for our more general set-up: (1) and (2) follow as there. For the proof of (3) observe that for $k^{-1} \in K$ and $S \in \mathcal{S}$

$$kS \subseteq kS_0K = S_0K,$$

and thus by (iii) $kS = (kS)_0 \in \mathcal{S}$. This implies statement (3):

$$(kS)_0|R = k^{|G:K|}(S|R) \text{ for } k \in K \text{ and } S, R \in \mathcal{S}.$$

For the proof of (4) let $x \in G$ and $(r, s) \in R \times S$ such that $Kr = Ks$, where as above $R, S \in \mathcal{S}$. We apply (ii) using the notation given there. Then

$$xr = s_{xr}k_{xr}a_{xr} \quad \text{and} \quad xs = s_{xs}k_{xs}a_{xs},$$

and $xrK = xsK$ yields

$$s_{xr}Ka_{xr} = s_{xs}Ka_{xs}.$$

This implies $s_{xr} = s_{xs}$ and also $a_{xr} = a_{xs}$ since $K \cap A = 1$. We get

$$(xr)_0(xs)_0^{-1} = xra_{xr}^{-1}(xsa_{xs}^{-1})^{-1} = xrs^{-1}x^{-1}$$

and thus

$$(xR)_0|(xS)_0 = x(R|S)x^{-1} \text{ for all } x \in G \text{ and } R, S \in \mathcal{S}.$$

Now the statements (4)–(6) follow as in the beginning of this section. As in the proof of the Theorem of Schur-Zassenhaus

$$R \sim S \iff R|S = 1$$

defines an equivalence relation on $\mathcal{S}$, and the existence of a complement follows as there, using the action of G and K on $\mathcal{S}/\sim$.

Let H_0, H_1 be as in (b). Then

$$A := U \cap H_0 = U \cap H_1$$

is a complement of K in U; and a left transversal of A in H_i $(i = 0, 1)$ is also a left transversal of U in G. Let S_0 be a fixed left transversal of A in H_0 and $\mathcal{S}$ be defined with respect to S_0 as before. For every $s \in S_0$ there exists a $k_s \in K$ such that $sk_s \in H_1$ $(k_s = 1$ if $s \in H_0 \cap H_1)$. Now

$$S_1 := \{sk_s \mid s \in S_0\}$$

is a left transversal of A in H_1 with $S_1 \subseteq S_0 K$, i.e. $S_1 \in \mathcal{S}$.

By (ii) we have $(L_i)_0 = S_i$ for every left transversal L_i of U in G that is contained in H_i. In particular $(xS_i)_0 = S_i$ for all $x \in H_i$. Hence, H_i fixes the equivalence class of $\mathcal{S}/\sim$ that contains S_i $(i = 0, 1)$.

Now as in the proof of 3.3.1 the transitive action of G on $\mathcal{S}/\sim$ implies that H_0 and H_1 are conjugate in G. $\square$

Exercises

Let G be a group and $\Phi(G)$ the intersection of all maximal subgroups of G.[15]

1.   Let N be an Abelian minimal normal subgroup of G. Then N has a complement in G, if and only if $N \not\leq \Phi(G)$.

2.   Let N be an Abelian normal subgroup of G such that $N \cap \Phi(G) = 1$. Then N has a complement in G.

3.   Let N_1, N_2 be normal subgroups of G. If N_1 has a complement L_i in G $(i = 1, 2)$ such that $N_2 \leq L_1$, then also $N_1 N_2$ has a complement in G.

4.   Let $p \in \pi(G)$ and K be an elementary Abelian normal p-subgroup of G such that

$$K = \langle K \cap Z(S) \mid S \in \mathrm{Syl}_p\, G \rangle.$$

Then $K = [K, G]\, C_K(G)$.

---

[15]$\Phi(G)$ is the **Frattini subgroup** of G; see 5.2.3 on p. 105.

Chapter 4

Permutation Groups

Let Ω be a set. A group G that acts faithfully on Ω is a **permutation group** on Ω. Every permutation group is isomorphic to a subgroup of S_Ω, and every subgroup of S_Ω is a permutation group on Ω.

The concept of a permutation group is not only interesting in its own right but also can be used to investigate and describe groups in general.

4.1 Transitive Groups and Frobenius Groups

In the following let G be a group that acts on the set Ω. Suppose that G also acts on another set Ω'. These two actions are **equivalent** if there exists a bijection $\rho\colon \Omega \to \Omega'$ such that

$$(\beta^x)^\rho = (\beta^\rho)^x \quad \text{for all } \beta \in \Omega, \ x \in G.$$

From now on we assume that G acts transitively on Ω. Let $\alpha \in \Omega$ be fixed. Set

$$U := G_\alpha \quad \text{and} \quad \Omega' := \{Ug \mid g \in G\}.$$

Then G acts by right multiplication on Ω' (3.1.1 (c) on page 56). As in the proof of 3.1.5 we obtain for $g \in G$

$$Ug = \{x \in G \mid \alpha^x = \alpha^g\}.$$

Hence the mapping

$$\rho\colon \Omega \to \Omega' \quad \text{with} \quad \alpha^g \mapsto Ug$$

77

is a bijection satisfying

$$((\alpha^g)^x)^\rho = (\alpha^{gx})^\rho = Ugx.$$

Thus:

4.1.1 *Let G act transitively on Ω and $\alpha \in \Omega$. Then this action is equivalent to the action of G on the set of right cosets of G_α by right multiplication.*

□

In this sense every transitive action of G can be understood as an action on the right cosets of a subgroup.[1] Hence, every statement about the transitive action of G on Ω can be reformulated as a statement about the internal structure of G.

The action of G on Ω is **regular** if, for every pair $(\alpha, \beta) \in \Omega \times \Omega$, there exists exactly one $g \in G$ such that $\alpha^g = \beta$. If N is a normal subgroup of G that acts regularly on Ω, then N is called a **regular normal subgroup** of G.

4.1.2 *The following statements are equivalent:*

(i)   *G acts regularly on Ω.*

(ii)   *G acts transitively on Ω and $G_\gamma = 1$ for some $\gamma \in \Omega$.*

Proof. The implication (i) $\Rightarrow$ (ii) holds by definition.

(ii) $\Rightarrow$ (i): Let $\alpha, \beta \in \Omega$ and $x, y \in G$ such that $\alpha^x = \alpha^y = \beta$; i.e., $xy^{-1} \in G_\alpha$. By 3.1.3 G_α is conjugate to G_γ and thus $x = y$. □

4.1.3 *Let G be a transitive Abelian permutation group on Ω. Then G acts regularly on Ω.*

Proof. $(G_\alpha)^g = G_\alpha$ for all $g \in G$ and $\alpha \in \Omega$ since G is Abelian. Hence G_α fixes every element in $\alpha^G = \Omega$. This gives $G_\alpha = 1$ and together with 4.1.2 the regularity of G. □

---

[1]Conversely, every such action is transitive.

4.1.4 Let $\alpha \in \Omega$ and N be a regular normal subgroup of G. For $\beta \in \Omega$ let $x_\beta \in N$ be the unique element of N such that $\alpha^{x_\beta} = \beta$. Then for all $\beta \in \Omega$ and $g \in G_\alpha$

$$(x_\beta)^g \; = \; x_{\beta^g}.$$

In particular, the action of G_α on $\Omega \setminus \{\alpha\}$ is equivalent to the action of G_α on $N^\#$ by conjugation.

Proof. We have $\beta^g = (\alpha^{x_\beta})^g = (\alpha^g)^{g^{-1} x_\beta g} = \alpha^{(x_\beta)^g}$.   □

We now introduce a class of permutation groups that will play an important role in later chapters, and whose internal structure is well understood.

Let G be a permutation group on Ω and $|\Omega| > 1$. Then G is a **Frobenius group** on Ω if

- G acts transitively on Ω;

- $G_\alpha \neq 1$ for any $\alpha \in \Omega$;

- $G_\alpha \cap G_\beta = 1$ for all $\alpha, \beta \in \Omega$, $\alpha \neq \beta$.

Let G be a Frobenius group on Ω, $\alpha \in \Omega$, and

$$H := G_\alpha.$$

The transitive action of G on Ω gives

$$\{H^g \mid g \in G\} \; = \; \{G_\beta \mid \beta \in \Omega\},$$

and $F := G \setminus \bigcup_{g \in G} H^g$ is the set of elements of G that do not have any fixed point in Ω. Let

$$K := F \cup \{1_G\}.$$

Then

$\mathcal{F}$
$$G^\# \; = \; K^\# \mathbin{\dot\cup} \dot{\bigcup_{g \in G}} (H^g)^\# \ {}^2$$

is a partition of $G^\#$.[3]

---

[2] $K^\# := F$.

[3] This partition is a **Frobenius partition** of G.

4.1.5 $|\Omega| = |K| = |G : H| \equiv 1 \pmod{|H|}$.

Proof. $\mathcal{F}$ implies

$$|K| = |G| - |G : H|(|H| - 1) = |G : H| = |\Omega|.$$

By our hypothesis $H \cap G_\beta = 1$ for all $\beta \in \Omega \setminus \{\alpha\}$. Hence, all orbits of H on $\Omega \setminus \{\alpha\}$ have length $|H|$ (3.1.5). This yields $|\Omega| \equiv 1 \pmod{|H|}$. $\quad\square$

The subgroup H is a **Frobenius complement** of G. Clearly all subgroups conjugate to H are also Frobenius complements of G. The set K is the corresponding **Frobenius kernel** of G.

The fundamental result about Frobenius groups is Frobenius's Theorem below, which we will not prove, apart from a special case.

4.1.6   Frobenius's Theorem.   *The Frobenius kernel of a Frobenius group is a normal subgroup.*

By this theorem a Frobenius group G is the semidirect product of a Frobenius complement H with the Frobenius kernel K. In particular, K acts transitively on Ω, so K is a regular normal subgroup of G.

For the proof of 4.1.6 it suffices to show that K is a subgroup of G since K is invariant under conjugation by elements of G. This can be done using character theory.[4] Up to now no purely group-theoretic proof is known for this result. But in the case $|H| \equiv 0 \pmod 2$ an elementary calculation with involutions gives the desired conclusion. We will do this on page 83, below.

Next we give an internal description of Frobenius groups, in the sense of the remark made at the beginning of this section:

4.1.7   *Let G be a group, H a nontrivial proper subgroup of G and $\Omega = \{Hg \mid g \in G\}$. Then the following statements are equivalent:*

(i)   *G is a Frobenius group on Ω with Frobenius complement H.*

(ii)   *$H \cap H^g = 1$ for all $g \in G \setminus H$.*

---

[4]See [46] and more recently [9], for example.

Proof. (i) $\Rightarrow$ (ii): For $g \in G \setminus H$ and $\alpha := H \in \Omega$ the element $\beta := \alpha^g \in \Omega$ is different from α, and $G_\beta = H^g$ (3.1.3). This gives $H \cap H^g = 1$.

(ii) $\Rightarrow$ (i): G acts transitively on Ω by right multiplication. Let $\alpha = Hg_1$ and $\beta = Hg_2$ be two different elements of Ω. Then $g := g_2 g_1^{-1} \in G \setminus H$ and

$$G_\alpha \cap G_\beta = H^{g_1} \cap H^{g_2} = (H \cap H^g)^{g_1} = 1. \qquad \square$$

We use 4.1.7 to give a second definition of Frobenius groups. A nontrivial proper subgroup of the group G is a **Frobenius complement** of G if

$$H \cap H^g = 1 \text{ for all } g \in G \setminus H,$$

and G is a **Frobenius group** (with respect to H), if G possesses such a Frobenius complement H. As before,

$$K := \left(G \setminus \bigcup_{g \in G} H^g \right) \cup \{1\}$$

is said to be the **Frobenius kernel** of G (with respect to H). By 4.1.7 such a group G is a Frobenius group on the set $\Omega := \{Hg \mid g \in G\}$.

This second definition seems to be more general than the version for permutation groups. But we will see in 8.3.7 on page 195 (using 4.1.6) that in a Frobenius group (in the above sense) all Frobenius complements are conjugate.

First two remarks, which do not need 4.1.6.

4.1.8 *Let G be a Frobenius group with Frobenius complement H and Frobenius kernel K.*

(a) *Let U be a subgroup of G such that $U \not\leq K$, and let $x \in G$ such that $H^x \cap U \neq 1$. Then either $U \leq H^x$ or U is a Frobenius group with Frobenius complement $H^x \cap U$ and Frobenius kernel $U \cap K$.*

(b) *Let H_0 be another Frobenius complement of G such that $|H_0| \leq |H|$. Then H_0 is conjugate to a subgroup of H.*

Proof. Since H is a Frobenius complement of G we get

$$\left| \bigcup_{g \in G} H^g \right| = |G : H|(|H| - 1) + 1 = |G| - |G : H| + 1,$$

and thus

(')
$$\left| \bigcup_{g \in G} H^g \right| > \frac{|G|}{2}.$$

(a) We may assume that $H = H^x$ and $U \cap H \neq U$. Now $H \neq H^u$ for $u \in U \setminus (H \cap U)$ and

$$(H \cap U) \cap (H \cap U)^u \leq H \cap H^u = 1.$$

Hence, $H \cap U$ is a Frobenius complement for U.

Let $g \in G$ such that $H^g \cap U \neq 1$. If $U \leq H^g$, then $H \cap H^g \neq 1$ and thus $H = H^g$, which contradicts $U \cap H \neq U$. Hence $1 \neq H^g \cap U \neq U$, and $H^g \cap U$ is also a Frobenius complement of U. From ('), applied to U and the two Frobenius complements $H \cap U$ and $H^g \cap U$, there exists $u_1 \in U$ such that

$$(H \cap U) \cap (H^g \cap U)^{u_1} \neq 1.$$

It follows that $H \cap H^{gu_1} \neq 1$ and thus $H^{gu_1} = H$ and $(H^g \cap U)^{u_1} = H \cap U$. We have shown that

$$\bigcup_{u \in U} (H \cap U)^u = \bigcup_{g \in G} (H^g \cap U).$$

Hence $K \cap U$ is the Frobenius kernel of U (with respect to $H \cap U$).

(b) Assume that $H_0 \not\leq H^x$ for all $x \in G$. Then (a) implies for $U := H_0$

$$m := |H_0^{\#} \cap K| \geq 1.$$

Because H_0 is a Frobenius complement of G and K is invariant under conjugation, we get

$$|G : H| \overset{4.1.5}{=} |K| \geq \left| \bigcup_{x \in G} (H_0^{\#} \cap K)^x \right| + 1 = m|G : H_0| + 1.$$

On the other hand, by our hypothesis $|H_0| \leq |H|$, and thus

$$|G : H_0| \geq |G : H|,$$

a contradiction.   □

Examples of Frobenius groups are:

- The dihedral groups D_{2n} of order $2n$, $n > 1$, n odd. Here the Frobenius complements are the subgroups of order 2.

- Let K be a finite field. Then the multiplicative group K^* acts by right multiplication on the additive group $K(+)$. The corresponding semidirect product $K^* \ltimes K(+)$ is a Frobenius group with Frobenius complement K^*.

The *proof* of 4.1.6 for the case $|H| \equiv 0 \pmod 2$ (Bender):

Let t be an involution in H and $g \in G \setminus H$. Then either

$$a := tt^g = [t, g]$$

is in K, or there exists $x \in G$ such that $1 \neq a \in H^x$. In the second case

$$a \in H^x \cap H^{xt} \cap H^{xt^g}$$

since $a^t = a^{-1} = a^{t^g}$, and we get $H^x = H^{xt} = H^{xt^g}$ and $t, t^g \in H^x$. But now $H^x = H$, which contradicts $t \in H$ and $t^g \notin H$. We have shown:

$$(*) \qquad\qquad tt^g \in K, \quad \text{if } g \in G \setminus H.$$

Let $\{g_1, \ldots, g_n\}$ be a transversal of H in G, $n := |G : H|$. Since

$$tt^{g_i} = tt^{g_j} \iff t^{g_i} = t^{g_j} \iff t^{g_i g_j^{-1}} = t \iff g_i g_j^{-1} \in H$$

the elements $tt^{g_1}, \ldots, tt^{g_n}$ are pairwise distinct. We get that

$$K = \{tt^{g_1}, \ldots, tt^{g_n}\}$$

since $|K| = n$. Conjugation with t gives

$$K = \{t^{g_1}t, \ldots, t^{g_n}t\}.$$

As mentioned above it suffices to show that K is a subgroup, i.e., to show that $KK \subseteq K$. For every $t^{g_i}t$ there exists g_s such that $t^{g_i}t = tt^{g_s}$. Hence

$$(tt^{g_i})(tt^{g_j}) = t(t^{g_i}t)t^{g_j} = t(tt^{g_s})t^{g_j} = t^{g_s}t^{g_j} = (tt^{g_j g_s^{-1}})^{g_s} \overset{(*)}{\in} K^{g_s} = K.$$

□

4.2   Primitive Action

As before G is a group acting on the set Ω. A nonempty subset $\Delta \subseteq \Omega$ is a **set of imprimitivity**, if for every $g \in G$:

$$\Delta^g \neq \Delta \quad \Rightarrow \quad \Delta^g \cap \Delta = \emptyset.^5$$

It is evident that also Δ^g is a set of imprimitivity.

For $\alpha \in \Omega$ and $H \leq G$ with $G_\alpha \leq H$ we set

$$\Delta := \alpha^H.$$

For all $g \in G \setminus H$

$$\Delta^g = \alpha^{Hg}$$

has an empty intersection with Δ, i.e., Δ is a set of imprimitivity.

We now assume that G acts transitively on Ω. Let $\alpha \in \Omega$ and Δ be a set of imprimitivity with $\alpha \in \Delta$. Then

$$\Delta = \alpha^H$$

where

$$H := G_\Delta := \{x \in G \mid \Delta^x = \Delta\}.$$

Thus, the sets of imprimitivity containing α correspond to the subgroups of G containing G_α.

Let Δ be a set of imprimitivity and $\Sigma := \{\Delta^g \mid g \in G\}$. The transitive action of G on Ω gives

$$\Omega = \overset{\cdot}{\underset{\Delta^g \in \Sigma}{\bigcup}} \Delta^g.$$

Hence, the action of G on Ω can be understood as composition of the transitive action of G on Σ and the transitive action of G_Δ on Δ.

The action of G on Ω is **imprimitive** if there exists a set of imprimitivity Δ such that

$$1 \neq |\Delta| \neq |\Omega|; \ ^6$$

otherwise the action is **primitive**.

In this section we discuss the primitive case, later in Section 4.4 the imprimitive case. As we have seen above:

---

[5] $\Delta^g := \{\alpha^g \mid \alpha \in \Delta\}$.
[6] Then also $1 \neq |\Sigma| \neq |\Omega|$.

4.2.1 *Let G act transitively on Ω. Then G is primitive on Ω if and only if G_α is a maximal subgroup of G ($\alpha \in \Omega$).*   □

4.2.2 *Let G be a primitive permutation group on Ω and $1 \neq N \trianglelefteq G$. Then N acts transitively on G. If in addition N is regular on Ω, then N is a minimal normal subgroup of G.*

Proof. Let $\alpha \in \Omega$. By 4.2.1 G_α is a maximal subgroup of G. If $N \leq G_\alpha$, then N acts trivially on $\Omega = \alpha^G$ (3.1.3 on page 58), which contradicts $N \neq 1$. Thus we have $G_\alpha < G_\alpha N = G$, and $\Omega = \alpha^G = \alpha^N$ follows.

Assume that N acts regularly on Ω. Then every normal subgroup $1 \neq M \trianglelefteq G$ with $M \leq N$ is also regular on Ω. It follows that

$$|N| = |\alpha^N| = |N| = |\Omega| = |\alpha^M| = |M|$$

and thus $N = M$.   □

For $n \in \mathbb{N}$ and $n \leq |\Omega|$ we set

$$\Omega^{(n)} := \{(\alpha_1, \ldots, \alpha_n) \in \Omega^n \mid \alpha_i \neq \alpha_j \text{ for } i \neq j\}.$$

We say that G is ***n*-fold transitive** on Ω if, for any two $(\alpha_1, \ldots, \alpha_n)$, $(\beta_1, \ldots, \beta_n) \in \Omega^{(n)}$, there exists $g \in G$ such that

$$\alpha_i{}^g = \beta_i \quad \text{for} \quad i = 1, \ldots, n,$$

i.e., the *componentwise* action of G on $\Omega^{(n)}$ defined by

$$(\alpha_1, \ldots, \alpha_n)^g := (\alpha_1{}^g, \ldots, \alpha_n{}^g) \quad (g \in G)$$

is transitive. Clearly n-fold transitivity implies m-fold transitivity for all $1 \leq m \leq n$. Suppose that G is $(n-1)$-fold transitive on Ω. Then G is n-transitive on Ω if and only if for $(\alpha_1, \ldots, \alpha_{n-1}) \in \Omega^{(n-1)}$ the stabilizer

$$G_{\alpha_1, \ldots, \alpha_{n-1}} := \bigcap_{i=1}^{n-1} G_{\alpha_i}$$

is transitive on $\Omega \setminus \{\alpha_1, \ldots, \alpha_{n-1}\}$.

4.2.3   *Let $\alpha \in \Omega$. Suppose that G acts transitively on Ω. Then G is 2-fold transitive on Ω if and only if*

$$G = G_\alpha \cup G_\alpha g \, G_\alpha$$

for $g \in G \setminus G_\alpha$.

Proof. By 4.1.1 Ω can be identified with the set of cosets $G_\alpha g$, $g \in G$. The transitivity of G_α on $\Omega \setminus \{\alpha\}$ implies, for $G_\alpha g \neq G_\alpha$,

$$G_\alpha g \, G_\alpha = G \setminus G_\alpha. \qquad \qquad \square$$

From 4.2.3 together with 4.2.1 we obtain:

4.2.4   *Every 2-fold transitive permutation group is primitive.* $\qquad \square$

Examples of n-fold transitive groups, resp. $(n-2)$-fold transitive groups, are the symmetric group S_n and the alternating group A_n. These groups will be introduced in the next section. It should be mentioned that apart from these two classes there are no n-transitive groups for $n \geq 6$.[7]

Here we only note:

4.2.5   *Let G be an n-fold transitive permutation group on Ω, $|\Omega| \geq 3$. Suppose that G contains a regular normal subgroup N. Then $n \leq 4$; more precisely:*

(a)   *For $n = 2$: N is an elementary Abelian p-group.*

(b)   *For $n = 3$: N is an elementary Abelian 2-group, or $N \cong C_3$ and $G \cong S_3$.*

(c)   *For $n = 4$: $N \cong C_2 \times C_2$ and $G \cong S_4$.*

Proof. Let $n \geq 2$ and $\alpha \in \Omega$. Then G_α is $(n-1)$-fold transitive on $\Omega \setminus \{\alpha\}$ and thus also $(n-1)$-transitive on $N^{\#}$ (4.1.4). This shows that for all $x, y \in N^{\#}$ there exists $g \in G_\alpha$ such that $x^g = y$. Hence, every element of $N^{\#}$ has the same order, and this order is a prime p (1.4.3). Now Cauchy's

---

[7]This follows from the classification of the finite simple groups.

Theorem implies that N is a p-group, and N is elementary Abelian since $Z(N) \neq 1$ (see 3.1.11 on page 61).

Let $n \geq 3$ and thus $3 \leq |N| = |\Omega|$. If $|N| = 3$ then $G \cong S_3$. Assume that $|N| \geq 4$, and let x_1, x_2, x_3 be three different elements of $N^{\#}$. Since G_α is 2-fold transitive on $N^{\#}$ there exists $g \in G_\alpha$ such that

$$x_1^g = x_1 \quad \text{and} \quad x_2^g = x_3.$$

If $p \geq 3$ then $x_1 \neq x_1^{-1}$; and for $x_2 := x_1^{-1}$ we get $x_2{}^g = x_2$, a contradiction. Thus, N is an elementary Abelian 2-group.

Let $n \geq 4$. Then $|\Omega| \geq 4$ and as seen above N is an elementary Abelian 2-group of order at least 4. Let $U = \langle x_1 \rangle \times \langle x_2 \rangle$ be a subgroup of order 4 of N. Assume that $U \neq N$. Choose $x_3 = x_1 x_2$ and $x_4 \in N \setminus U$. Since G_α is 3-fold transitive on $N^{\#}$ there exists $g \in G$ such that $x_1^g = x_1$, $x_2^g = x_2$ and $x_3^g = x_4$, which contradicts

$$x_3^g = (x_1 x_2)^g = x_1^g x_2^g = x_1 x_2 \in U.$$

This shows that $|\Omega| = 4$, and $n = 4$ follows. $\square$

4.3   The Symmetric Group

The **symmetric group** S_n **of degree** n is the group of all permutations of the set

$$\Omega := \{1, \ldots, n\}.$$

Then S_n has order $n!$ and acts by definition n-fold transitively on Ω. A permutation $z \in S_n$ is a **cycle of length** k (or a k-**cycle**) if there exist k different elements $\alpha_1, \ldots, \alpha_k \in \Omega$ such that

$$\alpha_i{}^z = \alpha_{i+1}, \quad \text{for } i = 1, \ldots, k-1, \quad \alpha_k{}^z = \alpha_1,$$

and

$$\beta^z = \beta \quad \text{for all } \beta \in \Omega \setminus \{\alpha_1, \ldots, \alpha_k\}.$$

We denote z by $(\alpha_1 \alpha_2 \ldots \alpha_k)$. Then for $g \in S_n$

(∗) $$g^{-1}(\alpha_1 \cdots \alpha_k) g = (\alpha_1 \cdots \alpha_k)^g = (\alpha_1{}^g \cdots \alpha_k{}^g).$$

A cycle $z' = (\beta_1 \cdots \beta_r)$ is **disjoint** from z, if

$$\{\beta_1, \ldots, \beta_r\} \cap \{\alpha_1, \ldots, \alpha_k\} = \varnothing.$$

In this case $zz' = z'z$. It is evident that every permutation can be written in a unique way as the product of pairwise disjoint and thus commuting cycles:

$$(**) \qquad x = (\alpha_{11}\cdots\alpha_{1k_1})(\alpha_{21}\cdots\alpha_{2k_2})\cdots(\alpha_{s1}\cdots\alpha_{sk_s}).$$

The cycles $(\alpha_{i1}\cdots\alpha_{ik_i})$ correspond to the orbits of $\langle x\rangle$ in its action on Ω and thus to a partition of Ω. After rearranging the cycles according to their lengths $k_1 \geq k_2 \geq \cdots \geq k_s$, the tuple $(k_1,\ldots,k_s)$ is called the **type** of x. The cycles of length 1 describe the fixed points of x in Ω. In the representation $(**)$ they are usually omitted.

4.3.1　*Two permutations of S_n are conjugate if and only if they have the same type.*

Proof. According to $(*)$ the conjugate of a k-cycle is a k-cycle. Hence, conjugate elements have the same type. Conversely, let x be as in $(**)$ and

$$x' = (\alpha'_{11}\cdots\alpha'_{1k_1})(\alpha'_{21}\cdots\alpha'_{2k_2})\cdots(\alpha'_{s1}\cdots\alpha'_{sk_s})$$

Then x and x' have the same type. Let $a \in S_n$ satisfying $a\colon \alpha_{ij} \mapsto \alpha'_{ij}$. Then $(*)$ implies $x^a = x'$. □

The 2-cycles of S_n are called **transpositions**. For $k \geq 2$ every k-cycle $(\alpha_1\cdots\alpha_k)$ is the product of $(k-1)$ transpositions:

$$(\alpha_1\cdots\alpha_k) = (\alpha_1\alpha_2)(\alpha_1\alpha_3)\cdots(\alpha_1\alpha_k).$$

Thus, $(**)$ shows that every permutation of S_n can be written as a product of transpositions t_i:[8]

$$x = t_1 t_2\cdots t_s.$$

In this representation of x the transpositions t_i are by no means uniquely determined – but in fact the number of factors (for a given element) is either always even or always odd.[9] Thus the mapping

$$\text{sgn}\colon x \mapsto (-1)^s$$

[8] 1 being written as the "empty product." Of course, for $n \geq 2$ also $1 = t^2$, t transposition.

[9] This property usually is proved in a beginner class in linear algebra, when determinants are introduced.

is well defined and is a homomorphism from S_n into the multiplicative group $\{1, -1\}$ ($\cong C_2$) of order 2. The kernel of this homomorphism is the **alternating group** A_n **of degree** n; it consists of all **even** permutations (the permutations in $S_n \setminus A_n$ are called **odd**). For $n \geq 2$ A_n is a normal subgroup of index 2 in S_n. For example, a k-cycle is an even permutation, if and only if k is odd.

4.3.2 A_n *is* $(n-2)$*-transitive on* Ω $(n \geq 3)$.

Proof. The tuple $T_1 := (3, 4, \ldots, n) \in \Omega^{(n-2)}$ can be mapped to any other tuple $T_2 \in \Omega^{(n-2)}$ by a permutation $x \in S_n$. Then also tx, $t = (12)$, maps T_1 to T_2, and either $x \in A_n$ or $tx \in A_n$. □

4.3.3 A_n *is the commutator subgroup of* S_n.

Proof. Let K be the commutator subgroup of S_n. Since $S_n = A_n = 1 = K$ for $n = 1$ we may assume that $n \geq 2$. Moreover, $K \leq A_n$ since $S_n/A_n \cong C_2$ (1.5.2).

Let t be a transposition of S_n. Then $\langle t \rangle K$ is a normal subgroup of S_n since S_n/K is Abelian. By 4.3.1 the transpositions of S_n are conjugate, and, as seen above, every element of S_n is a product of transpositions. Hence $S_n = \langle t \rangle K$ and $K = A_n$. □

Notice that $A_n = 1$ for $n = 2$ and $A_n = \langle (123) \rangle \cong C_3$ for $n = 3$. In the latter case

$$1 \leq A_3 \leq S_3$$

is a chief series (and also a composition series) for S_3 with cyclic chief factors.[10]

Let $n = 4$. The elements of order 2 in S_4 are either transpositions or of type (2,2). In the second case the elements are

$$t_1 = (12)(34), \quad t_2 = (13)(24), \quad t_3 = (14)(23).$$

The set

$$N := \{1, t_1, t_2, t_3\}$$

---

[10]The group table of S_3 is given on p. 3.

is a subgroup of A_4 isomorphic to $C_2 \times C_2$, and N is a regular normal subgroup of S_4. Let $d = (123)$. Then A_4 is the (internal) semidirect product of $\langle d \rangle$ with N and

$$t_1^d = t_3, \quad t_2^d = t_1, \quad t_3^d = t_2.$$

Hence

$$1 \leq N \leq A_4 \leq S_4$$

is a chief series of S_4 with Abelian chief factors, and

$$1 \leq \langle t_1 \rangle \leq N \leq A_4 \leq S_4$$

is a composition series with cyclic composition factors.

Later we will use the following description of S_4:

4.3.4 *Let G be a group of order 24 that is not 3-closed. Then either $G \cong S_4$ or $G/Z(G) \cong A_4$.* [11]

Proof. G acts on

$$\Omega := \mathrm{Syl}_3\, G$$

by conjugation. Since G is not 3-closed Sylow's Theorem gives $|\Omega| = 4$. Thus, there exists a homomorphism $\varphi \colon G \to S_4$ such that

$$\mathrm{Ker}\, \varphi = \bigcap_{S \in \Omega} N_G(S) =: N.$$

G/N is a subgroup of S_4 and $|N|$ a divisor of $\frac{24}{4} = 6$. If $|N| \in \{3,6\}$, then N and thus also G is 3-closed, a contradiction. The case $N = 1$ yields $G \cong S_4$, and the case $|N| = 2$ implies $N = Z(G)$ and $G/N \cong A_4$. $\qquad\square$

The subgroup of S_n fixing $n \in \Omega$ is the symmetric group S_{n-1} acting on $\{1, \ldots, n-1\}$. In this sense we will regard S_{n-1}, resp. A_{n-1}, as a subgroup of S_n. For example, S_4 is the semidirect product of S_3 with the regular normal subgroup N introduced above.

4.3.5   Theorem. *A_n is simple for $n \geq 5$.*

---

[11]In this case either $G \cong A_4 \times C_2$ or $G \cong \mathrm{SL}_2(3)$, see 8.6.10 on p. 219.

Proof. Let $n = 5$. Then $|A_n| = 60$, and A_n is not 5-closed since the number of 5-cycles in A_5 is larger than 4. Hence, 3.2.12 on page 68 shows that A_5 is simple.

Let $n \geq 6$ and N be a normal subgroup of A_n, $1 \neq N \neq A_n$. By induction on n we may assume that A_{n-1} (the stabilizer of $n \in \Omega$ in A_n) is simple. Moreover, A_n is 4-transitive and primitive on Ω (4.3.2 and 4.2.4). Thus, 4.2.1 and 4.2.2 imply that A_{n-1} is a maximal subgroup and N is a transitive normal subgroup of A_n. Now the simplicity of A_{n-1} shows that N is a regular normal subgroup, and the 4-transitivity of A_n together with 4.2.5 gives $n = 4$. This contradicts $n \geq 6$. $\qquad\square$

4.4 Imprimitive Groups and Wreath Products

Let G be a group and Ω a set on which G acts transitively and imprimitively. Then there exists a set of imprimitivity $\Delta \subseteq \Omega$ such that $1 \neq |\Delta| \neq |\Omega|$ and

$$\Sigma := \{\Delta^g \mid g \in G\}$$

is a partition of Ω. Fix $\alpha \in \Delta$, and set $U := G_\alpha$ and $H := G_\Delta$, where G_Δ is the stabilizer of Δ. Then

$$U < H < G.$$

We now describe the action of G on Ω by means of the action of G on Σ and H on Δ.

By 4.1.1 the action of G on Ω is equivalent to the action of G on the right cosets of U (in G) by right multiplication. Hence, we may assume that

$$\Omega = \{Ug \mid g \in G\}.$$

Then

$$\Delta^g = \{Uhg \in \Omega \mid h \in H\}.$$

Let S be a transversal of H in G. For every $x \in G$ and $s \in S$ there exist elements $f_x(s) \in H$ and $s_x \in S$ such that

$$sx = f_x(s) s_x,$$

and these elements $f_x(s)$ are uniquely determined by x and s. Hence, for $Uhs \in \Delta^s$:

(1) $(Uhs)x = Uhf_x(s)s_x.$

For $x, y \in G$ and $s \in S$ we obtain

$$f_{xy}(s)\, s_{xy} \;=\; s(xy) \;=\; f_x(s)\, s_x\, y \;=\; f_x(s)\, f_y(s_x)(s_x)_y.$$

It follows that

(2) $f_{xy}(s) = f_x(s)\, f_y(s_x),$ and

(3) $(s_x)_y = s_{xy}.$

Hence

$$s \mapsto s_x \quad (s \in S,\ x \in G)$$

defines an action of G on S, which is equivalent to the action of G on Σ.

Let

$$\widehat{H} := \underset{S}{\times}\, H$$

be the direct product of $|S|$ copies of H. We describe the elements of $\widehat{H}$ as functions from S in H

$$\widehat{H} = \{f \mid f\colon S \to H\},$$

where $f \in \widehat{H}$ is the "S-tuple" whose "s-th entry" is the element $f(s) \in H$. The multiplication is "componentwise", i.e., $(fg)(s) = f(s)g(s)$ for $f, g \in \widehat{H}$. The previously defined elements $f_x(s) \in H$ now give an element

$$f_x : s \mapsto f_x(s)$$

of $\widehat{H}$.

For $x \in G$ and $f \in \widehat{H}$ we define

(4) $f^x \in \widehat{H}$ such that $f^x(s) := f(s_{x^{-1}}),\ s \in S.$

Since

$$(f^x)^y(s) \;=\; f^x(s_{y^{-1}}) \;=\; f(s_{y^{-1}x^{-1}}) \;=\; f(s_{(xy)^{-1}}) \;=\; f^{xy}(s),$$

(4) defines an action of G on the group $\widehat{H}$. This action permutes the entries of the S-tuple f according to the action of G on S: $f(s)$ is the s_x-th entry of the S-tuple f^x.

Let $G \ltimes \widehat{H}$ be the semidirect product with respect to this action. For $(x, f) \in G \ltimes \widehat{H}$ and $Uhs \in \Omega$ $(h \in H,\ s \in S)$ we set:

(5) $(Uhs)^{(x,f)} := Uhf(s_x)s_x.$

Then
$$(x,f)\,(y,g) = (xy, f^y g) \text{ for } (x,f),(y,g) \in G \ltimes \widehat{H},$$

and
$$((Uhs)^{(x,f)})^{(y,g)} = (Uhf(s_x)s_x)^{(y,g)} \overset{(3)}{=} Uhf(s_x)g(s_{xy})s_{xy}$$
$$\overset{(4)}{=} Uhf^y(s_{xy})g(s_{xy})s_{xy} = (Uhs)^{(xy,f^y g)}.$$

Thus, (5) defines an action of $G \ltimes \widehat{H}$ on Ω. We denote this action by ρ' and the action of G on Ω by ρ.

4.4.1   *The mapping*
$$\eta\colon G \to G \ltimes \widehat{H} \quad \text{with} \quad x \mapsto (x, f_x{}^x)$$

is a monomorphism and $\rho = \eta\rho'$.

Proof. Evidently η is injective. For $x,y \in G$
$$x^\eta y^\eta = (x, f_x{}^x)(y, f_y{}^y) = (xy, f_x{}^{xy} f_y{}^y) = (xy, (f_x f_y{}^{x^{-1}})^{xy})$$
$$\overset{(2)(4)}{=} (xy, f_{xy}{}^{xy}) = (xy)^\eta.$$

Hence, η is a monomorphism.

For the proof of the second claim let $x \in G$ and $Uhs \in \Omega$ $(h \in H,\ s \in S)$. Then
$$(Uhs)^{x^\rho} = Uhsx \overset{(1)}{=} Uhf_x(s)s_x \overset{(4)}{=} Uhf_x{}^x(s_x)s_x$$
$$\overset{(5)}{=} (Uhs)^{(x,f_x{}^x)} = (Uhs)^{x^{\eta\rho'}}. \qquad \square$$

The group $G \ltimes \widehat{H}$ is a special case of a *wreath product*, which we define now.

We start with a quadruple (H,G,A,τ), where H and G are groups, A is a subgroup of G, and τ a homomorphism from A in $\operatorname{Aut} H$. We use the notation
$$h^a := h^{a^\tau}, \quad h \in H,\ a \in A.$$

Let S be a transversal of A in G. As above we define the $|S|$-fold direct product
$$\widehat{H} := \underset{S}{\times} H = \{f \mid f\colon S \to H\}$$

and, for every $(x, s) \in G \times S$, an element $(f_x, s_x) \in A \times S$ satisfying

$$sx = f_x(s) \, s_x.$$

As above, but with A in place of H, the equations (2) and (3) hold. In particular $s \mapsto s_x$ defines an action of G on S, which is equivalent to the action of G on the cosets Ag, $g \in G$. For $(x, f) \in G \times \widehat{H}$ let $f^x \in \widehat{H}$ be defined by

(6) $f^x(s) := f(s_{x^{-1}})^{f_x(s_{x^{-1}})}, \quad s \in S,$

where $f_x(s_{x^{-1}})$ acts on H with respect to τ.

Because of

$$(f^x)^y(s) = f^x(s_{y^{-1}})^{f_y(s_{y^{-1}})} = \left(f(s_{y^{-1}x^{-1}})^{f_x(s_{y^{-1}x^{-1}})}\right)^{f_y(s_{y^{-1}})}$$
$$\overset{(2)}{=} f(s_{(xy)^{-1}})^{f_{xy}(s_{(xy)^{-1}})} = f^{xy}(s)$$

(6) defines an action of G on the group $\widehat{H}$. Let

$$K_S := G \ltimes \widehat{H}$$

be the semidirect product with respect to this action. The index in K_S indicates that this definition might depend on the choice of the transversal S. But we show:

4.4.2   *Let S and $\widetilde{S}$ be two transversals of A in G. Then $K_S \cong K_{\widetilde{S}}$.*

Proof. For every $s \in S$ there exists a pair $(b_s, \widetilde{s}) \in A \times \widetilde{S}$ such that

(7) $\widetilde{s} = b_s s.$

For $(x, s) \in G \times S$ let $(\widetilde{f}_x, \widetilde{s}) \in \widehat{H} \times \widetilde{S}$ such that

$$\widetilde{s}x = \widetilde{f}_x(\widetilde{s}) \, \widetilde{s}_x.$$

Since

$$\widetilde{s}x = b_s sx = b_s f_x(s) s_x \overset{(7)}{=} b_s f_x(s) b_{s_x}^{-1} \widetilde{s}_x$$

we obtain

$$(*) \qquad\qquad \widetilde{f}_x(\tilde{s}) = b_s\, f_x(s)\, b_{s_x}^{-1}.$$

It is evident that the mapping

$$\beta\colon \underset{S}{\times} H \to \underset{\tilde{S}}{\times} H$$

defined by

$$(8) \qquad f^\beta(\tilde{s}) = f(s)^{b_s^{-1}}$$

is an isomorphism. Let

$$\psi\colon K_S \to K_{\tilde{S}} \quad \text{such that} \quad (x, f) \mapsto (x, f^\beta).$$

Then ψ is an isomorphism if and only if

$$(+) \qquad\qquad (f^x)^\beta = (f^\beta)^x, \quad \text{for all } f \in \widehat{H},\ x \in G.$$

Moreover

$$(f^x)^\beta(\tilde{s}) \overset{(8)}{=} f^x(s)^{b_s^{-1}} \overset{(6)}{=} f(s_{x^{-1}})^{f_x(s_{x^{-1}})b_s^{-1}},$$

and

$$(f^\beta)^x(\tilde{s}) = f^\beta(\tilde{s}_{x^{-1}})^{\tilde{f}_x(\tilde{s}_{x^{-1}})} \overset{(8)}{=} f(s_{x^{-1}})^{b_{s^{-1}}\tilde{f}_x(\tilde{s}_{x^{-1}})}.$$

Thus, $(+)$ follows from $(*)$. $\qquad\qquad\qquad\qquad\qquad\qquad\qquad\square$

Result 4.4.2 shows that the semidirect product

$$K := G \ltimes \widehat{H}$$

constructed from the quadruple (H, G, A, τ) is (up to isomorphism) independent of the choice of transversal. The group K is the **twisted wreath product** of G with H (with respect to $A \leq G$ and $\tau\colon A \to \operatorname{Aut} H$). If $A^\tau = 1$, then K is the **wreath product** of G with H.

Exercises

In the first three exercises G is a transitive permutation group on the set Ω.

1. Let N be a normal subgroup of G and Σ be the set of orbits of N on Ω. Then G acts transitively on Σ.[12]

2. (Witt, [101]) Let G be n-transitive on Ω and $\Sigma \subseteq \Omega$, $|\Sigma| \geq n$, and let P be a Sylow p-subgroup of $\bigcap_{\alpha \in \Sigma} G_\alpha$. Then $N_G(P)$ acts n-transitively on $C_\Omega(P)$.

3. Suppose that G is primitive on Ω and contains a transposition. Then $G = S_\Omega$.

In the following three exercises let G be a Frobenius group with Frobenius complement H and Frobenius kernel K.

4. If H has even order, then $Z(H) \neq 1$.

5. Suppose that every coset of H in G contains at least one element from K. Then K is a subgroup of G.

6. If H is a maximal subgroup of G, then K is an elementary Abelian p-group.[13]

7. Let p be a prime, $G := S_p$, and $P \in \mathrm{Syl}_p\, G$. Determine $N_G(P)$.

8. Let $x = (1 \cdots n) \in S_m$. Then $C_{S_m}(x) = R \times X$, where $R = \langle x \rangle$ and $X \cong S_{m-n}$ $(S_0 := 1)$.

9. Let $H, K \leq S_8$, $H = \langle (123)(456)(78) \rangle$ and $K = \langle (38) \rangle$. Determine the orbits of H, K and $\langle H, K \rangle$ on $\{1, \ldots, 8\}$.

10. Determine the class equation of A_7.

In the following three exercises let $G := S_n$, $n \geq 3$, and T be the conjugacy class of transpositions of G.

11. (a) $|T| = \frac{n(n-1)}{2}$ and $C_G(d) \cong C_2 \times S_{n-2}$ for $d \in T$.

    (b) $o(ab) \in \{1, 2, 3\}$ for all $a, b \in T$.

12. Let D be a conjugacy class of involutions of G such that

$$o(ab) \in \{1, 2, 3\} \quad \text{for all } a, b \in D.$$

    Then $D = T$, or $n = 6$ and $D = ((12)(34)(56))^G$, or $n = 4$ and $D = ((12)(34))^G$.

13. Let $\alpha \in \mathrm{Aut}\, G$ such that $T^\alpha = T$. Then α is an inner automorphism of G.

14. Let $\widetilde{G}$ be a group and $d_1, \ldots, d_m$ involutions in $\widetilde{G}$ $(m \geq 3)$ such that

    (i) $\widetilde{G} = \langle d_1, \ldots, d_m \rangle$,

---

[12]The action of G on Ω induces an action of G on the set of all subsets of Ω.
[13]For this exercise assume that K is a normal subgroup of G (see 4.1.6).

(ii) $\quad o(d_i d_j) = \begin{cases} 2 & \text{if } |i - j| \geq 2 \\ 3 & \text{if } |i - j| = 1 \end{cases}$,

(iii) $\quad \langle d_i, \ldots, d_m \rangle \cong S_{m-i+2}$ for $2 \leq i \leq m$.

Then for $D := d_1^G$ and $M := \langle d_2, \ldots, d_m \rangle$:

(a) $\quad |d_1^M| = m$, and M acts 2-transitively on d_1^M (by conjugation).

(b) $\quad a^b \in M$ for all $a, b \in d_1^M$ with $a \neq b$.

(c) $\quad D = d_1^M \cup (D \cap M)$.

(d) $\quad$ Let $a \in d_1^M$ and $b \in D$. Then
$$Mab = \begin{cases} M & a = b \\ Ma^b & \text{if} \quad b \in D \cap M \\ Ma & b \in d_1^M \setminus \{a\} \end{cases}.$$

(e) $\quad |\tilde{G} : M| = m + 1$, and d_1 acts as a transposition on $\{Mg \mid g \in \tilde{G}\}$.

15. Let $\tilde{G}$ be a group and $d_1, \ldots, d_{n-1}$ $(n \geq 2)$ involutions in $\tilde{G}$ such that

(i) $\quad \tilde{G} = \langle d_1, \ldots, d_{n-1} \rangle$ and

(ii) $\quad o(d_i d_j) = \begin{cases} 2 & \text{falls } |i - j| \geq 2 \\ 3 & \text{falls } |i - j| = 1 \end{cases}$.

Then there exists an isomorphism $\varphi \colon \tilde{G} \to S_n$ such that d_1^φ is a transposition of S_n.

16. There exist two subgroups isomorphic to S_5 in S_6 that are not conjugate.

17. $\operatorname{Aut} S_n = \operatorname{Inn} S_n$ or $n = 6$ and $|\operatorname{Aut} S_n / \operatorname{Inn} S_n| = 2$.

Chapter 5

p-Groups and Nilpotent Groups

As was mentioned in the introduction of Section 3.2, Sylow's Theorem directs attention to the p-subgroups of a finite group. In this chapter we will present some basic facts about p-groups (and more generally nilpotent groups), which will be used in later chapters.

In the second section we investigate p-groups that contain a cyclic maximal subgroup.

5.1   Nilpotent Groups

A group G is **nilpotent**, if every subgroup of G is subnormal in G.[1] It is evident that this property is equivalent to

$$U \; < \; N_G(U) \text{ for every subgroup } U < G.$$

As a direct consequence of 1.2.8 on page 14 and Cauchy's Theorem one obtains:

5.1.1   *Subgroups and homomorphic images of nilpotent groups are nilpotent. Maximal subgroups of nilpotent groups are normal and of prime index.*

$\square$

---

[1]This definition applies only to finite groups. See the footnote on p. 102.

5.1.2   *Let G be a group and Z a subgroup of $Z(G)$. Then G is nilpotent if and only if G/Z is nilpotent.*

Proof. One direction follows from 5.1.1. For the other direction let G/Z be nilpotent and $U \leq G$. Then $UZ/Z \lhd\lhd G/Z$ and thus $UZ \lhd\lhd G$ (1.2.8 on page 14). Since $Z \leq Z(G)$ also $U \lhd UZ$. Hence $U \lhd\lhd G$.   □

Result 3.1.10 on page 61 gives the most important class of nilpotent groups:

5.1.3   *p-Groups are nilpotent.*   □

Recall that $O_p(G)$ denotes the largest normal p-subgroup of a group G (3.2.2 on page 63), and G is p-closed if $O_p(G)$ is a Sylow p-subgroup of G.

5.1.4   Theorem.   *The following statements are equivalent:*

(i)   *G is nilpotent.*

(ii)   *For every $p \in \pi(G)$ G is p-closed.*

(iii) $G = \underset{p \in \pi(G)}{\times} O_p(G)$.

Proof. (i) $\Rightarrow$ (ii): Let $U := N_G(G_p)$, where $p \in \pi(G)$ and $G_p \in \mathrm{Syl}_p\, G$. Then the Frattini argument yields

$$N_G(U) = UN_{N_G(U)}(G_p) = U$$

since G_p is a Sylow p-subgroup of U. The definition of nilpotency gives $U = G$ and thus $G_p \lhd G$.

(ii) $\Rightarrow$ (iii): This follows from 1.6.5 on page 31.

(iii) $\Rightarrow$ (i): $Z(O_p(G)) \neq 1$ for $p \in \pi(G)$ by 3.1.11 on page 61. Hence also

$$Z(G) = \underset{p \in \pi(G)}{\times} Z(O_p(G)) \neq 1$$

(1.6.2 (a) on page 29). Let $\overline{G} := G/Z(G)$. Then 1.6.2 (c) implies

$$\overline{G} = \underset{p \in \pi(\overline{G})}{\times} \overline{O_p(G)} = \underset{p \in \pi(\overline{G})}{\times} O_p(\overline{G}).$$

By induction on $|G|$ we may assume that $\overline{G}$ is nilpotent. But then also G is nilpotent (5.1.2). □

From 1.6.2 (a) and 3.1.11 on page 61 we obtain:

5.1.5   *Let G be a nilpotent group and $N \neq 1$ a normal subgroup of G. Then $Z(G) \cap N \neq 1$.* □

This property characterizes nilpotent groups:

5.1.6   *The following statements are equivalent:*

(i)   *G is nilpotent.*

(ii)   *$Z(G/N) \neq 1$ for every proper normal subgroup $N \trianglelefteq G$.*

(iii)   *$[N, G] < N$ for every nontrivial normal subgroup $N \trianglelefteq G$.*

Proof. (i) $\Rightarrow$ (ii): This follows from 5.1.5 since factor groups of nilpotent groups are nilpotent.

(ii) $\Rightarrow$ (iii): Let $G \neq 1$ and $\overline{G} := G/Z(G)$. Then $Z(G) \neq 1$. Every factor group of G also satisfies (ii). Thus we may assume by induction on $|G|$ that either $\overline{N} = 1$ or $[\overline{N}, \overline{G}] < \overline{N}$. The first case gives $N \leq Z(G)$ and $[N, G] = 1 < N$. In the second case $[N, G] < N$ follows from 1.5.1 on page 24.

(iii) $\Rightarrow$ (i): Let $G \neq 1$ and M be a minimal normal subgroup of G. Since also $[M, G]$ is a normal subgroup of G (1.5.5 on page 26) we get $[M, G] = 1$ and thus

$$M \leq Z(G).$$

Let $\overline{G} := G/M$ and $M < N \trianglelefteq G$. Assume first that $[\overline{N}, \overline{G}] = \overline{N}$. Then $N = [N, G]M$ (1.5.1 on page 24) and

$$[N, G] = [[N, G], G].$$

Hence (iii) gives $[N, G] = 1$. But now $N = M$, which contradicts the choice of N.

We have shown that also $\overline{G}$ satisfies (iii). Thus, by induction on $|G|$ we may assume that $\overline{G}$ is nilpotent. But then G is nilpotent (5.1.2). □

The following is a useful property of nilpotent groups:

5.1.7   *Let G be nilpotent and N a maximal Abelian normal subgroup of G. Then $C_G(N) = N$.*

Proof. By way of contradiction we assume that $N < C_G(N) =: C$. Then C/N is a nontrivial normal subgroup of G/N. By 5.1.1 and 5.1.5

$$Z(G/N) \cap C/N \neq 1.$$

Let $N < U \leq G$ such that U/N is a cyclic subgroup of $Z(G/N) \cap C/N$. Then U is a normal subgroup, and U is Abelian (see 1.3.1 on page 16). This contradicts the maximality of N.   □

According to 5.1.6 (ii), resp. (iii), every nilpotent group G possesses a series of subgroups

$$1 = Z_0 \leq Z_1 \leq \cdots \leq Z_{i-1} \leq Z_i \leq \cdots \leq Z_{c-1} \leq Z_c = G$$

where $Z_i \trianglelefteq G$ and $Z_i/Z_{i-1} \leq Z(G/Z_{i-1})$ for $i = 1, \ldots, c$.[2]

Conversely, 5.1.2 shows that every group having such a **central series** is nilpotent.[3]

The length c of a shortest such central series of G is the **(nilpotent) class** of G, denoted by $c(G)$. For example, $c(G) = 1$ if $G \neq 1$ is Abelian; $c(G) \leq 2$ if $G/Z(G)$ is Abelian.

We conclude this section with two results about p-groups (of class 2) which we will need later.

5.1.8   *Let A and B be subgroups of the p-group G satisfying*

$$[A, B] \leq A \cap B \quad \text{and} \quad |[A, B]| \leq p.$$

Then

$$|A : C_A(B)| = |B : C_B(A)|.$$

---

[2]With 5.1.6 (ii) from the "bottom up" : $Z_1 := Z(G)$, $Z_2/Z_1 := Z(G/Z_1), \ldots$, and with 5.1.6 (iii) from the "top down": $G \geq [G, G] \geq [G, G, G] \geq \cdots$.

[3]An infinite group is **defined** to be **nilpotent**, if it possesses such a central series.

Proof. By 1.5.5 on page 26 $N := [A, B]$ is normal in $\langle A, B \rangle$. Hence, A/N and B/N centralize each other in $\langle A, B \rangle / N$. Moreover $N \leq Z(\langle A, B \rangle)$ since $|N| \leq p$ (3.1.11 on page 61). This gives

$$C_B(A) = C_B(AN) \trianglelefteq B.$$

Let

$$|B/C_B(A)| = p^n,$$

and $b_1, \ldots, b_n \in B$ such that

$$B = C_B(A)\langle b_1, \ldots, b_n \rangle.$$

A acts on $N b_i$ by conjugation ($i = 1, \ldots, n$). Let A_i be the kernel of this action. Then

$$C_A(B) = \bigcap_{i=1}^{n} A_i.$$

From $|N b_i| = |N| \leq p$ we get $|A/A_i| \leq p$ and thus with 1.6.4 on page 31

$$|A/C_A(B)| \leq p^n = |B/C_B(A)|.$$

The same argument with the roles of A and B reversed also gives

$$|B/C_B(A)| \leq |A/C_A(B)|.$$

Hence equality holds. $\square$

5.1.9 *Let P be a p-group and A a maximal Abelian subgroup of P. Suppose that $|P'| = p$. Then*

$$|P : A| = |A/Z(P)| \quad \text{and} \quad |P/Z(P)| = |A/Z(P)|^2.$$

In particular, all maximal Abelian subgroups of P have the same order.

Proof. The maximality of A implies

$$C_A(P) = Z(P), \quad C_P(A) = A;$$

and $|P'| = p$ implies $P' \leq Z(P)$. Hence $P' \leq A$. Thus, we can apply 5.1.8 with $B = P$. It follows that

$$|A/Z(P)| = |P/A|$$

and $|P/Z(P)| = |P/A|\,|A/Z(P)| = |A/Z(P)|^2$. $\square$

Exercises

1.   The dihedral group D_{2n} is nilpotent, if and only if n is a power of 2.

Let P be a p-group.

2.   Let M_1, M_2 be two different maximal subgroups of P. Then $P = M_1 M_2$ and $P/M_1 \cap M_2 \cong C_p \times C_p$.

3.   If P contains two different maximal subgroups that both are Abelian, then $P/Z(P)$ is Abelian.

4.   Let $U_1, \ldots, U_r$ be proper subgroups of P such that $P = U_1 \cup \cdots \cup U_r$. Then $r \geq p + 1$.

5.   Let A be a maximal Abelian normal subgroup of P. If $|A : C_A(x)| \leq p$ for all $x \in P$, then $P' \leq A$.

6.   Let $|P : C_P(x)| \leq p^2$ for all $x \in P$. Then P' is Abelian.

5.2   Nilpotent Normal Subgroups

Let G be a group and N a nilpotent normal subgroup of G. Then 5.1.4 (iii) shows:

$$N = \underset{p\in\pi(N)}{\times} O_p(N).$$

Since $O_p(N)$ is characteristic in N it is normal in G. Hence

$$O_p(N) \leq O_p(G).$$

The product of all nilpotent normal subgroups of G is a characteristic subgroup of G; it is called the **Fitting subgroup** of G and denoted by $F(G)$.[4]

As we have seen, $F(G)$ is contained in the product of the subgroups $O_p(G)$, $p \in \pi(G)$. On the other hand, by 5.1.3 each of the normal subgroups $O_p(G)$ is nilpotent and thus contained in $F(G)$. Thus:

5.2.1   (a) $F(G)$ is the largest nilpotent normal subgroup of G.

(b) $F(G) = \underset{p\in\pi(G)}{\times} O_p(G).$ □

---

[4]See [44].

For the following important property of the Fitting subgroup compare with 6.1.4 on page 123 and Section 6.5.

5.2.2   Let $C := C_G(F(G))$. Then

$$O_p(C/C \cap F(G)) = 1$$

for all $p \in \mathbb{P}$.

Proof. Let P be the inverse image of $O_p(C/C \cap F(G))$ in C. Then P is normal in G (1.3.2 on page 17), and P is nilpotent since $C \cap F(G) \leq Z(C)$ (5.1.2). Hence $P \leq F(G) \cap C$ and $O_p(C/C \cap F(G)) = 1$.   □

The intersection of all maximal subgroups of G is a characteristic subgroup of G; it is called the **Frattini subgroup** of G and denoted by $\Phi(G)$.[5] If $G = 1$ then $\Phi(G) = 1$ since G does not possess maximal subgroups. The crucial property of the Frattini subgroup is:

5.2.3   Let H be a subgroup of G such that $G = H\Phi(G)$. Then $G = H$.

Proof. If $G \neq H$, then there exists a maximal subgroup of G containing H and $\Phi(G)$. This contradicts $G = H\Phi(G)$.   □

5.2.4   Let N be a normal subgroup of G. Then $\Phi(G)N/N \leq \Phi(G/N)$.

Proof. The maximal subgroups of G/N are exactly the maximal subgroups of G that contain N.   □

An application of the Frattini argument gives:

5.2.5   (a) $\Phi(G)$ is nilpotent.

(b)   If $G/\Phi(G)$ is nilpotent, then G is nilpotent.

---

[5]See [45].

Proof. (a) By 5.1.4 (ii) it suffices to show that every Sylow p-subgroup P of $\Phi(G)$ is normal in $\Phi(G)$. The Frattini argument 3.2.7 yields $G = N_G(P)\Phi(G)$ and thus $G = N_G(P)$ by 5.2.3.

(b) By an argument similar to that of (a), we show that every Sylow p-subgroup P of G is normal in G. By 3.2.5 on page 65 $P\Phi(G)/\Phi(G)$ is a Sylow p-subgroup of the nilpotent group $G/\Phi(G)$ and thus normal in $G/\Phi(G)$ (5.1.4 (ii)). Hence $N := P\Phi(G)$ is normal in G. Now $P \in \mathrm{Syl}_p G$ implies

$$G \overset{3.2.7}{=} N_G(P)N = N_G(P)P\Phi(G) = N_G(P)\Phi(G),$$

and again $G = N_G(P)$. $\qquad\qquad\qquad\qquad\qquad\qquad\qquad\qquad\qquad\qquad$ $\square$

The last results of this section deal with the Frattini subgroup of a p-group. The first one is an observation that follows from 2.1.2:

5.2.6   *Let P be an elementary Abelian p-group. Then $\Phi(P) = 1$.* $\square$

5.2.7   *Let P be a p-group.*

(a) $P/\Phi(P)$ *is elementary Abelian.*

(b)   *If $|P/\Phi(P)| = p^n$, then there exist $x_1, \ldots, x_n \in P$ such that $P = \langle x_1, \ldots, x_n \rangle$.*

Proof. (a) In a nilpotent group every maximal subgroup is normal of index p. Hence (a) follows from 1.6.4 on page 31.

(b) Because of (a) and 2.1.8 (a) $|P/\Phi(P)| = p^n$ is generated by n elements $x_1\Phi(P), \ldots, x_n\Phi(P)$, $x_i \in P$. Hence $P = \langle x_1, \ldots, x_n \rangle\Phi(P) = \langle x_1, \ldots, x_n \rangle$ by 5.2.3. $\qquad\qquad\qquad\qquad\qquad\qquad\qquad\qquad\qquad\qquad\qquad\qquad$ $\square$

5.2.8   *Let P be a p-group. Then $\Phi(P)$ is the smallest normal subgroup of P that has an elementary Abelian factor group.*

Proof. Let $N \trianglelefteq P$ such that P/N is elementary Abelian. By 5.2.4 $\Phi(P)N/N \leq \Phi(P/N)$, so 5.2.6 shows that $\Phi(P) \leq N$. The result now follows from 5.2.7 (a). $\qquad\qquad\qquad\qquad\qquad\qquad\qquad\qquad\qquad\qquad\qquad$ $\square$

A p-group P is **special** if P is non-Abelian and

$$P' = \Phi(P) = Z(P) = \Omega(Z(P)).$$

If, in addition, $Z(P)$ is cyclic, then P is **extraspecial**.

From 5.1.9 we obtain:

5.2.9 *Let P be an extraspecial group and A a maximal Abelian subgroup of order p^n. Then $|P| = p^{2n-1}$.* $\qquad\square$

It should be pointed out that an extraspecial group is a central product of non-Abelian subgroups of order p^3.[6]

Exercises

Let G be a group.

1. $F(G/\Phi(G)) = F(G)/\Phi(G)$.

2. If $F(G)$ is a p-group, then $F(G/F(G))$ is a p'-group.

3. Suppose that G is nilpotent. Then the following statements are equivalent:

   (i) G is cyclic.

   (ii) G/G' is cyclic.

   (iii) Every Sylow p-subgroup of G is cyclic.

4. G is nilpotent if and only if every maximal subgroup of G is normal in G.

5. G is nilpotent if and only if for every noncyclic subgroup $U \leq G$:

   $$\langle x^U \rangle \neq U \text{ for all } x \in U.$$

6. $N \trianglelefteq G \;\Rightarrow\; \Phi(N) \leq \Phi(G)$.

7. Let $p \in \pi(G)$ such that $O_p(G) = 1$, and let N be a normal subgroup of G such that G/N is a p-group. Then $\Phi(G) = \Phi(N)$.

8. Let N be a normal subgroup of G such that G/N is nilpotent. Then there exists a nilpotent subgroup U of G such that $G = NU$.

9. Let P be a p-group. If $p = 2$, then

   $$\Phi(P) = \langle x^p \mid x \in G \rangle.$$

   Give a counterexample for $p \neq 2$.

---

[6]See Exercise 4 on p. 118.

5.3   p-Groups with Cyclic Maximal Subgroups

In this section we determine all the p-groups that contain a cyclic maximal subgroup.

Let P be a p-group and H a maximal subgroup of P. By 3.1.10 on page 61 H is normal in P and $P/H \cong C_p$. We first deal with the Abelian case:

5.3.1   *Let P be Abelian and H a cyclic maximal subgroup of P. Then either P is cyclic or $P = H \times C$, $C \cong C_p$.*

Proof. We may assume that P is not cyclic. Then H is a cyclic subgroup of maximal order in P. Hence, 2.1.1 on page 43 gives $P = H \times C$, $C \cong C_p$. □

We will treat separately the two cases

- H has a complement A in P.

- H has no complement in P.

In the first case P is the semidirect product of H with A and, as we have seen already, $A \cong C_p$. If in addition also H is cyclic, then the multiplication in P is completely determined by the action of A on H. Thus:

5.3.2   *Let P be a non-Abelian p-group and $H = \langle h \rangle$ a cyclic maximal subgroup of P, $|H| = p^n$. Suppose that H has a complement $A = \langle a \rangle$ in P. Then one of the following holds:*

(a) $p \neq 2$ and $h^a = h^{1+p^{n-1}}$ (if $a \in A$ is chosen suitably).

(b) $p = 2$ and $h^a = h^{-1}$.

(c) $p = 2$, $n \geq 3$ and $h^a = h^{-1+2^{n-1}}$.

(d) $p = 2$, $n \geq 3$ and $h^a = h^{1+2^{n-1}}$.

The four cases above describe four different isomorphism types of P.

Proof. The statements (a)–(d) follow from 2.2.6 on page 52. It remains to show that they describe different isomorphism types. This is only a problem for the cases (b), (c), and (d). In all these cases the involution

$$z := h^{2^{n-1}}$$

is in $Z(P)$. Moreover for $i \in \mathbb{N}$

$$(h^i)^a = \begin{cases} h^{-i}z^i & \text{in case (c)} \\ h^i z^i & \text{in case (d)} \end{cases}.$$

This gives

$$Z(P) = \begin{cases} \langle z \rangle & \text{(b)} \\ \langle z \rangle & \text{in case (c)} \\ \langle h^2 \rangle & \text{(d)} \end{cases}.$$

Hence it suffices to investigate the cases (b) and (c). In case (b) every element in $P \setminus H$ is an involution, while in case (c) $ha \in P \setminus H$ is an element of order 4. $\qquad\square$

In case (b) P is a dihedral group (see 1.6.9 on page 34). In case (c) P is said to be a **semidihedral group**.

We now turn to the *nonsemidirect* case and introduce the quaternion groups that arise here.

Let $3 \leq n \in \mathbb{N}$,

$$H = \langle h_1 \rangle \cong C_{2^{n-1}} \quad \text{and} \quad A = \langle a_1 \rangle \cong C_4.$$

Then A acts on H according to

$$h_1^{a_1} = h_1^{-1}.$$

In particular, $\langle a_1^2 \rangle$ acts trivially on H. Let P be the semidirect product AH with respect to this action. Then

$$\langle a_1^2 \rangle \langle h_1^{2^{n-2}} \rangle \quad (\cong C_2 \times C_2)$$

is a subgroup of $Z(P)$. Set

$$N := \langle a_1^2 h_1^{2^{n-2}} \rangle.$$

The group P/N (and every group isomorphic to P/N) is called a **quaternion group** of order 2^n and is denoted by Q_{2^n}.[7] Let

$$a = a_1 N \quad \text{and} \quad h = h_1 N.$$

Then

$$Q_{2^n} = \langle a, h \rangle, \text{ and}$$

$$o(h) = 2^{n-1}, \quad o(a) = 4, \quad h^a = h^{-1}, \quad h^{2^{n-2}} = a^2.$$

These relations determine the multiplication table of Q_{2^n}, i.e., all quaternion groups of a given order are isomorphic.

The quaternion group of order 8 can be found as a subgroup of $\mathrm{GL}(2, \mathbb{C})$, the group of all invertible 2×2-matrices over the field $\mathbb{C}$ of complex numbers, in the following way: Let

$$h = \begin{pmatrix} i & 0 \\ 0 & -i \end{pmatrix} \quad \text{and} \quad a = \begin{pmatrix} 0 & 1 \\ -1 & 0 \end{pmatrix}.$$

Then $h^2 = a^2 = \begin{pmatrix} -1 & 0 \\ 0 & -1 \end{pmatrix}$, and the subgroup $\langle h, a \rangle$ in $\mathrm{GL}(2, \mathbb{C})$ is a quaternion group.

The basic properties of a quaternion group $Q := Q_{2^n}$ are:

- $\langle h \rangle$ is a normal subgroup of index 2 in Q.

- $x^2 = h^{2^{n-2}}$ for every $x \in Q \setminus \langle h \rangle$.

- $Z(Q) = \langle h^{2^{n-2}} \rangle$.

- $Z(Q)$ is the unique subgroup of order 2 in Q.

- Every subgroup of Q is either a quaternion group or cyclic.

- If Q has order 8, then every subgroup is normal in Q.

We note:

5.3.3 Aut $Q_8 \cong S_4$.

---

[7]For $n > 2$ Q_{2^n} is also said to be a **generalized** quaternion group.

Proof. $A := \operatorname{Aut} Q_8$ acts on the set Ω of maximal subgroups of Q_8, and $|\Omega| = 3$. Hence, there exists a homomorphism

$$\varphi \colon A \to S_3.$$

Let x, y be elements of order 4 in Q_8 such that $y \notin \langle x \rangle$. Then

$$\Omega = \{\langle x \rangle, \langle y \rangle, \langle xy \rangle\}$$

and

$$x^y = x^{-1}, \quad y^x = y^{-1}, \quad x^2 = y^2.$$

These relations show that A contains an element that interchanges x and y. Thus $\operatorname{Im} \varphi$ contains all the transpositions of S_3, i.e.,

$$\operatorname{Im} \varphi = S_3.$$

We have

$$N := \operatorname{Inn} Q_8 \cong Q_8/Z(Q_8) \cong C_2 \times C_2$$

and $N \leq \operatorname{Ker} \varphi$. We first show:

(') $N = \operatorname{Ker} \varphi.$

Let $a \in \operatorname{Ker} \varphi$. Since $x^y = x^{-1}$ we may assume that $x^a = x$. If $y^a = y$ then $a = 1$, and if $y^a = y^{-1}$ then a is the inner inner automorphism induced by x. This shows (').

Now $A/N \cong S_3$ and thus $|A| = 24$. A subgroup of order 3 of $\operatorname{Im} \varphi$ is transitive on Ω, and thus on $N^\#$. This gives $Z(A) = 1$, and A is not 3-closed. Hence 4.3.4 on page 90 yields the conclusion. $\square$

To prove that only quaternion groups arise in the *nonsemidirect* case we need the following lemma:

5.3.4   *Let x, y be elements of the p-group P, and let*

$$[x, y] \in \Omega(Z(P)).$$

(a)   *If $p \neq 2$ then $(xy)^p = x^p y^p$.*

(b)   *If $p = 2$ then $(xy)^2 = x^2 y^2 [x, y]$ and $(xy)^4 = x^4 y^4$.*

Proof. Let $z := [x, y]$. Then $x^y = xz$ and $x^{y^i} = xz^i$ for $i \geq 1$. The hypothesis implies $z^p = 1$ and thus

$$(x^p)^y = (x^y)^p = x^p z^p = x^p.$$

Assume $p = 2$. Then

$$(xy)^2 = xy\,xy = xy^2 x[x, y] = x^2 y^2 z,$$

and $(xy)^4 = x^2 y^2 z x^2 y^2 z = x^4 y^4$.

Assume $p \neq 2$. Then

$$
\begin{aligned}
(xy^{-1})^p &= (xy^{-1})(xyy^{-2})(xy^2 y^{-3}) \cdots (xy^{p-1} y^{-p}) \\
&= x\,xz\,xz^2 \cdots xz^{p-1}\,y^{-p} \\
&= (x^p y^{-p})(z\,z^2 \cdots z^{p-1}).
\end{aligned}
$$

Since

$$z z^p \cdots z^{p-1} = z^{\frac{p(p-1)}{2}} = 1$$

(a) follows. □

As a corollary we have:

5.3.5   *Let P be a p-group and $p \neq 2$. Suppose that $P/Z(P)$ is Abelian.*[8]
Then

$$\Omega(P) = \{x \in P \mid x^p = 1\}.$$

Proof. Let $x, y \in P$ such that $x^p = y^p = 1$. Because $P/Z(P)$ is Abelian, the element $z := [x, y]$ is contained in $Z(P)$. Hence

$$1 = (x^p)^y = (x^y)^p = x^p z^p = z^p$$

and thus $z \in \Omega(Z(P))$ since $x^y = xz$. Now the result follows from 5.3.4 (a). □

Let P be a p-group with a cyclic maximal subgroup H, which does not have a complement in P. Then $1 \neq x^p \in H$ for every $x \in P \setminus H$, and $\Omega(H)$ is the unique subgroup of order p in P.

---

[8]That is, $c(P) \leq 2$.

5.3.6 *Let P be a non-Abelian p-group with a cyclic maximal subgroup H. Suppose that*

(∗) $$1 \neq x^p \in H \text{ for all } x \in P \setminus H.$$

Then P is a quaternion group; in particular $p = 2$.

Proof. Let $H = \langle h \rangle$, $o(h) = p^n$, and $z := h^{p^{n-1}}$. Then

$$\Omega(Z(P)) = \langle z \rangle.$$

Choose $x \in P \setminus H$ such that $o(x)$ is minimal. Since P is not cyclic we get $\langle x^p \rangle \leq \langle h^p \rangle$. Let $h_0 \in H$ such that

$$x^p = h_0^p.$$

The case $p \neq 2$ leads to a contradiction: Replacing x by a suitable power of x one gets from 2.2.6 (a) on page 52

$$h_0^x = h_0^{1+p^{n-1}}$$

and thus $[h_0^{-1}, x] \in \Omega(Z(P))$. Now 5.3.4 gives $(h_0^{-1}x)^p = h_0^{-p}x^p = 1$ which contradicts $h_0^{-1}x \in P \setminus H$.

Let $p = 2$. According to 2.2.6 we have to discuss the following cases:

(b) $h^x = h^{-1}$, i.e., $[h, x] = h^{-2}$.

(c) $n \geq 3$ and $h^x = h^{-1}z$, i.e., $[h, x] = h^{-2}z$.

(d) $n \geq 3$ and $h^x = hz$, i.e., $[h, x] = z$.

In case (d) we derive a contradiction. Namely, in this case for every power y of h

$$[y, x] \in \langle z \rangle.$$

5.3.4 (b) gives for $y := h_0^{-1}$

$$(h_0^{-1}x)^2 = h_0^{-2}x^2z = z$$

and thus $o(h_0^{-1}x) = 4$. The minimal choice of $o(x)$ yields $o(x) = 4$ and thus $x^2 = z$. It follows that $h_0^4 = 1$. On the other hand $(h^2)^x = h^2z^2 = h^2$

and $C_H(x) = \langle h^2 \rangle$. If $h_0 \in C_H(x)$ then $o(h_0^{-1}x) = 2$, which contradicts $h_0^{-1}x \in P \setminus H$. Hence $h_0 \in H \setminus \langle h^2 \rangle$ and

$$H = \langle h_0 \rangle \cong C_4,$$

which contradicts $n \geq 3$.

In the cases (b) and (c)

$$x^2 = (x^2)^x = (h_0^2)^x = h_0^{-2} = x^{-2}$$

and thus also $o(x) = 4$, i.e.,

$$x^2 = z.$$

Case (c) gives

$$(hx)^2 = hxhx = hx^{-1}hxx^2 = z^2 = 1,$$

which contradicts $hx \in P \setminus H$.

Hence we are in case (b), and $x^2 = z$ shows that P is quaternion group of order 2^{n+1}.   □

5.3.7   Theorem. *Let P be a p-group containing a unique subgroup of order p. Then either P is cyclic, or $p = 2$ and P is a quaternion group.*

Proof. By 2.1.7 on page 46 we may assume that P is not Abelian. Moreover, since also every subgroup U of P with $1 \neq U \neq P$ contains a unique subgroup of order p we may assume by induction on $|P|$ that U is cyclic, or $p = 2$ and U is a quaternion group. Let H be a maximal Abelian normal subgroup of P. Then H is cyclic and

$$C_P(H) = H$$

(5.1.7). Hence, we may regard $A := P/H$ as a subgroup of $\mathrm{Aut}\,H$ (3.1.9 on page 60). Let Q/H, $H \leq Q \leq P$, be a subgroup of order p in A. Then Q is non-Abelian and

$$1 \neq x^p \in H \quad \text{for all} \quad x \in Q \setminus H$$

since H contains the only subgroup of order p in P. Thus, $p = 2$ and Q is a quaternion group (5.3.6). In particular

$$h^a = h^{-1}$$

for every $a \in Q \setminus H$. Therefore the 2-group A contains only one subgroup of order 2. By 2.2.6 (c) $h^{\alpha^2} \neq h^{-1}$ for all $\alpha \in \operatorname{Aut} H$. It follows that $|A| = 2$ and $Q = P$. $\qquad\square$

As a corollary we have:

5.3.8 *Suppose that P is a p-group all of whose Abelian subgroups are cyclic. Then P is cyclic or a quaternion group.*

Proof. Let $U \leq P$ such that $|U| = p$. Because $Z(P)U$ is Abelian and thus cyclic, U is the unique subgroup of order p in $Z(P)U$ and thus $U \leq Z(P)$ since $Z(P) \neq 1$. Hence U is also unique in P, and the claim follows from 5.3.7. $\qquad\square$

In the following we will describe the *non-Abelian p-subgroups of order p^3.*[9]

Suppose that P contains an element h of order p^2. Then $H := \langle h \rangle$ is a cyclic maximal subgroup of P. This case was treated in 5.3.2 and 5.3.6:

If $p \neq 2$, then there exists $a \in P \setminus H$ such that $o(a) = p$ and $h^a = h^{1+p}$.

If $p = 2$, then P is a dihedral or quaternion group of order 8.

Suppose that P does not contain any element of order p^2; i.e.,

$(')$ $\qquad\qquad\qquad\qquad x^p = 1$ for all $x \in P$.

Then P is not isomorphic to one of the groups just considered. Since P is Abelian for $p = 2$ (Exercise 8 on page 10) we also have

$('')$ $\qquad\qquad\qquad\qquad p \neq 2$.

We now show that (up to isomorphism) P is uniquely determined by $(')$ and $('')$:

Let H be a subgroup of order p^2 in P and $a \in P \setminus H$. Then $P = \langle a \rangle H$ a semidirect product with

(1) $\qquad\qquad\qquad H \cong C_p \times C_p \quad \text{and} \quad \langle a \rangle \cong C_p$

(5.3.1). By 3.1.11 (a) on page 61 there exists $1 \neq z \in H$ such that $z^a = z$. Since P is non-Abelian we get $h^a \neq h$ for $h \in H \setminus \langle z \rangle$, and $[h, a] \in \langle z \rangle$ since

---

[9]These and other *small* groups were first determined by Hölder [69].

$P/\langle z \rangle$ is Abelian (5.3.1). After replacing z by a suitable power, we see that the action of a on H is given by

$$(2) \qquad H = \langle z \rangle \times \langle h \rangle, \quad z^a = z, \quad h^a = zh.$$

Hence (1) and (2) determine the isomorphism type of P.

It should be pointed out that the dihedral group of order 8 is also a semidirect product $\langle a \rangle H$ satisfying (1) and (2).

We conclude this section with two results that will be needed in Chapter 12.

5.3.9   *Let P be a p-group all of whose Abelian normal subgroups are cyclic. Then P is cyclic, or $p = 2$ and P is a quaternion group or a dihedral group of order > 8 or a semidihedral group.*

Proof. As in the proof of 5.3.7 let H be a maximal Abelian normal subgroup of P. Then $C_P(H) = H$ and by our hypothesis H is cyclic. We may assume that $H \neq P$. Then

$$(1) \qquad H \neq \Omega(H) = \Omega(Z(P)) \cap H$$

and P/H is Abelian. Hence, every subgroup containing H is normal in P. Let X be the maximal subgroup of H. We show:

($'$)   Let $a \in P \setminus H$ such that $a^p \in H$. Then $x^a = x^{-1}$ for all $x \in X$; in particular $p = 2$.

Assume that ($'$) does not hold. Then

$$(2) \qquad |H| \geq 2^3 \text{ if } p = 2,$$

and $H\langle a \rangle$ is not a quaternion group. Thus, we may choose a such that $o(a) = p$. Since a induces a nontrivial automorphism on H the result 2.2.6 on page 52 resp. 5.3.1 shows that

$$[H, a] = \Omega(H) \quad \text{and} \quad [X, a] = 1.$$

Assume $h \in H \setminus X$ and $o(ha) = p$. Then

$$o(h) = o(a^{-1}(ah)) \overset{5.3.4}{=} \begin{cases} p & \text{if } p \neq 2 \\ \leq 4 & \text{if } p = 2 \end{cases},$$

which contradicts (1), resp. (2). Hence, $\Omega(H\langle a \rangle)$ is contained in the non-cyclic Abelian group $X\langle a \rangle$. Therefore

$$C_p \times C_p \cong \Omega(H\langle a \rangle) \text{ char } H\langle a \rangle \trianglelefteq P,$$

which contradicts the hypothesis. Thus $(')$ is proved.

In 2.2.6 (b) Aut H has been described. It follows from there and $(')$ that $|P/H| = 2$. Now 5.3.2 gives the assertion; notice that in the case $P \cong D_8$ there exists a noncyclic Abelian normal subgroup in P. $\qquad \square$

5.3.10 *Let P be a 2-group and t an involution of P such that*

$$C_P(t) \cong C_2 \times C_2.$$

Then P is a dihedral or semidihedral group.[10]

Proof. Let H be a maximal Abelian normal subgroup of P. By 5.1.7

(1) $|P/H| \leq |\operatorname{Aut} H|$.

Because of 5.3.9 we may assume that H is not cyclic. Then H contains a subgroup isomorphic to $C_2 \times C_2$ (2.1.7).

Assume first that $t \in H$. Then $H \leq C_P(t)$ and thus $H \cong C_2 \times C_2$ and by (1) $|P/H| \leq 2$. The case $P = H$ gives $P = D_4$, and the case $|P| = 8$ gives $P = D_8$.[11]

Assume now that $t \notin H$. Then

(2) $C_2 \cong C_H(t) =: Z$.

The set

$$K := \{ [x, t] \mid x \in H \}$$

is a subgroup of H since H is Abelian. For $x, y \in H$

$$[x, t] = [y, t] \iff xZ = yZ.$$

---

[10]Conversely, dihedral and semidihedral groups have this property.
[11]Since otherwise $P = Q_8$ but Q_8 contains only one involution.

Hence

$$|H : K| = 2.$$

Moreover

$$[x, t]^t = (x^{-1}txt)^t = tx^{-1}tx = [x, t]^{-1},$$

and thus

(3) $k^t = k^{-1}$ for all $k \in K$.

In particular, the involutions of K are contained in $C_H(t)$. Hence, by (2) Z is the unique subgroup of order 2 in K, and K is cyclic (2.1.7). Since H is noncyclic there exists an involution $y \in H \setminus K$. Then also $[y, t]$ is an involution and thus

$$y^t = yz \quad \text{with} \quad \langle z \rangle := Z.$$

In the case $|H| = 4$, (1) yields $P = \langle y, t \rangle \cong D_8$.

In the case $|H| > 4$, we have $|K| \geq 4$, and there exists $k \in K$ such that $o(k) = 4$. It follows that $k^2 = z$ and

$$k^t \stackrel{(3)}{=} k^{-1} = kz.$$

This implies

$$(yk)^t = y^t k^t = yzkz = yk,$$

which contradicts (2) since $o(yk) = 4$.   □

Exercises

1.   Determine $\Omega(P)$ for all p-groups P of order p^3.

2.   Let A, B be two non-Abelian groups of order p^3, $Z(A) = \langle a \rangle$ and $Z(B) = \langle b \rangle$, and let $P := (A \times B)/\langle ab \rangle$. Then P is an extraspecial p-group.

3.   Let P be an extraspecial p-group of order p^3. Then

$$\text{Inn } P = \{\alpha \in \text{Aut } P \mid (xZ(P))^\alpha = xZ(P) \text{ for all } x \in P\}.$$

4.   Every extraspecial p-group is a central product of extraspecial p-groups of order p^3.

5.   Let P be a p-group and $Z_2(P')$ be the inverse image of $Z(P'/Z(P'))$ in P'. Then P' is cyclic if and only if $Z_2(P')$ is cyclic.

6. In the group $GL_2(3)$ of all invertible matrices over $\mathbb{F}_3$ let

$$P := \left\langle \begin{pmatrix} 0 & -1 \\ 1 & 0 \end{pmatrix}, \begin{pmatrix} -1 & 1 \\ 1 & 1 \end{pmatrix} \right\rangle.$$

Then $P \cong Q_8$ and $P \trianglelefteq GL_2(3)$.

7. $\operatorname{Aut} Q_{2^n}$ is a 2-group if and only if $n \geq 4$.

Chapter 6

Normal and Subnormal Structure

6.1 Solvable Groups

A group G is **solvable**,[1] if

$$U' \neq U \quad \text{for all subgroups } 1 \neq U \leq G.$$

Abelian and nilpotent groups are examples of solvable groups (5.1.1 on page 99). Thus, p-groups are solvable.

Let G be a dihedral group. Then G has a cyclic normal subgroup N of index 2 (1.6.9 on page 34). Hence, $U' \leq N$ for every subgroup $U \leq G$, and $U' = 1$ if $U \leq N$. This shows that dihedral groups are solvable.

Further examples of solvable groups are the symmetric groups S_3 and S_4. For S_3 this follows from the above since S_3 is a dihedral group. For S_4 a similar argument as for the dihedral groups, using the chief series of S_4 given in Section 4.3, yields the desired conclusion.

Groups containing a non-Abelian simple subgroup E cannot be solvable since $E = E'$; in particular, S_n is not solvable for $n \geq 5$ (4.3.5 on page 90).

By a classical Theorem of Burnside every group of order $p^a q^b$ $(q, p \in \mathbb{P})$ is solvable; we will prove this theorem in Section 10.2.

---

[1]For infinite groups *solvability* is defined differently; see the footnote on p. 123.

One of the most famous theorems in group theory, the theorem of Feit-Thompson [43], states that every group of odd order is solvable.[2]

6.1.1 *Subgroups and homomorphic images of solvable groups are solvable.*

Proof. For subgroups this is clear by the definition of solvability. Let φ be a homomorphism of the solvable group G. Let $1 \neq V \leq G^\varphi$, and let $U \leq G$ be of minimal order such that $U^\varphi = V$. By 1.5.1 on page 24

$$(U')^\varphi = V'.$$

Since $U' < U$ the minimality of U gives $V' < V$.   □

6.1.2 *A group G is solvable if and only if there exists a normal subgroup N of G such that N and G/N are solvable.*

Proof. One implication follows from 6.1.1. Let N be a normal subgroup of G such that N and G/N are solvable, and let $1 \neq U \leq G$. If $U \leq N$, then $U' < U$ by the solvability of N. If $U \not\leq N$, then $V := UN/N$ is a nontrivial subgroup of the solvable group G/N and

$$U'N/N \overset{1.5.1}{=} V' < V = UN/N.$$

Hence also in this case $U' < U$.   □

6.1.3 *Every minimal normal subgroup N of a solvable group G is an elementary Abelian p-group.*

Proof. 1 and N are the only characteristic subgroups of N (1.3.2 on page 17). Hence $N' = 1$, $|\pi(N)| = 1$ (2.1.6 on page 46) and $\Omega(N) = N$.   □

Result 6.1.3 shows that for a nontrivial solvable group G there exists $p \in \pi(G)$ such that $O_p(G) \neq 1$. Moreover, since also

$$G_1 := C_G(F(G))/C_G(F(G)) \cap F(G)$$

is solvable (6.1.1), we get $G_1 = 1$ by 5.2.2 on page 105. Thus (compare also with 6.5.8 on page 144) we have:

---

[2]See also [3].

6.1.4 *Let G be a solvable group. Then $C_G(F(G)) \leq F(G)$.* $\square$

In a group G the series

$$G^1 := G' \geq G^2 := (G^1)' \geq \cdots \geq G^i := (G^{i-1})' \geq \cdots$$

is the **commutator series** of G; notice that all subgroups in this series are characteristic in G.

6.1.5 **Theorem.** *For a group G the following statements are equivalent:*

(i) *G is solvable.*

(ii) *There exists $l \in \mathbb{N}$ such that $G^l = 1$.[3]*

(iii) *G possesses a normal series all of whose factors are Abelian.*

(iv) *G possesses a composition series all of whose factors have prime order.*

Proof. The implication (i) $\Rightarrow$ (ii) follows directly from the definition of solvability, and the implication (ii) $\Rightarrow$ (iii) is trivial. Assume (iii). Then we can extend this series to a composition series of G all of whose factors are of prime order[4] (see Section 1.8).

(iv) $\Rightarrow$ (i): Let $(A_i)_{i=0,\dots,a}$ be the composition series given in (iv). Then $N := A_{a-1}$ is a normal subgroup of G with cyclic factor group G/N. Since $(A_i)_{i=0,\dots,a-1}$ is a composition series of N we may assume by induction on $|G|$ that N is solvable. Now (i) follows from 6.1.2. $\square$

As in Section 5.1, where we considered nilpotent normal subgroups of arbitrary finite groups, we now investigate solvable normal subgroups.

6.1.6 *Let A and B be two solvable normal subgroups of the group G. Then the product AB is also a solvable normal subgroup of G.*

---

[3]An infinite group G is **defined** to be **solvable** if G satisfies this property.

[4]By the way, also to a chief series all of whose factors are elementary Abelian; compare with 6.1.3.

Proof. Since $AB/A \cong A/A \cap B$ the result follows from 6.1.1 and 6.1.2. $\square$

Result 6.1.6 shows that the product

$$S(G) := \prod_{\substack{A \trianglelefteq G \\ A \text{ solvable}}} A$$

is a (characteristic) solvable subgroup of G. Hence $S(G)$ is *the* largest solvable normal subgroup of G. In particular, the direct product of solvable groups is solvable.

Exercises

Let G be a group.

1. Let G be solvable. Then there exists a normal maximal subgroup in G.

2. Determine the commutator series of S_4.

3. G is solvable if one of the following holds:

    (a) $|G| = p^n q$ $(p, q \in \mathbb{P})$.

    (b) $|G| = pqr$ $(p, q, r \in \mathbb{P})$.

4. Determine all nonsolvable groups of order ≤ 100.

5. Let G be solvable. Suppose that all Sylow subgroups of G are cyclic. Then G' is Abelian.

6. Let G be solvable and $\Phi(G) = 1$. If G contains exactly one minimal normal subgroup N, then $N = F(G)$.

7. Suppose that every nontrivial homomorphic image of G contains a nontrivial cyclic normal subgroup (such a group G is **super-solvable**). Then $G/F(G)$ is Abelian.

8. (Carter, [36]) Let G be solvable. Then G contains exactly one conjugacy class of nilpotent subgroups A satisfying $N_G(A) = A$.[5]

9. Let G be solvable and $p \in \pi(G)$. Suppose that $N_G(P)/C_G(P)$ is a p-group for every p-subgroup P of G. Then G contains a normal p'-subgroup N such that G/N is a p-group. (Compare with 7.2.4 on page 170 and Exercise 6 on page 172.)

10. Suppose that every maximal subgroup of G is nilpotent. Then G is solvable.

---

[5]Such a subgroup A is a **Carter subgroup** of G.

11.   Let A and B be Abelian subgroups of G such that $G = AB$. Then G is solvable. (Do not use Exercise 5 on page 27.)

6.2   The Theorem of Schur-Zassenhaus

Let G be a group and K a normal subgroup of G such that

$$(|K|, |G/K|) = 1.$$

For Abelian K we have proved in 3.3.1 on page 73 that K has a complement in G and that all such complements are conjugate in G. The next theorem generalizes this result.

6.2.1   Theorem of Schur-Zassenhaus. *Let G a group and K a normal subgroup of G such that $(|K|,|G/K|) = 1$. Then K has a complement in G. If in addition K or G/K is solvable, then all such complements are conjugate in G.*[6]

Proof. Let $U \leq G$ and $N \trianglelefteq G$. Then

$$UK/K \cong U/U \cap K \quad \text{and} \quad (G/N)/(KN/N) \cong G/KN.$$

Hence, $U \cap K$ is a normal subgroup of U such that $(|U \cap K|, |U/U \cap K|) = 1$, and KN/N is a normal subgroup of G/N such that $(|KN/N|, |G/KN|) = 1$. Thus, the hypothesis is inherited by subgroups and factor groups of G. If in addition K or G/K is solvable, then by 6.1.1 also this property is inherited.

We now prove the existence of a complement by induction on $|G|$. Hence, we may assume that in all groups of order less than $|G|$ that satisfy the hypothesis, such complements exist. Moreover, we can assume that $1 \neq K < G$.

Let $p \in \pi(K)$, $P \in \mathrm{Syl}_p K$ and

$$U := N_G(P).$$

---

[6]If H is such a complement, then the factorization $G = KH$ shows that all complements are already conjugate under K.

First assume that $U \neq G$. Then by induction $U \cap K$ has a complement H in U. The Frattini argument yields

$$G = KU = K(U \cap K)H = KH.$$

Hence H is also a complement of K in G since $H \cap K = H \cap (U \cap K) = 1$. Assume now that $U = G$. Then P and thus also

$$N := Z(P) \overset{3.1.11}{\neq} 1$$

is a normal subgroup of G (1.3.2 on page 17). Let $\overline{G} := G/N$. By induction there exists $N \leq V \leq G$ such that $\overline{V}$ is a complement of $\overline{K}$ in $\overline{G}$. Then

$$V \cap K = N \quad \text{and} \quad G = KV.$$

Hence, a complement of N in V is also a complement of K in G. If $V \neq G$, then by induction such a complement exists. If $V = G$, then $\overline{K} = 1$ and K is Abelian. Now 3.3.1 on page 73 gives the desired complement.

Using the additional hypothesis that K or G/K is solvable we now show, again by induction on $|G|$, that all complements are conjugate in G. Let H and H_1 be two complements of K in G, and let N be a minimal normal subgroup of G that is contained in K. We set $\overline{G} := G/N$. Then $\overline{H}$ and $\overline{H_1}$ are complements of $\overline{K}$ in $\overline{G}$, and by induction there exists $g \in G$ such that

$$HN = (H_1 N)^g = H_1{}^g N.$$

Thus, H and $H_1{}^g$ are complements of N in HN. If $N \neq K$, then $HN \neq G$, and by induction the complements H and $H_1{}^g$ are conjugate in HN. Hence H and H_1 are conjugate in G.

Assume that $N = K$. If K is solvable, then N is a solvable minimal normal subgroup and thus Abelian (6.1.3 on page 122). Now the conclusion follows from 3.3.1 on page 73.

Assume that K is not solvable; i.e., $\overline{G} = G/K$ is solvable. Then there exists a normal subgroup $\overline{A}$ in $\overline{G}$ such that $K \leq A \trianglelefteq G$ and $\overline{G}/\overline{A}$ is a nontrivial p-group (6.1.5 on page 123). The Dedekind identity shows that $H \cap A$ and $H_1 \cap A$ are complements of K in A, and by induction they are conjugate in A. Thus we may assume (after suitable conjugation)

$$H \cap A = H_1 \cap A := D \trianglelefteq \langle H, H_1 \rangle$$

Since $H/D \cong G/A \cong H_1/D$ there exist $P \in \mathrm{Syl}_p H$ and $P_1 \in \mathrm{Syl}_p H_1$ such that

$$H = DP \quad \text{and} \quad H_1 = DP_1.$$

Moreover, $(|K|, |H|) = 1$ implies that P and P_1 are Sylow p-subgroups of $N_G(D)$. Hence by Sylow's Theorem there exists $g \in N_G(D)$ such that $P_1{}^g = P$, and $H_1{}^g = D^g P_1{}^g = DP = H$ follows. $\square$

The additional solvability hypothesis in the theorem of Schur-Zassenhaus is not really a loss of generality. Since $|K|$ and $|G/K|$ are coprime at least one of the groups K and G/K has odd order. But then the theorem of Feit-Thompson shows that one of these groups is solvable.

Result 6.2.1 will be used in Section 6.4. Next we give a corollary, which will be of importance for the discussion in Chapter 8.

6.2.2 *Let G be a group acting on the set Ω, and let K be a normal subgroup of G. Suppose that*

(1) $(|K|, |G/K|) = 1$,

(2) *K or G/K is solvable, and*

(3) *K acts transitively on Ω.*

Then for every complement H of K in G:

(a) *$C_\Omega(H) \neq \varnothing$, and*

(b) *$C_K(H)$ acts transitively on $C_\Omega(H)$.*

Proof. (a) Let $\beta \in \Omega$. By (3) $|\Omega|$ is a divisor of $|K|$ (3.1.5 on page 58), and $G = KG_\beta$ (Frattini argument) implies $G/K \cong G_\beta/K \cap G_\beta$. The theorem of Schur-Zassenhaus, applied to $K \cap G_\beta$ and G_β, gives a complement H_1 of $K \cap G_\beta$ in G_β. Then H_1 is also a complement of K in G such that $\beta \in C_\Omega(H_1)$. Now (a) follows since, according to Schur-Zassenhaus, all complements of K in G are conjugate to H_1.

(b) Let $\alpha, \beta \in C_\Omega(H)$ and $k \in K$ such that $\alpha^k = \beta$. Then H and H^k are two complements of $K \cap G_\beta$ in G_β. Again by Schur-Zassenhaus they are conjugate under G_β and thus even under $K \cap G_\beta$. Let $k' \in K \cap G_\beta$ such that $H^{kk'} = H$. Then $\alpha^{kk'} = \beta$ and

$$[kk', H] \le H \cap K = 1,$$

i.e., $kk' \in C_K(H)$. $\square$

If in 6.2.2 the complement H is a p-group (and this is the case in many applications), then (a) is a consequence of 3.1.7 on page 59 since $(|H|, |\Omega|) = 1$, and all complements are conjugate by Sylow's Theorem.

Exercises

1. Prove that the complements in the theorem of Schur-Zassenhaus are conjugate using the following hypothesis:

    $\operatorname{Aut} E / \operatorname{Inn} E$ is solvable for every simple group E. [7]

Let G be a group, $p \in \mathbb{P}$, $\pi \subseteq \mathbb{P}$ and $\pi' = \mathbb{P} \setminus \pi$.

2. If $G/\Phi(G)$ contains a nontrivial normal subgroup whose order is not divisible by p, then so does G.

3. Let A be a nilpotent π-subgroup of G and $q \in \pi'$. Let $\mathbb{M}_X(A)$ be the set of A-invariant q-subgroups $X \le G$ and $\mathbb{M}_X^*(A)$ the set of maximal elements of $\mathbb{M}_X(A)$ (with respect to inclusion). Suppose that

    $*$ $O_{\pi'}(C_G(A) \cap N_G(Q))$ acts transitively on $\mathbb{M}_{N_G(Q)}^*(A)$ for all $Q \in \mathbb{M}_G(A)$.

    Then the following hold:

    (a) Every nilpotent π-subgroup B of G containing A satisfies $*$ in place of A.

    (b) $\mathbb{M}_G^*(B) \subseteq \mathbb{M}_G^*(A)$, where B is as in (a).

6.3   Radical and Residue

In this section we present some of the arguments of Section 6.1 in a more general context. This allows us then to define further characteristic subgroups. In the following let $\mathcal{K}$ be always a class of groups that contains the

---

[7]This is Schreier's *conjecture*, which up to now only can be verified using the classification of the finite simple groups.

trivial group and with a given group also all its isomorphic images. For any group G

$$O^{\mathcal{K}}(G) := \bigcap_{\substack{A \trianglelefteq G \\ G/A \in \mathcal{K}}} A \quad \text{and} \quad O_{\mathcal{K}}(G) := \prod_{\substack{A \trianglelefteq G \\ A \in \mathcal{K}}} A$$

are characteristic subgroups of G. $O^{\mathcal{K}}(G)$ is the $\mathcal{K}$-**residue** and $O_{\mathcal{K}}(G)$ the $\mathcal{K}$-**radical** of G.

In general neither $O_{\mathcal{K}}(G)$ nor $G/O^{\mathcal{K}}(G)$ is in $\mathcal{K}$. For example, if $\mathcal{K}$ is the class of all cyclic groups, then $O^{\mathcal{K}}(G) = 1$ and $O_{\mathcal{K}}(G) = G$ for every Abelian group G (2.1.3 on page 45). On the other hand, if $O_{\mathcal{K}}(G) \in \mathcal{K}$ then $O_{\mathcal{K}}(G)$ is *the largest* normal subgroup of G contained in $\mathcal{K}$. Similarly, if $G/O^{\mathcal{K}}(G) \in \mathcal{K}$, then $O^{\mathcal{K}}(G)$ is *the smallest* normal subgroup of G with factor group in $\mathcal{K}$.

In this section we are interested in the classes:

- $\mathcal{A}$ of all Abelian groups,

- $\mathcal{N}$ of all nilpotent groups,

- $\mathcal{S}$ of all solvable groups,

- $\mathcal{P}$ of all p-groups ($p \in \mathbb{P}$),

- Π of all π-groups ($\pi \subseteq \mathbb{P}$).

Here G is a π-**group** if $\pi(G) \subseteq \pi \subseteq \mathbb{P}$.

Using the notation introduced earlier we get

$$O_{\mathcal{N}}(G) = F(G), \quad O_{\mathcal{S}}(G) = S(G), \quad O_{\mathcal{P}}(G) = O_p(G)$$

and

$$O^{\mathcal{A}}(G) = G'.$$

Let

$$O_\pi(G) := O_\Pi(G) \quad \text{and} \quad O^\pi(G) := O^\Pi(G).$$

Then for $\pi' := \mathbb{P} \setminus \pi$ also $O_{\pi'}(G)$ and $O^{\pi'}(G)$ are defined. In the case $\pi = \{p\}$ we write p, resp. p', in place of π, resp. π'; this gives the particularly important subgroups

$$O_p(G), \quad O^p(G), \quad \text{and} \quad O_{p'}(G), \quad O^{p'}(G).$$

If G is a quaternion group of order 8, then $O_{\mathcal{A}}(G) = G$ and thus $O_{\mathcal{A}}(G) \notin \mathcal{A}$. For all the other classes mentioned above

$$O_{\mathcal{K}}(G) \in \mathcal{K}.$$

For $\mathcal{K} = \mathcal{N}, \mathcal{S}, \Pi$ this assertion follows from 5.2.1, 6.1.6, and 1.1.6, respectively. In particular $O_p(G)$ resp. $O_{p'}(G)$ is the largest normal p-subgroup resp. p'-subgroup of G.

6.3.1   *Let $\mathcal{K} \in \{\mathcal{N}, \mathcal{S}, \Pi\}$. Then for every group G*

$$O_{\mathcal{K}}(G) = \langle A \mid A \trianglelefteq\trianglelefteq G, \;\; A \in \mathcal{K} \rangle;$$

in particular, $O_{\mathcal{K}}(G)$ is also the largest subnormal subgroup of G that is in $\mathcal{K}$.

Proof. We have to show that every subnormal subgroup $A \trianglelefteq\trianglelefteq G$ with $A \in \mathcal{K}$ is contained in $O_{\mathcal{K}}(G)$. For $A \trianglelefteq G$ this is obvious. Thus, we may assume that A is not normal in G. Hence, there exists a normal subgroup $N \trianglelefteq G$ such that

$$A \trianglelefteq\trianglelefteq N < G.$$

By induction on $|G|$ we may also assume that

$$A \leq O_{\mathcal{K}}(N).$$

$O_{\mathcal{K}}(N)$ is normal in G since it is characteristic in N. Moreover $\mathcal{K} \in \{\mathcal{N}, \mathcal{S}, \Pi\}$, and as mentioned above $O_{\mathcal{K}}(N) \in \mathcal{K}$. Hence $A \leq O_{\mathcal{K}}(N) \leq O_{\mathcal{K}}(G)$.   □

We call a class $\mathcal{K}$ of groups **closed**, if for every $X \in \mathcal{K}$:

- The homomorphic images of groups from $\mathcal{K}$ are in $\mathcal{K}$.

- The subgroups of groups from $\mathcal{K}$ are in $\mathcal{K}$.

- Direct products of groups from $\mathcal{K}$ are in $\mathcal{K}$.

All the above mentioned classes are closed. Similarly to 1.5.2 and 1.5.1 on page 24 one gets:

6.3.2   *Let $\mathcal{K}$ be a closed class. Then for every group G:*

(a) $G/O^{\mathcal{K}}(G) \in \mathcal{K}$.

(b) $(O^{\mathcal{K}}(G))^{\varphi} = O^{\mathcal{K}}(G^{\varphi})$ *for every homomorphism φ of G.*

Proof. (a) follows from 1.6.4 on page 31 and the definition of a closed class.

(b) Let φ be a homomorphism of G. Then $G^{\varphi}/(O^{\mathcal{K}}(G))^{\varphi}$ is in $\mathcal{K}$ since it is a homomorphic image of $G/O^{\mathcal{K}}(G)$, i.e.,

$$O^{\mathcal{K}}(G^{\varphi}) \leq (O^{\mathcal{K}}(G))^{\varphi}.$$

Now let ψ be a homomorphism of G^{φ} in a group $X \in \mathcal{K}$. Then $\varphi\psi$ is a homomorphism from G in X. This means that $O^{\mathcal{K}}(G) \leq \mathrm{Ker}\,(\varphi\psi)$ and thus $(O^{\mathcal{K}}(G))^{\varphi} \leq \mathrm{Ker}\,\psi$. It follows that

$$(O^{\mathcal{K}}(G))^{\varphi} \leq \bigcap_{\psi} \mathrm{Ker}\,\psi = O^{\mathcal{K}}(G^{\varphi}),$$

where ψ runs through all homomorphisms from G^{φ} in a group of $\mathcal{K}$.   □

Let $\mathcal{K}$ be a closed class of groups. In analogy to the definition of solvability we call a group G a $\mathcal{K}$-**group** if

$$O^{\mathcal{K}}(U) \neq U \quad \text{for all subgroups } 1 \neq U \leq G.$$

By $\widehat{\mathcal{K}}$ we denote the class of all $\mathcal{K}$-groups. Clearly $\mathcal{K} \subseteq \widehat{\mathcal{K}}$.

For example, by definition $\widehat{\mathcal{A}} = \mathcal{S}$ and $\widehat{\mathcal{N}} = \widehat{\mathcal{S}} = \mathcal{S}$.

With 6.3.2 in hand the same argument as in 6.1.1 shows that subgroups and homomorphic images of $\mathcal{K}$-groups are again $\mathcal{K}$-groups. The proof of 6.1.2 on page 122 gives:

6.3.3   *Let $\mathcal{K}$ be a closed class. Then $G \in \widehat{\mathcal{K}}$ if and only if there exists a normal subgroup N of G such that $N \in \widehat{\mathcal{K}}$ and $G/N \in \widehat{\mathcal{K}}$.*   □

From 6.3.3 we get as corollaries (also compare with 6.1.6 on page 123):

6.3.4   *Let $\mathcal{K}$ be a closed class. Then $O_{\widehat{\mathcal{K}}}(G) \in \widehat{\mathcal{K}}$.*   □

6.3.5 *Let $\mathcal{K}$ be a closed class. Then $\widehat{\mathcal{K}}$ is also closed.*   □

6.3.6 *Let $\mathcal{K}$ be a closed class. The following statements are equivalent:*

(i) $G \in \widehat{\mathcal{K}}$.

(ii)   *If $G^{(0)} = G$ and $G^{(i)} = O^{\mathcal{K}}(G^{(i-1)})$, for $i \geq 1$, then there exists $\ell \in \mathbb{N}$ such that $G^{(\ell)} = 1$.*

(iii)   *G has a composition series all of whose composition factors are in $\mathcal{K}$.*   □

For the class Π of π-groups we get $\Pi = \widehat{\Pi}$. A similar equality is no longer true for the class of π-*closed* groups:

Here a group G is π-**closed** if $G/O_\pi(G)$ is a π'-group; in other words if

$$O_\pi(G) = O^{\pi'}(G).^8$$

For example, the theorem of Schur-Zassenhaus 6.2.1 is a theorem about π-closed groups, $\pi := \pi(K)$. In its proof we have used the fact that subgroups and factor groups of π-closed groups are π-closed. Since also direct products of π-closed groups are π-closed the class of all π-closed groups is a closed class. We denote this class by Π_c.

Π_c contains all π-groups and all π'-groups. The groups in $\widehat{\Pi}_c$ are called π-**separable**.[9] Since Π_c is a closed class, π-separable groups—as explained earlier—possess the same formal properties as solvable groups.

Exercises

Let G be a group.

1.   Let $\mathcal{C}$ be the class of all groups H satisfying $C_H(F(H)) \leq F(H)$. Then the following hold:

 (a) $O_{\mathcal{C}}(G) \in \mathcal{C}$ and $G/O^{\mathcal{C}}(G) \in \mathcal{C}$.

 (b)   Let $N \trianglelefteq G$. If N and G/N are in $\mathcal{C}$, then also G is in $\mathcal{C}$.

---

[8]Compare with the remark following 3.2.2 on p. 63 about p-closed groups.

[9]Those who feel uncomfortable with this abstract definition may prefer the equivalent property given in 6.4.2.

6.4 π-Separable Groups

In this section we investigated π-separable groups ($\pi \subseteq \mathbb{P}$). Recall that a π-separable group G satisfies the following:

- Every nontrivial subgroup of G has a nontrivial π-closed factor group.

Since Abelian groups are π-closed we get:

6.4.1   *Solvable groups are π-separable.* $\square$

Hence, all results about π-separable groups are also results about solvable groups. We will present some of these results after we have introduced convenient notation. In the second part of this section we will characterize the solvable groups within the class of π-separable groups. It will turn out that solvable groups are characterized by the fact that a generalization of Sylow's Theorem holds for them. This was proved by P. Hall and became the starting point for a today highly developed theory of solvable groups; we refer the reader the book of Doerk-Hawkes [8].

6.4.2   *A group G is π-separable if and only if G possesses a series*

$$1 = A_0 < A_1 < \cdots < A_{i-1} < A_i < \cdots < , A_n = G$$

of characteristic subgroups A_i $(i = 1, \ldots, n)$ such that every factor A_i/A_{i-1} is a π-group or a π'-group.

Proof. Let G be π-separable. Then also $O^{\Pi_c}(G)$ is π-separable,[10] and

$$\overline{G} := G/O^{\Pi_c}(G)$$

is π-closed. Hence $\overline{G}/O_\pi(\overline{G})$ is a π'-group. As $O^{\Pi_c}(G)$ is characteristic in G, the characteristic subgroups of $O^{\Pi_c}(G)$ and the inverse images of characteristic subgroups of $\overline{G}$ are characteristic in G. Now induction on $|G|$ yields the desired series for G.

Conversely, let $(A_i)_{i=0,\ldots,n}$ be as in the statement of 6.4.2 and $G^{(i)}$ as in 6.3.6. Then $G^{(1)}$ is contained in A_{n-1}, and more generally $G^{(i)} \leq A_{n-i}$; in particular $G^{(n)} = 1$. Now 6.3.6 implies the G is π-separable. $\square$

---

[10]Π_c is the class of π-closed groups.

It is evident that a group is π-separable if and only if it is π'-separable.
For a group G the subgroup $O_{\pi'\pi}(G)$ is defined by

$$O_{\pi'\pi}(G)/O_{\pi'}(G) := O_{\pi}(G/O_{\pi'}(G))$$

and $O_{\pi'\pi\pi'}(G) \leq G$ by

$$O_{\pi'\pi\pi'}(G)/O_{\pi'\pi}(G) := O_{\pi'}(G/O_{\pi\pi'}(G)).$$

Continuing in this way one gets a series of characteristic subgroups

$$1 \leq O_{\pi'}(G) \leq O_{\pi'\pi}(G) \leq O_{\pi'\pi\pi'}(G) \leq O_{\pi'\pi\pi'\pi}(G) \leq \cdots,$$

which terminates in G if and only if G is π-separable.
For the particularly important case $\pi = \{p\}$ this series is

$$1 \leq O_{p'}(G) \leq O_{p'p}(G) \leq O_{p'pp'}(G) \leq O_{p'pp'p}(G) \leq \cdots.$$

6.4.3   *Let G be a π-separable group and $O_{\pi'}(G) = 1$. Then*

$$C_G(O_{\pi}(G)) \leq O_{\pi}(G).$$

Proof. Let $C := C_G(O_{\pi}(G))$ and $K := C \cap O_{\pi}(G)$ $(= Z(O_{\pi}(G)))$. Then C/K is a π-separable normal subgroup of G/K satisfying $O_{\pi}(C/K) = 1$ since $O_{\pi}(G)$ is the largest normal π-subgroup of G. Let $K \leq A \leq C$ such that $A/K = O_{\pi'}(C/K)$. Then the theorem of Schur-Zassenhaus gives a complement H of K in A, and

$$A = KH = K \times H$$

since $A \leq C$. This implies $H = O_{\pi'}(A)$ and thus $H \leq O_{\pi'}(G) = 1$. Hence $O_{\pi'}(C/K) = 1 = O_{\pi}(C/K)$ and $C = K$.   □

The following result consists of some frequently used consequences of 6.4.3:

6.4.4   *Let G be p-separable for $p \in \pi(G)$ and P a Sylow p-subgroup of $O_{p'p}(G)$.*

(a) $C_G(P) \leq O_{p'p}(G)$; *in particular:*

$$O_{p'}(G) = 1 \quad \Rightarrow \quad C_G(O_p(G)) \leq O_p(G).$$

(b)   Let U be a P-invariant p'-subgroup of G. Then U is contained in $O_{p'}(G)$.

(c)   If G has Abelian Sylow p-subgroups, then $G = O_{p'pp'}(G)$.

Proof. (a) Because of 3.2.8 (a) on page 66 we may assume that $O_{p'}(G) = 1$. Then $P = O_p(G)$, and the conclusion follows from 6.4.3.

(b) Again we may assume that $O_{p'}(G) = 1$. Then $P = O_p(G)$ and thus $[U, P] \leq U \cap P = 1$. Now (a) gives $U \leq O_{p'}(G) = 1$.

(c) Let $P \leq S \in \mathrm{Syl}_p G$, S Abelian. Then $S \leq C_G(P)$ and thus by (a) $S = P$.   □

A π-subgroup H of the group G is a **Hall π-subgroup** of G if

$$\pi(|G : H|) \subseteq \pi'.$$

For example, for $p \in \mathbb{P}$ the Hall p-subgroups of G are the Sylow p-subgroups of G. As for Sylow subgroups we denote by $\mathrm{Syl}_\pi G$ the set of Hall π-subgroups of G.

In contrast to the case $\pi = \{p\}$, where one has Sylow's Theorem, in general Hall π-subgroups do not always exist. For example, the alternating group A_5 possesses Hall $\{2,3\}$-subgroups but not Hall $\{3,5\}$- and $\{2,5\}$-subgroups (3.2.12 on page 68).

With the same argument as in 3.2.2 on page 63, resp. 3.2.5 on page 65, we get:

• Let $\mathrm{Syl}_\pi G \neq \varnothing$. Then

$$O_\pi(G) = \bigcap_{H \in \mathrm{Syl}_\pi G} H.$$

• Let $H \in \mathrm{Syl}_\pi G$ and $N \trianglelefteq G$. Then

$$N \cap H \in \mathrm{Syl}_\pi N \quad \text{and} \quad NH/N \in \mathrm{Syl}_\pi G/N.$$

6.4.5   *Every π-separable group contains Hall π-subgroups.*

Proof. Let $G \neq 1$ be a π-separable group and $N \neq 1$ a normal subgroup of G. Since also G/N is π-separable we may assume by induction on $|G|$ that G/N contains a Hall π-subgroup

$$H/N \quad (N \leq H \leq G).$$

If $O_\pi(G) \neq 1$, we choose $N := O_\pi(G)$. Then H is a Hall π-subgroup of G.

Assume that $O_\pi(G) = 1$. Then $1 \neq O_{\pi'}(G)$, and we choose

$$N := O_{\pi'}(G).$$

Now N is a normal π'-subgroup of G and $\pi(H/N) \subseteq \pi$. The theorem of Schur-Zassenhaus gives a complement H_1 of N in H. This complement is a Hall π-subgroup of G.   □

We say that the π-**Sylow Theorem** holds in G, if every π-subgroup of G is contained in a Hall π-subgroup of G and all Hall π-subgroups are conjugate in G.

6.4.6 *Let G be a π-separable group satisfying the following:*

(∗) *Every π-section or every π'-section of G is solvable.*[11]

Then the π-Sylow Theorem holds in G.

Proof. Let U be a π-subgroup and H a Hall π-subgroup of G (6.4.5). It suffices to show that U is contained in a conjugate of H. We prove this by induction on $|G|$.

Obviously, we may assume that $G \neq 1$. Let $1 \neq N \trianglelefteq G$ and $\overline{G} := G/N$. By induction there exists $\overline{g} \in \overline{G}$ such that $\overline{U}^{\overline{g}} \leq \overline{H}$ and $(UN)^g = U^g N \leq HN$. Since we are allowed to replace U by any conjugate we may assume

$$U \leq HN.$$

We now proceed as in 6.4.5: If $O_\pi(G) \neq 1$ we choose $N := O_\pi(G)$. Then $HN = H$ and $U \leq H$. Assume that $O_\pi(G) = 1$. Then $O_{\pi'}(G) \neq 1$ and we choose $N := O_{\pi'}(G)$. The π-subgroup U is a complement of N in NU. Since also $H \cap NU$ is such a complement (1.1.11) the theorem of Schur-Zassenhaus shows that U is conjugate to the subgroup $H \cap NU$ of H.   □

Since solvable groups are π-separable 6.4.6 implies:

---

[11] A π-**section** is a section that is a π-group.

6.4.7   *The π-Sylow Theorem holds in solvable groups for every $\pi \subseteq \mathbb{P}$.* □

Result 6.4.7 in fact characterizes the solvable groups. To show this we need two lemmata.

6.4.8   *Let H, K be subgroups of the group G such that*

$$(|G : H|, |G : K|) = 1.$$

Then $G = HK$ and $|G : H \cap K| = |G : H| \, |G : K|$.

Proof. By 1.1.6

$$n := \frac{|G|}{|HK|} = \frac{|G| \, |H \cap K|}{|H| \, |K|}.$$

Hence n is a divisor of $|G : H|$ and $|G : K|$. Since these two integers are coprime we get $n = 1$ and $G = HK$. Now

$$|G : H| \, |G : K| = \frac{|G|^2}{|H| \, |K|} \overset{1.1.6}{=} \frac{|G|^2}{|G| \, |H \cap K|} = |G : H \cap K|. \qquad \square$$

6.4.9   *Let H_1, H_2 and H_3 be solvable subgroups of the group G such that*

$$G = H_1 H_2 = H_1 H_3 \quad \text{and} \quad (|G : H_2|, |G : H_3|) = 1.$$

Then G is solvable.

Proof. If $H_1 = 1$ then $G = H_2$ is solvable. Assume that $H_1 \neq 1$. Let A be a minimal normal subgroup of H_1. Then A is a p-group (6.1.3 on page 122). Since $(|G : H_2|, |G : H_3|) = 1$ we may assume that p does not divide $|G : H_2|$. Thus H_2 contains a Sylow p-subgroup of G. By Sylow's Theorem there exists $g \in G$ such that $A \leq H_2{}^g$. Since $G = H_2 H_1$ we may assume that $g \in H_1$. It follows that $A^{g^{-1}} = A \leq H_2$ and

$$N := \langle A^G \rangle = \langle A^{H_1 H_2} \rangle = \langle A^{H_2} \rangle \leq H_2.$$

Hence, N is a solvable normal subgroup of G. Since G/N satisfies the hypothesis we may assume by induction on $|G|$ that G/N is solvable. But then also G is solvable (6.1.2). □

Let G be a group. A set $\mathcal{S}$ of Sylow subgroups of G is a **Sylow system** of G, if

- $|\mathcal{S} \cap \mathrm{Syl}_p G| = 1$ for all $p \in \pi(G)$, and

- $PQ = QP$ for all $P, Q \in \mathcal{S}$.

Let $\mathcal{S}$ be a Sylow system of G. Then for every nonempty $\mathcal{S}_0 \subseteq \mathcal{S}$ a repeated application of 1.1.5 and 1.1.6 shows that the group

$$\prod_{P \in \mathcal{S}_0} P$$

is a Hall π-subgroup of G, where

$$\pi = \{p \in \pi(G) \mid (\mathrm{Syl}_p G) \cap \mathcal{S}_0 \neq \varnothing\}.$$

Suppose that $\pi(G) = \{p, q\}$, then $PQ = QP = G$ for every $P \in \mathrm{Syl}_p G$ and $Q \in \mathrm{Syl}_q G$. Thus, each such pair is a Sylow system of G. Moreover, a theorem of Burnside, mentioned earlier, shows that G is solvable; we will prove this theorem in Section 10.2.

The following theorem, the characterization of solvable groups announced above, shows in general that the existence of Sylow systems is equivalent to solvability. The proof of the implication (v) $\Rightarrow$ (i) requires the theorem of Burnside.

6.4.10   Theorem (P. Hall).[12] *Let G be a group. The following statements are equivalent:*

(i)   *G is solvable.*

(ii)   *G is π-separable for every set of primes π.*

(iii)   *G contains a Hall π-subgroup for every set of primes π.*

(iv)   *G contains a Hall p'-subgroup for every prime p.*

(v)   *G possesses a Sylow system.*

Proof. (i) $\Rightarrow$ (ii): 6.4.1.

(ii) $\Rightarrow$ (iii): 6.4.5.

(iii) $\Rightarrow$ (iv): Trivial.

---

[12]See [63], [65], [66].

(iv) $\Rightarrow$ (v): For $p \in \pi(G)$ let H_p be a Hall p'-subgroup of G, and for $\varnothing \neq \pi \subseteq \pi(G)$ let

$$H_\pi := \bigcap_{p \in \pi} H_p.$$

First we show:

(') H_π is a Hall π'-subgroup of G.

This follows from 6.4.8 using induction on $|\pi|$. For $|\pi| = 1$ there is nothing to prove. Assume that $|\pi| \geq 2$ and let $p \in \pi$ and $\sigma := \pi \setminus \{p\}$. Then

$$H_\pi = H_\sigma \cap H_p.$$

Since by induction H_σ is a Hall σ'-subgroup we are allowed to apply 6.4.8 with respect to the subgroups H_σ and H_p. This gives (').

In particular, for $p_i \in \pi(G)$,

$$P_i := \bigcap_{p \in \pi(G) \setminus \{p_i\}} H_p$$

is a Sylow p_i-subgroup of G. Let $p_i, p_j \in \pi(G)$. Then P_i, P_j are Sylow subgroups of

$$H := \bigcap_{p \in \pi(G) \setminus \{p_i, p_j\}} H_p,$$

and by (') H is a Hall $\{p_i, p_j\}$-subgroup of G. This gives $P_i P_j = P_j P_i = H$. Hence $\{P_i \mid p_i \in \pi(G)\}$ is a Sylow system of G.

(v) $\Rightarrow$ (i): If $|\pi(G)| = 1$, then G is a p-group and thus solvable. If $|\pi(G)| = 2$, then the solvability of G follows from 10.2.1 on page 276. We now assume that $|\pi(G)| \geq 3$ and that $\{P_1, \ldots, P_n\}$ is a Sylow system of G. For $i \in \{1, 2, 3\}$ let

$$H_i = \prod_{j \neq i} P_j.$$

Then $|G : H_1|$, $|G : H_2|$, and $|G : H_3|$ are pairwise coprime and

$$G = H_1 H_2 = H_1 H_3 = H_2 H_3.$$

Moreover, since $\{P_1, \ldots, P_n\} \setminus \{P_i\}$ is a Sylow system of H_i we may assume by induction on $|G|$ that H_1, H_2, and H_3 are solvable. Hence 6.4.9 shows that G is solvable. $\square$

We conclude this section with a property of π-separable groups, which we will refer to in Chapter 12. For $\pi = \{p\}$ this property is a consequence of Baer's Theorem (6.7.6 on page 160) and therefore also true for groups which are not p-separable.

6.4.11   *Let G be a π-separable group and A a π-subgroup of G. Then the following statements are equivalent:*

(i) $A \not\leq O_\pi(G)$.

(ii)   *There exists $x \in O_{\pi\pi'}(G)$ such that $x \in \langle A, A^x \rangle$ and $\langle A, A^x \rangle$ is not a π-group.*

Proof. If $A \leq O_\pi(G)$, then $\langle A, A^x \rangle \leq O_\pi(G)$ for all $x \in G$. This shows (ii) $\Rightarrow$ (i).

(i) $\Rightarrow$ (ii): Let $\overline{G} := G/O_\pi(G)$. Assume that $\langle A, A^x \rangle$ is a π-group for all $x \in O_{\pi\pi'}(G)$. Then

$$[O_{\pi'}(\overline{G}), \overline{A}] = 1$$

and by 6.4.3 $\overline{A} = 1$, which contradicts $A \not\leq O_\pi(G)$. Hence, there exists $x \in O_{\pi\pi'}(G)$ such that $\langle A, A^x \rangle$ is not a π-group. Let

$$G_1 := \langle A, A^x \rangle \quad (\leq O_{\pi\pi'}(G)A).$$

Then $A \not\leq O_\pi(G_1)$ and $A^x \not\leq O_\pi(G_1)$. If $G_1 < G$, then induction on $|G|$ yields (ii). If $G = G_1$, then (ii) is obvious. $\square$

Exercises

1.   Let G be a p-separable group, $p \in \pi(G)$. Suppose that for all $q \in \pi(G)$ and $S \in \mathrm{Syl}_q G$:
$$\mathrm{Syl}_p N_G(S) \subseteq \mathrm{Syl}_p G.$$
Then $G = O_{p'p}(G)$.

2.   (Example for 6.4.11) Let $G := S_5$ and $A := \langle (1\,2) \rangle \leq G$. Then there exists $\pi \subseteq \pi(G)$ such that $A \not\leq O_\pi(G)$ and

$$\langle A, A^x \rangle \text{ is } \pi\text{-subgroup for all } x \in G.$$

3.   Let G be a p-separable group. The p-**length** $\ell_p(G)$ of G is defined recursively by:

$$\ell_p(G) := 0, \text{ if } G = O_{p'}(G), \text{ and}$$
$$\ell_p(G) := 1 + \ell_p(G/O_{p'p}(G)), \text{ if } G \neq O_{p'}(G).$$

Show that $\ell_p(G) \leq c(P)$, $P \in \mathrm{Syl}_p G$.

6.5 Components and the Generalized Fitting Subgroup

The concepts mentioned in the title of this section came up around 1970 in the course of the classification of the finite simple groups. They are examples of how the essence of a new development is reflected by appropriate concepts and that these concepts—as a language—contribute to the success of this development.[13]

A group $K \neq 1$ is **quasisimple** if K is perfect and $K/Z(K)$ is simple. Clearly, for every subnormal subgroup N of a quasisimple group K either

$$N \leq Z(K) \quad \text{or} \quad N = K.$$

This implies that nontrivial homomorphic images of quasisimple groups are quasisimple.

Let G be a group. A subgroup K of G is a **component** of G, if K is quasisimple and subnormal in G. The first of these two properties is an internal property of K, while the second one describes the embedding of K in G. Therefore components K are endowed with similar inheritance properties as subnormal subgroups in general:

- If $K \leq U \leq G$, then K is a component of U.

- If $K \not\leq N \trianglelefteq G$, then KN/N is a component of G/N.

- If K is a component of a subnormal subgroup of G, then K is a component of G.

Minimal subnormal subgroups of G are simple groups. Hence, those which in addition are non-Abelian are components of G.

6.5.1 *Let Z and E be subgroups of G such that $Z \leq Z(G)$ and EZ/Z is a component of G/Z. Then E' is a component of G.*

Proof. Since $Z \leq Z(G)$ we have $E' = (EZ)'$, and since $EZ \trianglelefteq\trianglelefteq G$ also $E' \trianglelefteq\trianglelefteq G$. Moreover, 1.5.3 on page 25 shows that E' is perfect. Let N be a normal subgroup of E' and $\overline{G} := G/Z$. Then either

$$\overline{N} = \overline{E} = \overline{E}' \quad \text{or} \quad \overline{N} \leq Z(\overline{E}).$$

---

[13]Namely the classification of the finite simple groups.

The first case gives $N \leq E' \leq NZ$ and thus $N(Z \cap E') = E'$. Hence $N = E'$ since E' is perfect. The second case gives $[E', N] \leq Z$ and

$$[E', N, E'] = 1 = [N, E', E'].$$

The Three-Subgroups Lemma (1.5.6 on page 26) yields $[E', E', N] = 1$ and thus $[E', N] = 1$, again since E' is perfect. Hence $N \leq Z(E')$, and E' is quasisimple. $\square$

6.5.2 *Let K be a component of G and U a subnormal subgroup of G. Then $K \leq U$ or $[U, K] = 1$.*

Proof. Obviously, $U = G$ implies $K \leq U$. Moreover, as mentioned earlier, $K = G$ implies either $U = K$ or $[U, K] = 1$. Thus, we may assume that there exist proper normal subgroups N, M of G such that

$$K \leq N < G \quad \text{and} \quad U \leq M < G.$$

In particular

$$U_1 := [U, K] \leq N \cap M$$

and $K \leq N_N(U_1) =: G_1$ (1.5.5 on page 26). Thus, K is a component of G_1, and U_1 is subnormal (in fact normal) in G_1. By induction on $|G|$, applied to G_1, we get

$$[U_1, K] = 1 \quad \text{or} \quad K \leq U_1.$$

The first case gives

$$1 = [U, K, K] = [K, U, K]$$

and then using the Three-Subgroups Lemma

$$1 = [K, K, U] = [K', U] = [K, U].$$

The second case gives $K \leq M$ since $[K, U] \leq M$, and the conclusion follows by induction on $|G|$, now applied to M. $\square$

6.5.3 *Let K_1 and K_2 be components of G. Then either $K_1 = K_2$ or $[K_1, K_2] = 1$. In particular, products of components are subgroups of G.*

Proof. In the case $[K_1, K_2] \neq 1$ 6.5.2 implies $K_1 \leq K_2$ and by symmetry also $K_2 \leq K_1$.   □

We now define two characteristic subgroups of G.

$E(G)$:   the subgroup generated by the components of G,

$F^*(G)$:= $F(G) E(G)$.

$F^*(G)$ is the **generalized Fitting subgroup** of G. Notice that by 6.5.2

$$[F(G), E(G)] = 1.$$

Let N be a minimal normal subgroup of G. By 1.7.3 on page 38 either N is Abelian and $N \leq F(G)$, or N is product of components and $N \leq E(G)$. Hence:

6.5.4 $F^*(G)$ *contains every minimal normal subgroup of G; in particular* $F^*(G) \neq 1$ *if $G \neq 1$.*   □

6.5.5   (a)   *Let K be a component of G such that $Z(K) = 1$. Then $\langle K^G \rangle$ is a minimal normal subgroup of G; in particular, $\langle K^G \rangle$ is the direct product of the components conjugate to K.*

(b)   *Let $F(G) = 1$. Then $E(G)$ is the product of the minimal normal subgroups of G.*

Proof. (a) By 6.5.3 $\langle K^G \rangle$ is the central product of the components K^g, $g \in G$, and thus by 1.6.7 a direct product since K is simple. Let N be a minimal normal subgroup of G contained in $\langle K^G \rangle$. Then 1.6.3 on page 30 shows that at least one of the factors K^g is in N. But then $N = \langle (K^g)^G \rangle = \langle K^G \rangle$.

(b) By 6.3.1 on page 130 $Z(K) \leq F(G)$ for every component K of G. Now the hypothesis $F(G) = 1$ implies that all the components of G are simple, and (b) follows from (a).   □

In general one gets:

6.5.6   *Let $E(G) \neq 1$ and $K_1, \ldots, K_n$ be the components of G. Set*

$$Z := Z(E(G)), \quad Z_i := Z(K_i), \quad E_i := K_i Z/Z \quad (i = 1, \ldots, n).$$

(a) $E(G)$ is the central product of $K_1, \ldots, K_n$, in particular
      $Z = Z_1 \cdots Z_n$.

(b) $Z_i = Z \cap K_i$ and $E_i \cong K_i/Z_i$ $(i = 1, \ldots, n)$.

(c) $E(G)/Z = E_1 \times \cdots \times E_n$.

Proof. Let $Z_0 = \prod\limits_{i=1}^{n} Z_i$. By 6.5.3 $E(G)$ is the product of the normal
subgroups $K_1 Z_0, \ldots, K_n Z_0$ and

$$K_i Z_0 \cap \prod_{i \neq j} K_k Z_0 = Z_0.$$

Now 1.6.2 and 1.6.7 imply $Z_0 = Z$ and (a)–(c).   □

6.5.7   *Let L be a subnormal subgroup of G.*

(a)   *If $L \leq F^*(G)$, then $L = (L \cap F(G))(L \cap E(G))$.*

(b) $F^*(L) = F^*(G) \cap L$.

(c) $E(L) C_{E(G)}(L) = E(G)$. *In particular, $E(L)$ is normal in $E(G)$.*

Proof. Every component of L is also a component of G and $F(L) \leq F(G)$
(6.3.1 on page 130). Now apply 6.5.2, 6.5.6 (a) and $[F(G), E(G)] = 1$.   □

The following is the fundamental property of $F^*(G)$. It generalizes 6.1.4:

6.5.8   Theorem. *Let G be a group. Then $C_G(F^*(G)) \leq F^*(G)$.*

Proof. Let $L := C_G(F^*(G))$, $Z := Z(L)$, and $\overline{L} := L/Z$. It suffices to show
that $F^*(\overline{L}) = 1$ since this implies $\overline{L} = 1$ and $L \leq Z \leq F^*(G)$.
By 6.5.7 $F^*(L) \leq F^*(G)$ and thus

$$F^*(L) = Z.$$

Hence, the inverse image of $F(\overline{L})$ in L is a nilpotent normal subgroup of L
(5.1.2 on page 100) and thus contained in $F(L)$. It follows that $F(\overline{L}) = 1$.

If $F^*(\overline{L}) \neq 1$, then $\overline{L}$ contains a component $\overline{E}$, $Z < E \leq L$. But then E' is a component of L (6.5.1), which contradicts $F^*(L) = Z$. $\qquad\square$

Exercises

Let G be a group.

1. Describe $F^*(C_G(E(G)))$ and $F^*(C_G(F(G)))$.

2. Let t be an involution of G and E a component of $C_G(t)$. Then E normalizes every component of G.

3. Let $\operatorname{Aut} E / \operatorname{Inn} E$ be solvable for every component E of G, and let $F(G) = 1$. Then
$$E(C_G(t)) \leq E(G)$$
for every involution t of G.

4. Let K be a subgroup of G. Suppose that for every $g \in G$:
$$K \text{ is a component of } \langle K, K^g \rangle.$$
Then K is a component of G (compare with 6.7.4 on page 159).

6.6   Primitive Maximal Subgroups

In this section we investigate embedding properties of maximal subgroups. Let G be a group and M a maximal subgroup of G, and let N be a normal subgroup of G. If $N \leq M$, then M/N is a maximal subgroup of G/N. Hence, we may assume—possibly after substituting for G a suitable factor group—that no nontrivial normal subgroup of G is contained in M. Then M satisfies:

$(*)$ $\qquad\qquad\qquad 1 \neq N \trianglelefteq M \quad \Rightarrow \quad M = N_G(N).$

$(**)$ $\qquad\qquad\qquad 1 \neq N \trianglelefteq G \quad \Rightarrow \quad G = MN.$

Since the embedding property $(*)$ will also be central in later investigations we call a proper—not necessarily maximal—subgroup M of G **primitive**, if M satisfies $(*)$.

Recall that point stabilizers of primitive permutation groups are primitive maximal subgroups. Conversely, the action of G on the right cosets of a primitive maximal subgroup M (by right multiplication) or on the conjugates of M is faithful and primitive.

We start with two elementary properties of primitive subgroups:

6.6.1   *Let M be a primitive subgroup and N a normal subgroup of G such that $M \cap N \neq 1$. Then $C_G(N) = 1$.*

Proof. $1 \neq N \cap M \trianglelefteq M$ and the primitivity of M give

$$C_G(N) \leq C_G(N \cap M) \leq M.$$

Thus $C_G(N) = 1$ since $C_G(N)$ is a normal subgroup of G.   □

6.6.2   *Let M be a primitive subgroup of G. Then no nontrivial subnormal subgroup of G is contained in M. In particular $M \cap F(G) = 1$.*

Proof. Assume by way of contradiction that there exists a subnormal subgroup $L \neq 1$ of G such that $L \leq M$. Without loss we may further assume that L is a minimal subnormal subgroup of G. Then $L \leq F^*(G)$, and from 6.6.1, applied to $N = F^*(G)$, we get

$$1 = Z(F^*(G)) \quad (= Z(F(G))\, Z(E(G))).$$

In particular $F(G) = 1$ (5.1.5 on page 101). It follows that L is a component of G. Now 6.5.6 on page 143 shows that $\langle L^M \rangle$ is a normal subgroup of $E(G)$. Hence, the primitivity of M yields first $E(G) \leq M$ and then $E(G) = 1$. This contradicts $L \neq 1$.   □

6.6.3   *Let M be a primitive subgroup, $p \in \pi(M)$, and N a normal subgroup of G. Suppose that $M \cap N = 1$ and $O_p(M) \neq 1$.*

(a) $p \notin \pi(N)$.

(b)   *For every $q \in \pi(N)$ there exists a unique M-invariant Sylow q-subgroup of N.*

(c)   If $|\pi(N)| \geq 2$, then M is not a maximal subgroup of G.

Proof. (a) For $P := O_p(M)$ the primitivity of M gives

$$M = N_G(P).$$

In particular, P is a Sylow p-subgroup of NP since $N \cap M = 1$ (3.1.10 on page 61). This implies $p \notin \pi(N)$ (3.2.5).

(b) PN acts on $\Omega := \mathrm{Syl}_q N$ by conjugation, and by Sylow's Theorem N is a transitive normal subgroup of PN. Hence 6.2.2 on page 127 applies to PN and Ω. It follows that $C_\Omega(P) \neq \varnothing$ and $C_N(P)$ is transitive on $C_\Omega(P)$. Now $C_N(P) \leq M \cap N = 1$ gives $|C_\Omega(P)| = 1$; in particular $C_\Omega(P) = C_\Omega(M)$ since P is normal in M.

(c) According to (b) there exists an M-invariant $Q \in \mathrm{Syl}_q N$. Since Q is a proper subgroup of N we get $M < QM < NM \leq G$. $\square$

6.6.4   *Let M be a primitive subgroup and N a normal subgroup of G such that*

$$M \cap F^*(N) \neq 1.$$

Then $F(G) = 1$ and $F^*(N) = F^*(G) = E(G)$. In particular, every minimal normal subgroup of G is contained in N.

Proof. Note that $F^*(N) \leq F^*(G)$ (6.5.7). Hence 6.6.1 implies

$$Z(F(G)) \leq C_G(F^*(N)) = 1,$$

and thus $F(G) = 1$ by 5.1.5 on page 101. In particular $F^*(N) = E(N)$, and

$$F^*(G) \overset{6.5.7}{=} C_{F^*(G)}(E(N))\, E(N).$$

Another application of 6.6.1 gives $F^*(G) = E(N) = F^*(N)$. $\square$

In the following let M be a primitive maximal subgroup of G. Then

$$G = F^*(G)\, M.$$

For the rest of this section we will investigate this factorization. The results will be collected in the theorem of O'Nan-Scott.

We distinguish three cases:

(F1) $F(G) = F^*(G)$, $F(G)$ is the unique minimal normal subgroup of G,[14] and M is a complement of $F(G)$ in G.

(F2) G contains exactly two minimal normal subgroups N_1 and N_2. These normal subgroups are non-Abelian, i.e.,

$$F^*(G) = N_1 \times N_2 = E(G).$$

(F3) $F^*(G)$ is a non-Abelian minimal normal subgroup of G.

6.6.5   *Suppose that G contains a primitive maximal subgroup M. Then either (F1), (F2), or (F3) holds.*

Proof. Let N_1 be a minimal normal subgroup of G. Then

$$('):\qquad\qquad\qquad G = N_1 M.$$

From 6.6.1 we get $C_G(N_1) \cap M = 1$. If N_1 is Abelian, then $(')$ implies $N_1 = C_G(N_1)$ and thus $N_1 = Z(F^*(G))$. It follows that $N_1 = F^*(G)$. Hence (F1) holds since M is a complement of N_1 by 6.6.2.

We may assume now that no minimal normal subgroup of G is Abelian. Then $F(G) = 1$, and $E(G)$ is the product of the minimal normal subgroups of G (6.5.5 (b)). If N_1 is the only minimal normal subgroup of G, then (F3) holds.

Assume that there exists another minimal normal subgroup N_2 of G. Then

$$N := N_1 N_2 = N_1 \times N_2,$$

and $N \cap M \neq 1$ by $(')$. As $N = F^*(N)$ we get $E(G) = N$ from 6.6.4. Hence, N_1 and N_2 are the only minimal normal subgroups of G (1.6.3 (b) on page 30), and (F2) holds.   □

We now discuss the three cases (F1), (F2), and (F3) separately.

6.6.6   *Suppose that (F1) holds. Let $p \in \pi(M)$ such that $O_p(M) \neq 1$. Then all primitive maximal subgroups of G are conjugate.*[15]

---

[14]Thus $F(G)$ is elementary Abelian.

[15]In particular, this holds for solvable groups G.

Proof. Set $P := O_p(M)$ and $F := F^*(G)$. Then

$$M = N_G(P) \quad \text{and} \quad FP \trianglelefteq G,$$

and by 6.6.3 (a)

$$\mathrm{Syl}_p\, M \subseteq \mathrm{Syl}_p\, G.$$

Let H be another primitive maximal subgroup of G. Then also H is a complement of F; in particular $|H| = |M|$. According to Sylow's Theorem there exists $g \in G$ such that $P \leq H^g$. This implies

$$P = H^g \cap FP \trianglelefteq H^g,$$

and thus $H^g = N_G(P) = M$. $\qquad\qquad\square$

6.6.7 *Suppose that* (F2) *holds. Then there exists an M-isomorphism* $\alpha \colon N_1 \to N_2$ *such that*

$$M \cap F^*(G) = \{xx^\alpha \mid x \in N_1\}.[16]$$

Proof. Let $D := M \cap F^*(G)$. Then 6.6.1 implies

$$D \cap N_1 = 1 = D \cap N_2.$$

Since $G = N_i M$ we get $F^*(G) = N_i D$. Hence, for every $x_1 \in N_1$ there exists a unique $x_2 \in N_2$ such that $x_1 x_2 \in D$, and the mapping

$$\alpha \colon N_1 \to N_2, \quad x_1 \mapsto x_2$$

is an isomorphism. Moreover, this isomorphism commutes with the conjugation by elements of M since N_1, N_2, and D are M-invariant. $\qquad\square$

We now start the discussion of case (F3) and begin with the remark (compare with 1.7.1 (b) on page 36):

6.6.8 *Let F be a minimal normal subgroup of G and M a proper subgroup of G such that $G = FM$.*

---

[16]Thus, $M \cap F^*(G)$ is a "diagonal" of $N_1 \times N_2$.

(a)   Suppose that U is a proper M-invariant subgroup of F. Then UM is
      a proper subgroup of G.

(b) M is a maximal subgroup of G if and only if $F \cap M$ is the unique
      maximal M-invariant subgroup of F.

Proof. (b) follows from (a). For the proof of (a) we assume, by way of
contradiction, $G = UM$. Then U is normal in G, and thus F is not a
minimal normal subgroup of G.   □

In case (F3) $F^*(G)$ is a *non-Abelian* minimal normal subgroup of G. We
investigate the following situation:

$\mathcal{F}$ F is a non-Abelian minimal normal subgroup of G;

       M is a maximal subgroup of G such that $G = FM$;

       K is a component of F;

       $M_0 := N_M(K)$;

       $G_0 := KM_0$;

       $\overline{G_0} := G_0/C_{G_0}(K)$.

Then K is a non-Abelian simple group, and F is the direct product of the
conjugates of K. In fact, since $G = FM$ these conjugates are already
conjugate under M. This also shows that $K \not\leq M$ since $F \not\leq M$.

Note that $\overline{K}$ $(\leq \overline{G_0})$ is isomorphic to K and thus is a minimal normal
subgroup of
$$\overline{G_0} = \overline{K}\,\overline{M_0}.$$

6.6.9   *Suppose that $\mathcal{F}$ holds.*

(a) M_0 *is a maximal subgroup of* G_0.

(b)   *Let* $\overline{M_0} \neq \overline{G_0}$. *Then* $\overline{M_0}$ *is a primitive maximal subgroup of* $\overline{G_0}$.

(c) $\overline{M_0} \cap \overline{K} \in \{\overline{M \cap K}, \overline{K}\}$.

Proof. (a) We apply 6.6.8 (b) to G_0 and M_0 (with K in place of F). Then it suffices to show that every proper M_0-invariant subgroup V of K is contained in $K \cap M_0$.

Let $U := \langle V^M \rangle$. Then either $UM = M$ or $UM = G$. In the first case we have $V \le K \cap M_0$. In the second case we derive a contradiction, as follows. Note that now U is normal in G. The minimality of F gives $F = U$. On the other hand, $V^x \le K^x \ne K$ for every $x \in M \setminus M_0$. It follows that $U = V C_F(K) = F$ (6.5.3) and $V \trianglelefteq F$. But then $V = K$, a contradiction.

(b) The maximality of $\overline{M_0}$ follows from (a) since $\overline{M_0} \ne \overline{G_0}$. To show the primitivity of $\overline{M_0}$ let $N \le M_0$ such that $\overline{N} \trianglelefteq \overline{G_0}$. Then $[N, K] = 1$ since $\overline{K} \not\le \overline{M_0}$. As the mapping

$$K \to \overline{K} \text{ with } x \mapsto \overline{x}$$

is an N-isomorphism we get $N \le C_{G_0}(K)$ and thus $\overline{N} = 1$.

(c) Let $V := \{x \in K \mid \overline{x} \in \overline{M_0} \cap \overline{K}\}$. Then $\overline{V} = \overline{M_0} \cap \overline{K}$ and V is an M_0-invariant subgroup of K, which contains $M_0 \cap K$. Now the conclusion follows from (a) and 6.6.8. $\qquad\square$

6.6.10 *Suppose that $\mathcal{F}$ holds.*

(a) If $K \cap M \ne 1$, then $\overline{M_0}$ is a primitive maximal subgroup of $\overline{G_0}$.

(b) If $K \cap M = 1$, then (b1) or (b2) holds:

  (b1) $\overline{K} \le \overline{M_0} = \overline{G_0}$.

  (b2) $\overline{M_0} \cap \overline{K} = 1$, and $\overline{M_0}$ is a primitive maximal subgroup of $\overline{G_0}$.

Proof. (a) This follows from 6.6.9 (b) if $\overline{M_0} \ne \overline{G_0}$. Assume that $\overline{M_0} = \overline{G_0}$. Then $\overline{K} \le \overline{M_0}$, and $M \cap K = M_0 \cap K$ is K-invariant. This contradicts the simplicity of K.

(b) This follows from 6.6.9 (c) and (b). $\qquad\square$

The following conjecture can be verified using the classification of the finite simple groups.

Schreier's Conjecture:
Let E be a simple group. Then $\operatorname{Aut} E / \operatorname{Inn} E$ is solvable.

With the help of this conjecture one can show that case (b2) in 6.6.10 does
not occur:

According to 3.1.9 on page 60, $\overline{G}_0$ can be identified with a subgroup of
$\operatorname{Aut} K$ such that $\overline{K} = \operatorname{Inn} K$. Then

$$\overline{G}_0/\overline{K} \leq \operatorname{Aut} K/\operatorname{Inn} K.$$

It follows that $\overline{G}_0/\overline{K}$ and thus also $\overline{M}_0 \cong \overline{M}_0\overline{K}/\overline{K}$ is solvable. Moreover,
$|\pi(\overline{K})| \geq 2$ since K is a non-Abelian simple group. Hence 6.6.3 (c) shows
that $\overline{M}_0$ is not maximal, which contradicts (b2).

We denote the set of components of the group X by $\mathcal{K}(X)$. Let F be as in
$\mathcal{F}$ and N a normal subgroup of F. Then (see 1.7.5)

$$N = \underset{E\in\mathcal{K}(N)}{\times} E \quad \text{and} \quad F = N \times \left(\underset{E\in\mathcal{K}(F)\backslash\mathcal{K}(N)}{\times} E \right).$$

For $E \in \mathcal{K}(F)$ let

$$\pi_E \colon F \to E$$

be the projection of F onto E (we write $\pi_E(x)$ for the image of $x \in F$).

6.6.11   *Suppose that $\mathcal{F}$ holds and*

$$1 = K \cap M \neq F \cap M.$$

*Then there exist normal subgroups $N_1,\ldots,N_r$ of F such that the following
hold:*

(a) $F = N_1 \times \cdots \times N_r$, *and M acts transitively on* $\{N_1,\ldots,N_r\}$.

(b) $F \cap M = \overset{r}{\underset{i=1}{\times}}(N_i \cap M)$.

(c)   *For every $E \in \mathcal{K}(N_i)$ the mapping*

$$N_i \cap M \to E \quad \text{with} \quad x \mapsto \pi_E(x)$$

   is a $N_M(E)$-isomorphism $(i = 1,\ldots,r)$.[17]

(d) $\overline{M}_0 = \overline{G}_0$.

---

[17]Thus, $N_i \cap M$ is a *diagonal* of the direct product $N_i = \underset{E\in\mathcal{K}(N_i)}{\times} E$.

Proof. Let $D := F \cap M$ and

$$F_0 := \underset{E \in \mathcal{K}(F)}{\times} \pi_E(D).$$

Since $D \neq 1$ also $F_0 \neq 1$. Moreover $F_0 \not\leq M$ since $K \cap M = 1$ and all components of F are conjugate under M. Hence 6.6.8 (b) gives $F_0 = F$ and thus

(1) $\pi_E(D) = E$ for all $E \in \mathcal{K}(F)$.

Choose $a \in D^{\#}$ such that the number of components $E \in \mathcal{K}(F)$ with $\pi_E(a) \neq 1$ is minimal, and let N be the product of these components, i.e.,

$$\mathcal{K}(N) = \{ E \in \mathcal{K}(F) \mid \pi_E(a) \neq 1 \}.$$

Set

$$C := D \cap N;$$

note that $a \in C^{\#}$ and $C \trianglelefteq D$. Then $1 \neq \pi_E(C) \trianglelefteq \pi_E(D)$ for $E \in \mathcal{K}(N)$. Now (1) and the simplicity of E give $\pi_E(C) = E$. By the minimal choice of a the mapping $\pi_E|_C$ is injective. Hence we obtain:

(2) For every $E \in \mathcal{K}(N)$ the mapping

$$C \to E \quad \text{with} \quad x \mapsto \pi_E(x)$$

is a D-isomorphism.

That this isomorphism commutes with the action of D follows from the fact that $C^D = C$ and $E^D = E$.

We now show:

(3) Let $d \in D$ and $c \in C$ such that

$$\pi_{E_0}(d) = \pi_{E_0}(c) \quad \text{for some } E_0 \in \mathcal{K}(N).$$

Then $[N, dc^{-1}] = 1$.

For the proof of (3) let $x \in C$. By (2)

$$\pi_{E_0}(x^d) = \pi_{E_0}(x)^d = \pi_{E_0}(x)^{\pi_{E_0}(d)} = \pi_{E_0}(x)^{\pi_{E_0}(c)} = \pi_{E_0}(x^c),$$

and thus $x^d = x^c$ since $\pi_{E_0}|_C$ is injective; in particular $[C, f] = 1$ where $f := dc^{-1}$. It follows that for all $E \in \mathcal{K}(N)$

$$1 = \pi_E([C, f]) = [\pi_E(C), \pi_E(f)] \overset{(2)}{=} [E, \pi_E(f)],$$

and thus $\pi_E(f) = 1$ since $Z(E) = 1$. This implies (3).

Of course, the statements (2) and (3) also hold with N^m ($m \in M$) in place of N.

Let $E_0 \in \mathcal{K}(N) \cap \mathcal{K}(N^m)$ and $d := a^m$. By (2) there exists $c \in C$ such that $\pi_{E_0}(d) = \pi_{E_0}(c)$. Hence (3), applied to N and N^m, gives $[NN^m, dc^{-1}] = 1$. Together with $dc^{-1} \in NN^m$ we get that $d = c \in N \cap N^m$, and the minimality of a yields $N = N^m$. We have shown:

(4) $N \cap N^m = 1$ or $N = N^m$ for $m \in M$. In particular $N_M(E) \leq N_M(N)$ for all $E \in \mathcal{K}(N)$.

The second part of (4) implies that the mapping in (2) is an $N_M(E)$-isomorphism for all $E \in \mathcal{K}(N)$. With this remark (a) and (c) follow from (2) and (4), where the N_i are the conjugates of N. Moreover, (3) first gives

$$D = (D \cap N) \times C_D(N)$$

and then after repeated application

$$D = \underset{m \in M}{\times} (N^m \cap D).$$

This is (b). Finally (d) follows from (c). $\square$

The results 6.6.5 – 6.6.11 now yield:

6.6.12   Theorem of O'Nan-Scott.[18] *Let M be a primitive maximal subgroup of G. Then one of the following holds:*

(a)   *$F^*(G) = F(G)$, and $F(G)$ is the unique minimal normal subgroup of G.*

(b)   *$F(G) = 1$, and $F^*(G) = N_1 \times N_2$; here N_1 and N_2 are the only minimal normal subgroups of G. There exists an isomorphism $\alpha\colon N_1 \to N_2$ such that $F^*(G) \cap M = \{xx^\alpha \mid x \in N_1\}$.*

---

[18]See [81] and [35].

(c) $F(G) = 1$, and $F^*(G)$ is the unique minimal normal subgroup of G.
Moreover one of the following holds, where the notation is as in $\mathcal{F}$:

(c1) $\overline{M}_0$ is a primitive maximal subgroup of $\overline{G}_0$ (and $K \cap M \neq 1$).[19]

(c2) $\overline{M}_0 = \overline{G}_0$ and $M \cap F = 1$.

(c3) $\overline{M}_0 = \overline{G}_0$, $1 = K \cap M \neq F \cap M$, and F is as described in
6.6.11.   □

We now give an example for each of the cases that arise in the theorem of
O'Nan-Scott.

Case (a): $G = S_3$ and $M = S_2$ ($\leq G$) or $G = S_4$ and $M = S_3$ ($\leq G$).
In general (a) holds for every solvable group G as long as $\Phi(G) = 1$ and G
contains exactly one minimal normal subgroup.

In the other cases $F^*(G)$ is the direct product of its components. For the
cases (b), (c1), and (c3) let

$$K \cong A_5 \quad \text{and} \quad H := K \times K$$

and $t \in \operatorname{Aut} H$ such that

$$(k_1, k_2)^t = (k_2, k_1) \quad \text{for all } (k_1, k_2) \in H.$$

Using these data we construct a group G (and a primitive maximal subgroup
M) such that $F^*(G) = H$.

Case (b): $G := H$ and $M := \{(k, k) \mid k \in K\}$.

Case (c1): $G = \langle t \rangle H$, the semidirect product of H with $\langle t \rangle$, and M_1 is a
maximal subgroup of K. Let $M_2 := \{(k_1, k_2) \mid k_1, k_2 \in M_1\}$ and $M :=
M_2 \langle t \rangle$.

Case (c3): G is as in the example for (c1) but $M_2 := \{(k, k) \mid k \in K\}$ and
$M := M_2 \langle t \rangle$.

Case (c2): In the alternating group $M := A_6$ the stabilizer

$$M_0 := \{x \in A_6 \mid 6^x = 6\}$$

is a subgroup isomorphic to A_5. Let G be the twisted wreath product

$$(A_5, \ M, \ M_0, \ \tau),$$

---

[19] $K \cap M \neq 1$ follows if one uses Schreier's conjecture; see above.

where $\tau\colon M_0 \to \operatorname{Aut} A_5$ describes the action by conjugation, i.e., $\operatorname{Im}\tau = \operatorname{Inn} A_5$ (see Section 4.4). Then G is the semidirect product of a normal subgroup

$$\widehat{A}_5 = A_5 \times A_5 \times A_5 \times A_5 \times A_5 \times A_5$$

with M, and $F^*(G) = \widehat{A}_5$. Moreover, M is primitive since A_6 is simple, and 1 and K are the only M_0-invariant subgroups of K since M_0 acts on the first component $K \cong A_5$ of $\widehat{A}_5$ as $\operatorname{Inn} A_5$. Now 6.6.8 shows that M is a maximal subgroup of G (also compare with the proof of 6.6.9 (a)).

Exercises

Let G be a solvable group.

1. Let U be a primitive group and N a minimal normal subgroup of G. Set $\overline{G} := G/N$. Then $\overline{U} = \overline{G}$ or $\overline{U}$ is a primitive subgroup of $\overline{G}$.

2. Let U_1 and U_2 be primitive subgroups of G such that $|U_1| \le |U_2|$. Then U_1 is conjugate to a subgroup of U_2.

6.7   Subnormal Subgroups

In this last section of the chapter we present two theorems of Wielandt about subnormal subgroups. In particular, corollary 6.7.6 (Baer's Theorem) is a frequently used result.

6.7.1   Theorem (Wielandt [98]). *Let G be a group and A and B subnormal subgroups of G. Then also $\langle A, B \rangle$ is subnormal in G.*

Proof. Let G be a minimal counterexample[20] and $\mathcal{S}$ the set of all subnormal subgroups of G. Then there exist $A, B \in \mathcal{S}$ such that $\langle A, B \rangle \notin \mathcal{S}$. We fix B and choose $A \in \mathcal{S}$ maximal such that $\langle A, B \rangle \notin \mathcal{S}$. It follows:

(1)   If $A < X \in \mathcal{S}$ then $\langle X, B \rangle \in \mathcal{S}$.

---

[20]This means: We assume that the theorem is false. Then there exist groups G that satisfy the hypothesis but not the conclusion of the theorem. Among these groups we choose G such that $|G|$ is minimal.

If $A \trianglelefteq G$, then 1.2.8 on page 14 implies $AB/A \trianglelefteq\trianglelefteq G/A$ and thus $\langle A, B \rangle = AB \in \mathcal{S}$. Hence

(2) $\qquad\qquad\qquad A$ is not normal in G.

Since $A \in \mathcal{S}$ there exist subgroups X and G_1 of G such that

(3) $\qquad\qquad A \trianglelefteq X \trianglelefteq\trianglelefteq G_1 \trianglelefteq G$ and $A \neq X$, $G_1 \neq G$.

Clearly $A^b \trianglelefteq\trianglelefteq G_1$ for every $b \in B$ since G_1 is normal in G. The minimality of G gives
$$A \leq \langle A^B \rangle \trianglelefteq\trianglelefteq G_1;$$
in particular $\langle A^B \rangle \in \mathcal{S}$. If $A < \langle A^B \rangle$ then (1) implies
$$\langle A, B \rangle = \langle \langle A^B \rangle, B \rangle \trianglelefteq\trianglelefteq G,$$
which is not the case. Thus, we have $\langle A^B \rangle = A$, i.e.,
$$B \leq N_G(A).$$

Again by (1)
$$G_2 := \langle X, B \rangle \trianglelefteq\trianglelefteq G.$$
If $G_2 \neq G$ then as above the minimality of G gives $\langle A, B \rangle \trianglelefteq\trianglelefteq G_2$ and $\langle A, B \rangle \in \mathcal{S}$. Thus, we have
$$G = G_2 = \langle X, B \rangle \leq N_G(A),$$
which contradicts (2). $\qquad\qquad\qquad\qquad\qquad\qquad\qquad\qquad \square$

The following lemma gives a typical property of subnormal subgroups (compare with 3.2.6 on page 66):

6.7.2 Let Σ be a set of subnormal subgroups of the group G satisfying $\Sigma^G = \Sigma$, and let Σ_0 be a proper subset of Σ. Then there exists $X \in \Sigma \setminus \Sigma_0$ such that $\langle \Sigma_0 \rangle^X = \langle \Sigma_0 \rangle$.

Proof. By 6.7.1, $\langle \Sigma_0 \rangle$ is subnormal in G. Since we may assume that $\langle \Sigma_0 \rangle \neq G$ there exists a proper normal subgroup G_1 of G containing $\langle \Sigma_0 \rangle$. Hence, the claim follows by induction on $|G|$, applied to G_1, provided
$$\Sigma_1 := \{ U \in \Sigma \,|\, U \leq G_1 \} \neq \Sigma_0.$$

Assume that $\Sigma_0 = \Sigma_1$. Clearly $(\Sigma_1)^G = \Sigma_1$ since $G_1^G = G_1$ and $\Sigma^G = \Sigma$. Hence $\langle \Sigma_0 \rangle = \langle \Sigma_1 \rangle$ is normal in G. $\square$

Central in the proof of the next theorem is the fact that a certain set of subgroups contains a unique maximal element.[21] Such uniqueness results are frequently used tools in the investigation of finite groups. As the uniqueness result used here[22] is also needed in Chapter 12, we formulate it separately:

6.7.3   *Let A be a subgroup of the group G and $\mathcal{U}$ a nonempty set of subgroups of G. For $U \in \mathcal{U}$ set*

$$\Sigma_U := \{A^g \mid g \in G,\ A^g \trianglelefteq\trianglelefteq U\}.$$

Suppose that for all $U, \tilde{U} \in \mathcal{U}$:

(1) $A \in \Sigma_U$.

(2) $\{B \in \Sigma_{\tilde{U}} \mid B \le U\} \subseteq \Sigma_U$.

(3)   *There exists $\hat{U} \in \mathcal{U}$ such that $N_G(\langle \Sigma_U \cap \Sigma_{\tilde{U}} \rangle) \le \hat{U}$.*

Then $\mathcal{U}$ contains a unique maximal element.

Proof. Set

$$\Sigma := \bigcup_{U \in \mathcal{U}} \Sigma_U.$$

From (2) we obtain

$$\Sigma_U = \{B \in \Sigma \mid B \le U\} \text{ for } U \in \mathcal{U}.$$

By way of contradiction we assume that there exist two different maximal elements U_1 and U_2 of $\mathcal{U}$. In addition, we choose these maximal elements such that

$$\Sigma_0 := \Sigma_{U_1} \cap \Sigma_{U_2}$$

is maximal. According to (3) $N_G(\langle \Sigma_0 \rangle)$ is contained in a maximal element U_3 of $\mathcal{U}$. The definition of Σ_{U_i} shows that $\langle \Sigma_{U_i} \rangle \trianglelefteq U_i$, and the maximality of U_i and (3) give

(*) $U_i = N_G(\langle \Sigma_{U_i} \rangle), \quad i = 1, 2, 3;$

---

[21] With respect to inclusion.

[22] One version of this result Wielandt calls the *Zipper Lemma*—according to the method used in the proof; see [99], p. 586.

in particular $\Sigma_{U_1} \neq \Sigma_{U_2}$.

Let $i \in \{1, 2\}$ such that $\Sigma_0 \subsetneqq \Sigma_{U_i}$. Then by 6.7.2 there exists $X \in \Sigma_{U_i} \setminus \Sigma_0$ with $\langle \Sigma_0 \rangle^X = \langle \Sigma_0 \rangle$. It follows that $X \in \Sigma_{U_3}$, and thus $\Sigma_0 \subsetneqq \Sigma_{U_3} \cap \Sigma_{U_i}$ and $U_i = U_3$ by the maximal choice of Σ_0. Hence, we can choose notation such that $U_2 = U_3$ and $U_1 \neq U_3$. Then $\Sigma_0 = \Sigma_{U_1}$ and by $(*)$

$$U_1 = N_G(\langle \Sigma_{U_1} \rangle) \leq U_3,$$

i.e., $U_1 = U_3$, a contradiction. $\square$

6.7.4   Theorem (Wielandt [98]). *Let A be a subgroup of the group G. Suppose that*
$$A \trianglelefteq\trianglelefteq \langle A, A^g \rangle \quad \text{for all } g \in G.$$
Then A is subnormal in G.

Proof. Note that the hypothesis also holds for all conjugates A^x, $x \in G$:

$$A \trianglelefteq\trianglelefteq \langle A, A^{gx^{-1}} \rangle \quad \Rightarrow \quad A^x \trianglelefteq\trianglelefteq \langle A^x, A^g \rangle.$$

We now proceed by induction on $|G|$ and assume that A is not subnormal in G. Let $\mathcal{U}$ be the set of all proper subgroups of G that contain A; in particular $\langle A, A^g \rangle \in \mathcal{U}$ for all $g \in G$ since A is not subnormal in G. Let $U \in \mathcal{U}$. By induction on $|G|$ we may assume that every subgroup of

$$\Sigma_U := \{ A^x \mid A^x \leq U, \ x \in G \}$$

is subnormal in U. Moreover for $\Sigma_0 \subseteq \Sigma_U$ and $A \in \Sigma_0$

$$A \trianglelefteq\trianglelefteq \langle \Sigma_0 \rangle.$$

Hence $\langle \Sigma_0 \rangle$ is not normal in G. It follows that $N_G(\langle \Sigma_0 \rangle) \in \mathcal{U}$, and $\mathcal{U}$ satisfies the hypothesis of 6.7.3. Thus there exists a maximal subgroup M of G that contains $\langle A, A^g \rangle$ for all $g \in G$. Hence

$$A \trianglelefteq\trianglelefteq \langle \Sigma_M \rangle \trianglelefteq G$$

and thus $A \trianglelefteq\trianglelefteq G$ which contradicts our assumption. $\square$

Every subgroup of a nilpotent group is subnormal. This gives the following corollary:

6.7.5 *Let A be a subgroup of G. Suppose that $\langle A, A^g \rangle$ is nilpotent for every $g \in G$. Then A is subnormal in G; in particular $A \leq F(G)$.* □

Since p-groups are nilpotent we get as another corollary:

6.7.6 Baer's Theorem [24]. *Let x be a p-element of G. Suppose that $\langle x, x^g \rangle$ is a p-subgroup for every $g \in G$. Then $x \in O_p(G)$.* □

For the prime 2 we obtain:

6.7.7 *Let t be an involution of G that is not in $O_2(G)$. Then there exists an element $y \in G^{\#}$ of odd order such that $y^t = y^{-1}$.*

Proof. By 6.7.6 there exists $g \in G$ such that $\langle t, t^g \rangle$ is not a 2-subgroup. Then 1.6.9 on page 34 shows that $d := tt^g$ is not a 2-element. Hence there exists $1 \neq y \in \langle d \rangle$ of odd order, and again by 1.6.9 $y^t = y^{-1}$. □

The following lemma, which is similar to 6.7.6, will be needed in Chapters 10 and 11.

6.7.8 Matsuyama's Lemma [80]. *Let Z, Y be subgroups of G and $p \in \pi(G)$. Suppose that*

$$\langle (Z^g)^Y \rangle \text{ is a p-subgroup for all } g \in G.^{23}$$

Then there exists a Sylow p-subgroup P of G such that

$$\langle Z^g \mid g \in G, \ Z^g \leq P \rangle$$

is normalized by Y.[24]

Proof. Let $\mathcal{M}$ be the set of all Y-invariant p-subgroups Q of G that have the following property:

$$Z \leq Q \quad \text{and} \quad Q = \langle Z^g \mid g \in G, \ Z^g \leq Q \rangle.$$

---

[23]This implies $Z^g \trianglelefteq\trianglelefteq \langle Z^g, Y \rangle$ for all $g \in G$.
[24]On p. 169 this subgroup is denoted by $\mathrm{wcl}_G(Z, P)$.

By our hypothesis $\langle Z^Y \rangle$ is contained in $\mathcal{M}$; in particular $\mathcal{M}$ is nonempty. Let Q be a maximal element of $\mathcal{M}$ and

$$Q \leq P \in \mathrm{Syl}_p\, G.$$

We set

$$\Sigma := \{Z^g \mid g \in G,\ Z^g \leq P\} \quad \text{and} \quad \Sigma_0 := \{Z^g \mid g \in G,\ Z^g \leq Q\}.$$

Then $Q = \langle \Sigma_0 \rangle$, and the claim follows if $\Sigma_0 = \Sigma$.

Thus, we may assume that $\Sigma_0 \subset \Sigma$. Since all subgroups of P are subnormal in P we can apply 6.7.2. Hence, there exists $Z^g \in \Sigma \setminus \Sigma_0$ such that $Z^g \leq N_G(Q)$. Then also $\langle Z^{gY} \rangle \leq N_G(Q)$ since $Q^Y = Q$. It follows that $Q \langle Z^{gY} \rangle \in \mathcal{M}$, which contradicts the maximal choice of Q. $\qquad\square$

Exercises

Let G be a group.

1. Let $H \trianglelefteq\trianglelefteq G$. Then $H \cap S \in \mathrm{Syl}_p\, H$ for all $p \in \mathbb{P}$ and $S \in \mathrm{Syl}_p\, G$.

2. Let H be a solvable subgroup of G such that

$$S \cap H \in \mathrm{Syl}_p\, H \text{ for all } p \in \mathbb{P} \text{ and } S \in \mathrm{Syl}_p\, G.$$

   Then H is subnormal in G.

Let D be a conjugacy class of p-elements of G, $p \in \mathbb{P}$.

3. If $\langle D \rangle$ is not a p-group, then there exist $x, y \in D$ such that $x \neq y$ and x is conjugate to y in $\langle x, y \rangle$.

4. Let $E \subseteq D$ and $|E|$ be maximal satisfying

   $(*)$ $\quad E$ is a conjugacy class of $\langle E \rangle$.

   Then $\langle E \rangle \trianglelefteq\trianglelefteq G$.

5. Let $G = \langle D \rangle$, $E \subseteq D$ and $|E|$ be maximal satisfying

   $(**)$ $\quad E \neq D$ and E is a conjugacy class of $\langle E \rangle$.

   Then the set of all $U \leq G$ with $E \subseteq U$ and $U = \langle U \cap D \rangle$ contains a unique maximal element.

6. (Baumann, [25]) Let $G = \langle D \rangle$ and $D \subseteq U_1 \cup \cdots \cup U_r$ for proper subgroups $U_1, \ldots, U_r$ of G. Then $r \geq p + 1$.

Chapter 7

Transfer and p-Factor Groups

7.1   The Transfer Homomorphism

To search for nontrivial proper normal subgroups is often the first step in the investigation of a finite group. For example, if the group G has such a normal subgroup N, then in proofs by induction one frequently gets information about N and G/N, allowing one to derive the desired result for G (e.g., 6.1.2 on page 122).

Since normal subgroups are kernels of homomorphisms it is suggestive to construct homomorphisms of G in order to find normal subgroups. The difficulty then is to decide whether the kernel of such a homomorphism is a nontrivial and proper subgroup of G.

In the following let P be a subgroup of G. In this chapter we define a homomorphism τ from G into the Abelian group P/P', whose kernel and image can be described by means of p-elements if P is a Sylow p-subgroup of G. This is in the spirit of the philosophy mentioned earlier, that the structure of a group be deduced from its p-structure.

If G is non-Abelian, then clearly Ker τ is nontrivial since $G/$ Ker τ is Abelian. Hence, either G contains a proper nontrivial normal subgroup or $G = $ Ker τ. In the second case the description of Ker τ in terms of the conjugacy of p-elements in G will yield information concerning the structure of G.

Let

$$\overline{P} := P/P'$$

be the commutator factor group of P and

$$P \to \overline{P} \quad \text{with} \quad x \mapsto \overline{x}$$

the natural epimorphism to the *Abelian* group $\overline{P}$.

Let $\mathcal{S}$ be the set of transversals of P in G. For $R, S \in \mathcal{S}$ let

$$R|S := \prod_{\substack{(r,s) \in R \times S \\ Pr = Ps}} \overline{rs^{-1}} \quad (\in \overline{P}).$$

(Compare with the definition on page 71.) Since the factors are elements of the Abelian group $\overline{P}$ this product does not depend on their ordering. As in Section 3.3 for $R, S, T \in \mathcal{S}$ the following properties hold:

(1) $$(R|S)^{-1} = S|R$$

(2) $$(R|S)\,(S|T) = R|T.$$

We investigate the action of G on $\mathcal{S}$ by right multiplication:

$$S \xmapsto{g \in G} Sg.$$

Then

(3) $$Rg \,|\, Sg = R|S$$

and

(4) $$Rg|R = Sg|S.$$

For the proof of (4) note that

$$\begin{aligned}
(Rg|R)\,(Sg|S)^{-1} &= (Rg|R)\,(R|Sg)\,(R|Sg)^{-1}\,(Sg|S)^{-1} \\
&= (Rg|R)\,(R|Sg)\,((R|Sg)\,(Sg|S))^{-1} \\
&\overset{(2)}{=} (Rg|Sg)\,(R|S)^{-1} \overset{(3)}{=} 1.
\end{aligned}$$

7.1.1   Transfer Homomorphism. *Let $S \in \mathcal{S}$. The mapping*

$$\tau_{G \to P} : G \to \overline{P} \quad \text{with} \quad g \mapsto Sg|S$$

is a homomorphism that is independent of the choice of $S \in \mathcal{S}$.

Proof. The independence of the choice of S follows from (4). For $x, y \in G$

$$Sxy|S \overset{(2)}{=} (Sxy|Sy)(Sy|S) = ((Sx)y|Sy)(Sy|S) \overset{(3)}{=} (Sx|S)(Sy|S).$$

Hence $\tau_{G \to P}$ is a homomorphism. $\qquad\qquad\qquad\qquad\qquad\qquad\square$

Next we want to calculate the transfer $x^{\tau_{G \to P}}$ for $x \in G$. To do so we study the action of $\langle x \rangle$ by right multiplication on the set $\Omega := \{\, Pg \mid g \in G\}$. Let $\Omega_1, \ldots, \Omega_k$ be the $\langle x \rangle$-orbits of Ω and $Pg_i \in \Omega_i$. Then there exists a divisor n_i of $o(x)$ such that $\langle x^{n_i} \rangle$ is the kernel of the action of $\langle x \rangle$ on Ω_i.

For $i = 1, \ldots, k$:

- $n_i = |\Omega_i|$ and $\displaystyle\sum_{i=1}^{k} n_i = |G : P|$;

- $\Omega_i = \{Pg_i, Pg_i x, \ldots, Pg_i x^{n_i - 1}\}$;

- $Pg_i x^{n_i} = Pg_i$, and thus $g_i x^{n_i} g_i^{-1} \in P$.

In particular

$$S := \overset{\cdot}{\underset{i=1,\ldots,k}{\bigcup}} \{g_i x^j \mid j = 0, \ldots, n_i - 1\}$$

is an element of $\mathcal{S}$ and satisfies

$$Sx \cap Pg_i x^j = \begin{cases} \{g_i x^j\} & \text{for} \quad j = 1, \ldots, n_i - 1 \\ \{g_i x^{n_i}\} & \text{for} \quad j = 0. \end{cases}$$

Hence

(5) $$\qquad\qquad\qquad x^{\tau_{G \to P}} = \overline{\prod_{i=1}^{k} g_i x^{n_i} g_i^{-1}}.$$

We now set

$$P^* := \langle\, y^{-1} y^g \mid y, y^g \in P,\ g \in G \,\rangle.$$

Note here that $y^{-1} y^g = [y, g]$ and thus

$$P' \leq P^* \leq P \cap G'.$$

With this notation we get:

7.1.2 $\qquad (x^{\tau_{G \to P}})\, \overline{P^*} = \overline{x}^{|G:P|}\, \overline{P^*} \ \text{for } x \in P.$

Proof. For each factor in (5) we have

$$g_i x^{n_i} g_i^{-1} = x^{n_i}(x^{-n_i} g_i x^{n_i} g_i^{-1}) \in x^{n_i} P^*,$$

and thus

$$x^{\tau_{G \to P}} \equiv \overline{x}^{\sum_i n_i} = \overline{x}^{|G:P|} \pmod{\overline{P^*}}. \qquad \square$$

Now let π be a nonempty set of primes and let P be a Hall π-subgroup of G. Then $PG'/G' = O_\pi(G/G')$ and by 2.1.6 on page 46

$$G/G' = PG'/G' \times O_{\pi'}(G/G').$$

We denote the inverse image of $O_{\pi'}(G/G')$ in G by $G'(\pi)$. Then $G'(\pi)$ is the smallest normal subgroup of G having an Abelian π-factor group.[1] Since $G = PG'(\pi)$ we get

(6) $P \cap G'(\pi) = P \cap G'$ and $P/P \cap G' \cong G/G'(\pi).$

7.1.3   Theorem. *Let P be a Hall π-subgroup of G. Then*

$$P^* = P \cap G'(\pi) = P \cap G', \quad \text{and} \quad P/P^* \cong G/G'(\pi).$$

More precisely: $\operatorname{Ker} \tau_{G \to P} = G'(\pi)$ *and* $\overline{P} = \overline{P^*} \times \operatorname{Im} \tau_{G \to P}.$

Proof. Let $\tau := \tau_{G \to P}$. Note that $(|P|, |G : P|) = 1$ since P is a Hall π-subgroup. Hence 7.1.2 implies

$$\langle x^\tau \overline{P^*} \rangle = \langle \overline{x} \rangle \overline{P^*}$$

for all $x \in P$ (1.4.3 (b)). This gives $\overline{P \cap \operatorname{Ker} \tau} \leq \overline{P^*}$ and then

$$P \cap \operatorname{Ker} \tau \leq P^* \quad \text{and} \quad \overline{P} = \overline{P^*} \operatorname{Im} \tau$$

since $P' \leq P^*$.

Conversely, $G'(\pi) \leq \operatorname{Ker} \tau$ since τ is a homomorphism into the Abelian π-group $\overline{P}$. It follows that

$$P^* \leq P \cap G' \overset{(6)}{=} P \cap G'(\pi) \leq P \cap \operatorname{Ker} \tau,$$

---

[1] In the terminology of 6.3, $G'(\pi) = O^{\mathcal{K}}(G)$, where $\mathcal{K}$ is the class of Abelian π-groups.

and thus $P^* = P \cap \mathrm{Ker}\,\tau = P \cap G'(\pi) = P \cap G'$. Hence

$$|G/G'(\pi)| \geq |G/\mathrm{Ker}\,\tau| = |\mathrm{Im}\,\tau| \geq |P/P^*| = |P/P \cap G'| \overset{(6)}{=} |G/G'(\pi)|,$$

and this implies $\mathrm{Ker}\,\tau = G'(\pi)$ and $|\mathrm{Im}\,\tau| = |P/P^*|$. Since $\overline{P} = \overline{P^*}\,\mathrm{Im}\,\tau$ we get $\overline{P} = \overline{P^*} \times \mathrm{Im}\,\tau$. $\square$

As a corollary one gets:

7.1.4 *Let P be a Hall π-subgroup of G and $P \neq P^*$. Then $G \neq O^\pi(G)$.*
$\square$

The importance of 7.1.3 and 7.1.4 lies mainly in the fact that the subgroup $P \cap G'$ [2] can be calculated in P, provided one knows which of the elements of P are conjugate in G. For $\pi = \{p\}$ —according to Alperin's Fusion Theorem [20]—this conjugation takes place in the normalizers of certain nontrivial p-subgroups. In a very special case this result has long been known. Thus:

7.1.5 **Burnside's Lemma** ([4], p. 155). *Let P be a Sylow p-subgroup of G and A_1, A_2 normal subsets of P.[3] If A_1 and A_2 are conjugate in G, then they are already conjugate in $N_G(P)$.*

Proof. Let $g \in G$ such that $A_1^g = A_2$. Then $P \leq N_G(A_1)$ implies $P^g \leq N_G(A_1^g) = N_G(A_2)$. Hence, P and P^g are two Sylow p-subgroups of $N_G(A_2)$; in particular they are conjugate in $N_G(A_2)$. Let $z \in N_G(A_2)$ such that $P^{gz} = P$. Then $y := gz \in N_G(P)$ and $A_1{}^y = A_2$. $\square$

If P is an *Abelian* Sylow p-subgroup, then 7.1.5 can be applied to all subsets of P; in particular

$$x, x^g \in P,\ g \in G \quad \Rightarrow \quad x^g = x^y \quad \text{for some } y \in N_G(P).$$

This implies $P^* = \{x^{-1}x^y \mid y \in N_G(P),\ x \in P\}$, and 7.1.3 gives:

7.1.6 **Theorem.** *Let P be an Abelian Sylow p-subgroup of G and $H := N_G(P)$. Then $P \cap G' = P \cap H'$ and*

$$P/P \cap H' \cong G/G'(p) \cong H/H'(p). \qquad \square$$

---

[2]That is called the **focal** subgroup of P in G.
[3]That is, $A_i = A_i^x$ for all $x \in P$.

Let Z and P be subgroups of G, $Z \leq P$. Then Z is **weakly closed** in P (with respect to G), if

$$Z^g \leq P, \ g \in G \ \Rightarrow \ Z^g = Z.$$

7.1.7 Let $P \in \mathrm{Syl}_p\, G$ and let Z be a subgroup of $Z(P)$ that is weakly closed in P. Suppose that $y \in P$ and $g \in G$ such that $y^g \in P$. Then there exists $g' \in N_G(Z)$ with $y^g = y^{g'}$.

Proof. Note that $y^g \in P \cap P^g$ and thus $\langle Z, Z^g \rangle \leq C_G(y^g)$. By Sylow's Theorem there exists $c \in C_G(y^g)$ such that $\langle Z^g, Z^c \rangle$ is a p-group. It follows that

$$\langle Z^{gh}, Z^{ch} \rangle = \langle Z^g, Z^c \rangle^h \leq P$$

for some $h \in G$, again by Sylow's Theorem. Since Z is weakly closed in P we get $Z^{gh} = Z^{ch} = Z$ and thus

$$g' := gc^{-1} \in N_G(Z).$$

Now $c \in C_G(y^g)$ gives $y^{g'} = y^g$. $\square$

From 7.1.7 we conclude, using 7.1.3:

7.1.8 Grün's Theorem [62]. *Let P be a Sylow p-subgroup of G and Z a subgroup of $Z(P)$ that is weakly closed in P. Set $H := N_G(Z)$. Then $P \cap G' = P \cap H'$ and*

$$P/(P \cap G') \cong G/G'(p) \cong H/H'(p).$$

In particular

$$G \neq O^p(G) \ \Longleftrightarrow \ H \neq O^p(H).$$ $\square$

We conclude this section with an elementary remark about weakly closed subgroups:

7.1.9 *Let P be a Sylow p-subgroup of G and Z a subgroup of P that is normal in $N_G(P)$. Then the following two statements are equivalent:*

(i)   *Z is weakly closed in P with respect to G.*

(ii) $Z \leq R \in \mathrm{Syl}_p\, G \;\;\Rightarrow\;\; Z \trianglelefteq R.$

Proof. (i) $\Rightarrow$ (ii): If $Z \leq R = P^{g^{-1}}$, $g \in G$, then $Z^g \leq P$ and thus $Z^g = Z$. Hence $Z^R = Z^{P^{g^{-1}}} = Z$.

(ii) $\Rightarrow$ (i): Let $Z^g \leq P$. Since (ii) also holds for all conjugates of Z we have $Z^g \trianglelefteq P$. By 7.1.5 there exists $y \in N_G(P)$ such that $Z^y = Z^g$, and the hypothesis implies $Z^g = Z^y = Z$. $\square$

Let Z and P be subgroups of G, $Z \leq P$. The subgroup

$$\mathrm{wcl}_G(Z, P) := \langle\, Z^g \mid g \in G,\; Z^g \leq P \,\rangle$$

is said to be the **weak closure** of Z in P (with respect to G).[4]

It is evident that the weak closure $\mathrm{wcl}_G(Z, P)$ is normal in $N_G(P)$ and weakly closed in P. In particular, one gets a result similar to that of 7.1.9 (ii):

$$\mathrm{wcl}_G(Z, P) \leq R \in \mathrm{Syl}_p\, G \;\;\Rightarrow\;\; \mathrm{wcl}_G(Z, P) = \mathrm{wcl}_G(Z, R).$$

7.2   Normal p-Complements

A normal subgroup N of the group G is a **normal p-complement** of G if G is the semidirect product of N with a Sylow p-subgroup of G. This is equivalent to

$$O_{p'}(G) = N = O^p(G).$$

In other words, the existence of a normal p-complement is equivalent to G being p'-closed. As we have seen in Section 6.3, also subgroups and factor groups of groups with a normal p-complement have a normal p-complement.

From 7.1.6 one gets:

7.2.1   Theorem (Burnside ([4], S. 327)). *Let P be a Sylow p-subgroup of G. Suppose that $N_G(P) = C_G(P)$. Then G has a normal p-complement.*

---

[4]Compare 6.7.8 on p. 160.

Proof. Let $H := N_G(P)$. Then $P \leq Z(H)$, so P is Abelian. By 3.3.1 on page 73 there exists a complement A of P in H. Now $P \leq Z(H)$ gives $H = P \times A$ and thus $H' \cap P = 1$, and the claim follows from 7.1.6.　　　□

If P in 7.2.1 is cyclic and p is the smallest prime divisor of $|G|$, then $N_G(P) = C_G(P)$ by 3.1.9 on page 60 and 2.2.5 (a) on page 51. Thus, we have the following corollary:

7.2.2　*Suppose that the Sylow p-subgroups of G are cyclic, where p is the smallest prime divisor of $|G|$. The G has a normal p-complement.*　　□

The following observation is used in the proof of the next theorem.

7.2.3　*Let G be the semidirect product of the normal subgroup N with the subgroup P. Let Z be a subgroup of P and let $g \in G$ such that $Z^g \leq P$. Then there exists $x \in P$ with $Z^g = Z^x$. In particular, every normal subgroup of P is weakly closed in P.*

Proof. The element g can be written $g = yx$ with $y \in N$ and $x \in P$ since $G = NP$. Then $Z^g \leq P$ implies $Z^y \leq P$. This shows that for all $z \in Z$

$$[z, y] = z^{-1}y^{-1}zy \in N \cap P = 1$$

and thus $y \in C_G(Z)$. Now $Z^g = Z^x$ follows.　　　□

7.2.4　**Normal p-Complement Theorem of Frobenius** [47]. *Let P be a Sylow p-subgroup of G. Suppose that for every nontrivial p-subgroup U of P, $N_G(U)$ has a normal p-complement. Then G has a normal p-complement.*

Proof. Obviously, G has a normal p-complement if $P = 1$. Thus, we may assume that $P \neq 1$. Then also

$$Z := Z(P) \neq 1.$$

By hypothesis $H := N_G(Z)$ has a normal p-complement; in particular $O^p(H) \neq H$. We show:

$(')$ Z is weakly closed in P.

Using $(')$ and Grün's Theorem we get $O^p(G) \neq G$. Since the hypothesis is inherited by subgroups we may assume by induction on $|G|$ that $O^p(G)$ has a normal p-complement K. Then $K \trianglelefteq G$ and G/K is a p-group. Hence, K is also a normal p-complement of G.

For the proof of $(')$ it suffices to show the implication

$$Z \leq R \in \mathrm{Syl}_p\, G \quad \Rightarrow \quad Z \trianglelefteq R$$

(see 7.1.9). Hence, we assume that there exists $R \in \mathrm{Syl}_p\, G$ such that $Z \leq R$ and $Z \ntrianglelefteq R$. In addition, we choose R such that

$$S := N_R(Z)$$

is maximal. Let $S \leq T \in \mathrm{Syl}_p\, N_G(Z)$. Since $S < R$ and $T \in \mathrm{Syl}_p\, G$ we also have $S < T$ and thus by 3.1.10 on page 61

$$S < N_R(S) \quad \text{and} \quad S < N_T(S).$$

Let $M := N_G(S)$ and $N_T(S) \leq T_1 \in \mathrm{Syl}_p\, M$. Then the maximality of S shows that Z is normal in T_1. Since by our hypothesis M has a normal p-complement 7.2.3 implies that Z is weakly closed in T_1 with respect to M. But then by 7.1.9 Z is normal in every Sylow p-subgroup of M in which it is contained. Hence $Z \trianglelefteq N_R(S)$, which contradicts $S < N_R(S)$, and $(')$ is proved.   □

For $p \neq 2$ the preceding theorem was improved considerably by a result of Thompson. In 9.4.7 on page 255 we will give a version of Thompson's Normal p-Complement Theorem. It turns out—for odd primes p—that G has a normal p-complement if $N_G(U)$ has one, where U is a certain characteristic subgroup of P.[5]

Exercises

Let G be a group and P a subgroup of G.

1.   Let $P \leq Z(G)$. Then $x^{\tau_{G \to P}} = x^{|G:P|}$ for every $x \in G$.

2.   If P is an Abelian Hall subgroup of G, then $P \cap G' \cap Z(G) = 1$.

---

[5]$U = W(P)$ in the notation of 9.4.

3. Suppose that all Sylow subgroups of G are Abelian. Then $G' \cap Z(G) = 1$.

4. Let P be an Abelian Sylow p-subgroup of G. Then G has a factor group isomorphic to $Z(N_G(P)) \cap P$.

5. Let P be a Hall subgroup of G such that $N_G(P) = C_G(P)$. Then P has a normal complement in G.

6. Suppose that $N_G(P)/C_G(P)$ is a p-group for every nontrivial p-subgroup P of G. Then G has a normal p-complement.

7. (Iwasawa [71]) Suppose that every proper subgroup of G is nilpotent. Then G is solvable.[6]

8. If G contains a nilpotent Hall π-subgroup, $(\pi \subseteq \pi(G))$, then the π-Sylow theorem holds in G.

9. Let $S \in \mathrm{Syl}_2\, G$ and $S = H\langle a \rangle$ be as in 5.3.2 (d) on page 108. Then $G \neq O^2(G)$.

10. Let $G = O^2(G)$. Suppose that G has dihedral or semidihedral Sylow 2-subgroups. Then all involutions in G are conjugate.

11. Let G be a perfect group with (generalized) quaternion groups of order at least 16 as Sylow 2-subgroups. Then $C_G(t)$ is nonsolvable for every involution t of G.

12. Prove Frobenius's Theorem 4.1.2 for Frobenius groups with solvable Frobenius complements.

Let p and q be two different odd primes. We denote the multiplicative group of the field $\mathbb{Z}/p\mathbb{Z}$ by $\mathbb{Z}_p^*$ and set $\mathbb{Z}_p^* = \{\overline{1}, \ldots, \overline{p-1}\}$, where $\overline{z} = z + p\mathbb{Z}$. In addition, set

$$R := \{1, \ldots, \tfrac{p-1}{2}\},$$
$$S := \{1, \ldots, \tfrac{q-1}{2}\},$$
$$F(z,p) := \{r \in R \mid (-rz + p\mathbb{Z}) \cap R \neq \varnothing\},$$
$$F(z,q) := \{s \in S \mid (-sz + q\mathbb{Z}) \cap S \neq \varnothing\},$$
$$M := \{(a,b) \in R \times S \mid -\tfrac{q-1}{2} \leq bp - aq \leq \tfrac{p-1}{2}\}.$$

13. Let $H := \{\overline{1}, \overline{p-1}\} \leq \mathbb{Z}_p^*$ and $\overline{R} = \{\overline{x} \mid x \in R\}$. Then for all $\overline{x} \in \mathbb{Z}_p^*$:

    (a) $\overline{R}$ is a transversal of H in $\mathbb{Z}_p^*$.

    (b) $\overline{x}^{\tau_{\mathbb{Z}_p^*} \to H} = \overline{x}^{\frac{p-1}{2}} = (\overline{-1})^{|F(x,p)|}$.

    (c) $\overline{x}$ is a square in $\mathbb{Z}_p^*$, if and only if $|F(x,p)|$ is even.

---

[6]Compare with Exercise 10 on p. 124.

14. (a) $|M| = |F(q,p)| + |F(p,q)|$.

(b)   The mapping

$$\varepsilon \colon R \times S \to R \times S \quad \text{with} \quad (a,b) \mapsto (\tfrac{p+1}{2} - a, \tfrac{q+1}{2} - b)$$

is an involutionary bijection on $R \times S$ such that

i. $M^\varepsilon = M$,

ii. $y^\varepsilon \neq y$ for all $y \in (R \times S) \setminus M$.

(c) $\frac{p-1}{2} \frac{q-1}{2} \equiv |F(q,p)| + |F(p,q)| \pmod 2$.

15. Prove Gauß' Quadratic Reciprocity Law using Exercises 13 and 14.

Chapter 8

Groups Acting on Groups

The action of a group A on a set G is described by a homomorphism

$$\pi \colon A \to S_G;$$

see Section 3.1. Suppose that G is not only a set but also a group. Then $\operatorname{Aut} G \leq S_G$, and we say that π describes the action of A on the *group* G if $\operatorname{Im} \pi$ is a subgroup of $\operatorname{Aut} G$. In other words, in this case the action of A on G not only satisfies $\mathcal{O}_1$ and $\mathcal{O}_2$ but also

$$\mathcal{O}_3 \qquad\qquad (gh)^a = g^a h^a \text{ for all } g, h \in G \text{ and } a \in A.$$

The action by conjugation is the most important example for an action that also satisfies $\mathcal{O}_3$. For example, if A a subgroup and G a normal subgroup of a group H, then A acts by conjugation on the group G. In fact, in the semidirect product $A \ltimes_\pi G$ the action described by π is the conjugation of A on G (page 34).

In this chapter it is sometimes convenient (or even necessary) to use this semidirect product, as it allows us to apply, for example, Sylow's Theorem or the Theorem of Schur-Zassenhaus. We then simply write AG in place of $A \ltimes_\pi G$.

8.1 Action on Groups

Let A be a group that acts on the group G. First we introduce some notation that coincides with earlier notion if A and G are embedded in their semidirect product.

For $U \subseteq G$ and $B \subseteq A$

$$N_B(U) := \{b \in B \mid U^b = U\},$$

$$C_B(U) := \{b \in B \mid u^b = u \text{ for all } u \in U\},$$

$$C_U(a) := \{u \in U \mid u^a = u\} \quad (a \in A),$$

$$C_U(B) := \bigcap_{b \in B} C_U(b).$$

$C_G(A)$ is the subgroup of **fixed points** of A in G, and $C_A(G)$ is the kernel of the action of A on G. With respect to

$$g \overset{aC_A(G)}{\longmapsto} g^a \quad (g \in G, \ a \in A)$$

the factor group $A/C_A(G)$ acts faithfully on G.

We also use the commutator notation in this slightly more general situation:

$$[g, a] := g^{-1}g^a \quad (g \in G, \ a \in A),$$

$$[U, a] := \langle [g, a] \mid g \in U \rangle \quad (a \in A, \ U \subseteq G),$$

$$[U, B] := \langle [U, a] \mid a \in B \rangle \quad (B \subseteq A).$$

Similarly we define $[a, g] := g^{-a}g$, $[a, U]$ and $[B, U]$. The commutator relations given in Section 1.5 also hold in this more general context:

$$[U, B]^a = [U^a, B^a] \quad (a \in A),$$

$$[A, G] = [G, A],$$

$$U \leq C_G(A) \iff [U, A] = 1.$$

In particular, the Three-Subgroups Lemma is at hand:

$$[X, Y, Z] = [Y, Z, X] = 1 \quad \Rightarrow \quad [Z, X, Y] = 1,$$

where X, Y, Z now can be subgroups of G or A. Result 1.5.4 on page 25 now reads:

$$[gx, a] = [g, a]^x [x, a] \quad (g, x \in G, \ a \in A).$$

From this one gets that $[G, A]$ is an A-invariant normal subgroup of G.

8.1.1   *Let U be an A-invariant subgroup of G. Then*

$$[G, A] \leq U \iff (Ug)^a = Ug \text{ for all } g \in G,\ a \in A.$$

Proof. For $a \in A$ and $g \in G$

$$(Ug^{-1})^a = Ug^{-1} \iff Ug^{-a} = Ug^{-1} \iff [g, a] \in U. \qquad \square$$

8.1.2   *Let N be an A-invariant normal subgroup of G.*

(a)   *If A acts trivially on G/N, then $[G, A] \leq N$.*

(b)   *If A acts trivially on N, then A also acts trivially on $G/C_G(N)$.*

(c)   *If A acts trivially on N and G/N, then $[G, A] \leq Z(N)$ and $A' \leq C_A(G)$.*

Proof. (a) follows from 8.1.1.

(b) Let $[N, A] = 1$. Then

$$[N, A, G] = 1 = [G, N, A],$$

and the Three-Subgroups Lemma gives $[A, G, N] = 1$.

(c) From (a) and (b) we get

$$[G, A] \leq N \cap C_G(N) = Z(N),$$

and thus $[G, A, A] = 1 = [A, G, A]$. Again the Three-Subgroups Lemma yields the desired conclusion $[A', G] = [A, A, G] = 1$. $\qquad \square$

8.1.3   *Let A be a p-group. Then there exists an A-invariant Sylow p-subgroup of G.*

Proof. Let $A \leq \widehat{P} \in \operatorname{Syl}_p AG$. Then $P := \widehat{P} \cap G$ is the desired Sylow p-subgroup of G (3.2.5 on page 65). $\qquad \square$

8.1.4   *Let A be a p-group.*

(a)   If $p \in \pi(G)$, then $C_G(A) \neq 1$.

(b)   If G p-group, then $[G, A] < G$.

Proof. (a) By 8.1.3 there exists an A-invariant Sylow p-subgroup P of G. Hence, P is a normal subgroup of the semidirect product AP. Since AP is a p-group (a) follows from 3.1.11 (a) on page 61.

(b) This is 5.1.6 (iii) on page 101.   □

8.1.5   *Let K be an A-composition factor of G that is a p-group. Then $[K, O_p(A)] = 1$.*

Proof. The p-group $B := O_p(A)$ acts on the p-group K, so by 8.1.4 $C_K(B) \neq 1$. Since K is an A-composition factor and $C_K(B)$ is A-invariant we get $C_K(B) = K$.   □

Assume that G allows a direct decomposition

$$G = E_1 \times \cdots \times E_n$$

that is *invariant* under A, i.e.,

$$E_i{}^a \in \{E_1, \ldots, E_n\} \text{ for all } a \in A \text{ and } i \in \{1, \ldots, n\}.$$

Under the additional hypothesis that A acts transitively on $\{E_1, \ldots, E_n\}$ we compare the fixed-point groups $C_G(A)$ and $C_{E_i}(N_A(E_i))$.

Let

$$E \in \{E_1, \ldots, E_n\} \quad \text{and} \quad B := N_A(E),$$

and let S be a transversal for the cosets of B in A. Then

$$(+) \qquad\qquad G = \langle E^A \rangle = \underset{s \in S}{\times}\, E^s.$$

Under the above hypotheses the following hold:

8.1.6   (a) $C_G(A) = \{ \prod_{s \in S} e^s \mid e \in C_E(B) \}$.

$(b)^1$ *If B acts trivially on E and $P \leq E$ such that $\langle P^E \rangle = E$, then*

$$G = \langle C_G(A), \prod_{s \in S} P^s \rangle.$$

Proof. (a) Let $g \in G$ and

$$F := \Big\{ \prod_{s \in S} e^s \mid e \in C_E(B) \Big\}.$$

As S is a transversal, for every $(s,a) \in S \times A$ there exists a unique $(b(s,a), s_a) \in B \times S$ such that

$$sa = b(s,a)s_a.$$

Note here that the mapping $s \mapsto s_a$ is a bijection on S.

Let $g = \prod_{s \in S} e^s \in F$. Then for every $a \in A$

$$g^a = \prod_{s \in S} e^{sa} = \prod_{s \in S} e^{b(s,a)s_a} = \prod_{s \in S} e^{s_a} = g$$

since $e \in C_G(B)$. Thus $F \leq C_G(A)$.

Let $g \in C_G(A)$. By $(+)$ g has the unique representation

$$g = \prod_{s \in S} e_s \quad (e_s \in E^s).$$

For all $a \in A$

$$\prod_{s \in S} e_s = g = g^a = \prod_{s \in S} e_s{}^a,$$

and the uniqueness of the representation gives

$$\{e_s \mid s \in S\} = \{e_s{}^a \mid s \in S\}.$$

Let $s_0 \in B \cap S$ and $e := e_{s_0}$. Then $e^b = e$ for all $b \in B$ and $g = \prod_{s \in S} e^s \in F$, so $C_G(A) \leq F$.

(b) From (a) we get

$$C_G(A) = \Big\{ \prod_{s \in S} e^s \mid e \in E \Big\},$$

---

[1]This will be needed in Chapter 9.

and for $s \in S$

$$\langle (P^s)^{C_G(A)} \rangle = \langle (P^s)^{E^s} \rangle = \langle P^E \rangle^s = E^s.$$

This is (b).   □

We conclude this section with some remarks about cyclic operator groups.

8.1.7   Let $A = \langle a \rangle$ be cyclic. Then for $x, y \in G$

$$[x, a] = [y, a] \iff xy^{-1} \in C_G(a).$$

In particular, $|G : C_G(a)|$ is the number of commutators $[x, a]$, $x \in G$.[2]

Proof. $x^{-1}x^a = y^{-1}y^a \iff yx^{-1} = y^a x^{-a} \iff yx^{-1} = (yx^{-1})^a$
$\iff yx^{-1} \in C_G(a)$.   □

8.1.8   Let $A = \langle a \rangle$ such that $[G, a^2] = 1$, and let G be of odd order. Then

$$\{ x \in G \mid x^a = x^{-1} \} = \{ [x, a] \mid x \in G \},$$

and every coset of $C_G(a)$ in G contains exactly one commutator $[x, a]$.

Proof. Since $[G, a^2] = 1$ for every commutator $[x, a]$

$$[x, a]^a = (x^{-1}x^a)^a = x^{-a}x^{a^2} = x^{-a}x = [x, a]^{-1}.$$

The conclusion now follows from 8.1.7, if we can show that every coset of $C_G(a)$ contains at most one x such that $x^a = x^{-1}$.

Let x and xf, $f \in C_G(a)$, be two such elements. Then

$$x^a = x^{-1}, \quad (xf)^a = f^{-1}x^{-1} \quad \text{and} \quad f^a = f.$$

This implies

$$f^{-1}x^{-1} = (xf)^a = x^a f^a = x^{-1}f,$$

---

[2]This number is equal to $|a^S|$ in the semidirect product $S := \langle a \rangle G$.

so $f^x = f^{-1}$ and thus $f^{x^2} = f$. As x has odd order, we get $\langle x^2 \rangle = \langle x \rangle$. Hence $f = f^{-1}$ and thus $f = 1$ since f has odd order. $\square$

The operator group A acts **fixed-point-freely** on G if

$$C_G(A) = 1.$$

Similarly, the element $a \in A$ acts fixed-point-freely on G if $C_G(a) = 1$.

From 8.1.4 we get:

8.1.9 *Let A be a p-group. Suppose that A acts fixed-point-freely on G. Then G is a p'-group.* $\square$

Now 8.1.8 implies:

8.1.10 *Let a be a fixed-point-free automorphism of G of order 2. Then for all $x \in G$*

$$x^a = x^{-1}.$$

In particular G is Abelian.[3] $\square$

Also for arbitrary $p \in \mathbb{P}$ the existence of a fixed-point-free automorphism of order p has consequences for the structure of G. A theorem of Thompson shows in this case that G is nilpotent. We postpone the proof of this theorem and further discussion of fixed-point-free action to Section 9.5 since another fundamental theorem of Thompson (9.4.7 on page 255) is needed for this. Here we only remark that fixed-point-free automorphisms "behave well" with respect to induction:

8.1.11 *Let a be a fixed-point-free automorphism of G.*

(a) $G = \{[x, a] \mid x \in G\} = \{x^{-1}x^a \mid x \in G\}$.

(b) *For every $p \in \pi(G)$ there exists an a-invariant* [4] *Sylow p-subgroup of G.*

(c) *Let N be an a-invariant normal subgroup of G. Then a acts fixed-point-freely on G/N.*

---
[3]This is Exercise 10 on p. 10.
[4]a-invariant $= \langle a \rangle$-invariant.

Proof. 8.1.7 is (a). For the proof of (b) let $P \in \mathrm{Syl}_p\, G$ and $g \in G$ such that $P^a = P^g$. By (a) there exists $x \in G$ such that $g = x^{-1}x^a$. Now (b) follows since

$$(P^{x^{-1}})^a = P^{ax^{-a}} = P^{gx^{-a}} = P^{x^{-1}x^a x^{-a}} = P^{x^{-1}}.$$

(c) Let $(xN)^a = xN$ for some $x \in G$, so $x^{-1}x^a \in N$. Then (a), applied to $(N, a|_N)$, shows that there exists $y \in N$ such that $x^{-1}x^a = y^{-1}y^a$. This implies

$$yx^{-1} = y^a x^{-a} = (yx^{-1})^a$$

and thus $x = y$ and $xN = yN = N$. $\qquad\qquad\qquad\qquad\qquad\qquad\qquad$ □

Frobenius groups provide examples for fixed-point-free action:

8.1.12 *Let G be the semidirect product of the nontrivial subgroup H with the normal subgroup K. Then the following statements are equivalent:*

(i) *G is a Frobenius group with Frobenius complement H and Frobenius kernel K.*

(ii) *$C_K(h) = 1$ for all $h \in H^{\#}$.*[5]

Proof. The factorization $G = HK$ together with 4.1.7 on page 80 implies:

$$\text{(i)} \iff H \cap H^x = 1 \text{ for all } x \in K^{\#}.$$

On the other hand, since $H \cap K = 1$ we get for all $h \in H^{\#}$ and $x \in K^{\#}$:

$$h^x \in H \cap H^x \iff x^{-1}h^{-1}xh = [x,h] \in H \cap K \iff x \in C_K(h)^{\#}.$$

Now the equivalence of (i) and (ii) follows. $\qquad\qquad\qquad\qquad\qquad\qquad$ □

Exercises

Let A be a group acting on the group G, and let AG be the semidirect product of A with G.

---

[5]That is, h acts fixed-point-freely on K (with respect to conjugation).

1. Let $\varphi \colon A \to S_G$ and $\rho \colon G \to S_G$ be the homomorphisms describing the action of A on G and the action of G on the set G by right multiplication, respectively. Suppose that A acts faithfully on G. Then

    $$AG \cong A^{\varphi} G^{\rho}.$$

2. Let G be solvable, A nilpotent and N an A-invariant normal subgroup of G. Suppose that A acts fixed-point-freely on G. Then A acts fixed-point-freely on G/N (compare with Exercise 8 on page 124).

Let $[G, A; 1] := [G, A]$ and $[G, A; n] := [[G, A; n-1], A]$ for $n \geq 2$. Then A acts **nilpotently** on G if there exists an $n \in \mathbb{N}$ such that $[G, A; n] = 1$.

3. Let A and G be p-groups. Then A acts nilpotently on G.

4. A acts nilpotently on G if and only if A is a subnormal subgroup of AG.

5. Let A_1 and A_2 be two normal subgroups of A. If A_1 and A_2 act nilpotently on G, them also $A_1 A_2$ acts nilpotently on G.

6. Let $C_A^*(G)$ be the subgroup generated by all subnormal subgroups of A that act nilpotently on G. Then $C_A^*(G)$ acts nilpotently on G.

In the next two exercises G acts by conjugation on G, and $C_G^*(G)$ is the subgroup defined in Exercise 6.

7. $C_G^*(G) = F(G)$.

8. Let $\mathcal{F}$ be the set of all normal subgroups N of G satisfying $C_G^*(N) \leq N$. Then

    $$F^*(G) = \bigcap_{N \in \mathcal{F}} N.$$

8.2 Coprime Action

As in Section 8.1 let A be a group that acts on the group G. The action of A on G is **coprime** if

(1) $(|A|, |G|) = 1$,

(2) A or G is solvable.[6]

---

[6] Again we want to emphasize that the theorem of Feit-Thompson mentioned earlier shows that (1) implies (2) since at least one of the groups A and G has odd order.

In the semidirect product AG the subgroup A is a complement of the normal subgroup G, so in the case of coprime action the hypothesis of the theorem of Schur-Zassenhaus is satisfied. Hence, every subgroup of order $|A|$ in AG is conjugate to A.

A first consequence of this conjugacy-property is:[7]

8.2.1 *Suppose that the action of A on G is coprime. Let U be an A-invariant subgroup of G and $g \in G$ such that $(Ug)^A = Ug$. Then there exists $c \in C_G(A)$ such that $Ug = Uc$.*

Proof. $U^A = U$ and $(Ug)^A = Ug$ imply $g^a g^{-1} \in U$ for all $a \in A$. In the semidirect product AG we get $a^{-1} g a g^{-1} \in U$ and

$$A^{g^{-1}} \le AU.$$

Hence, A and $A^{g^{-1}}$ are complements of U in AU, and by the theorem of Schur-Zassenhaus (6.2.1 on page 125) they are conjugate in AU. Thus, there exists $u \in U$ such that $A^u = A^{g^{-1}}$. For $c := ug$ this gives

$$c \in N_{AG}(A) \cap Ug,$$

and $[A, c] \le A \cap G = 1$. $\qquad\qquad\qquad\qquad\qquad\qquad\qquad\qquad\qquad$ $\square$

For the special case that A is a p-group the proof of 8.2.1 does not require the theorem of Schur-Zassenhaus but follows from 3.1.7 on page 59 (with $\Omega := Ug$).

8.2.2 *Let N be an A-invariant normal subgroup of G. Suppose that the action of A on N is coprime.*

(a) $C_{G/N}(A) = C_G(A)N/N$. [8,9]

(b)   *If A acts trivially on N and G/N, then A acts trivially on G.*[10]

---

[7] A similar statement is also true for left cosets.

[8] In general only $C_G(A)N/N \le C_{G/N}(A)$.

[9] Compare with 3.2.8 (a) on p. 66.

[10] Compare with 8.1.2.

Proof. (b) is a consequence of (a), and (a) follows from 8.2.1 with $U := N$.

□

The next result, another important consequence of the theorem of Schur-Zassenhaus, is in the spirit of Sylow's Theorem. As for 8.2.1, the proof is elementary if A is a p-group.

8.2.3   *Let p be a prime divisor of $|G|$. Suppose that the action of A on G is coprime.*

(a)   *There exists an A-invariant Sylow p-subgroup of G.*

(b)   *The A-invariant Sylow p-subgroups of G are conjugate under $C_G(A)$.*

(c)   *Every A-invariant p-subgroup is contained in an A-invariant Sylow p-subgroup of G.*

Proof. The semidirect product AG acts on the set $\Omega := \mathrm{Syl}_p\, G$ by conjugation, and by Sylow's Theorem G is transitive on Ω. Hence, (a) and (b) follow from 6.2.2 on page 127 with (G, A) in place of (K, A).

(c) Let U be a maximal A-invariant p-subgroup of G. We show that U is a Sylow p-subgroup of G.

Assume that $U \notin \mathrm{Syl}_p\, G$. Then U is not a Sylow p-subgroup of $G_1 := N_G(U)$ (3.2.6 on page 66). As G_1 is A-invariant, there exists an A-invariant $T \in \mathrm{Syl}_p\, G_1$ by (a). But $U < T$, which contradicts the maximality of U.   □

The intersection $O_p(G)$ of all Sylow p-subgroups of G is the largest normal p-subgroup of G. An analogue statement is true in the situation of 8.2.3.

8.2.4   *Suppose that the action of A on G is coprime. Let $p \in \pi(G)$. Then the intersection of all A-invariant Sylow p-subgroups of G is the largest A-invariant p-subgroup of G that is normalized by $C_G(A)$.*

Proof. By 8.2.3 (a), (b) there exists $S \in \mathrm{Syl}_p\, G$ such that $S^A = S$, and

$$\{P \in \mathrm{Syl}_p\, G \mid P^A = P\} = \{S^c \mid c \in C_G(A)\}.$$

Hence, the intersection of these Sylow p-subgroups is $C_G(A)$-invariant.

Any A-invariant p-subgroup U is contained in an A-invariant Sylow p-subgroup of G (8.2.3 (c)). If in addition U is normalized by $C_G(A)$, then—as seen above—U is contained in every A-invariant Sylow p-subgroup and thus in their intersection. $\qquad\qquad\qquad\qquad\qquad\qquad\qquad\qquad\qquad$ □

8.2.5 *Suppose that the action of A on G is coprime. Let P be an A-invariant Sylow p-subgroup of G. If H is a subgroup of G that is invariant under A and $C_G(A)$, then $P \cap H$ is a Sylow p-subgroup of H.*

Proof. By 8.2.3 (a), (c) there exists an A-invariant Sylow p-subgroup R of H such that $P \cap H \leq R$ and an A-invariant Sylow p-subgroup S of G such that $R \leq S$; i.e.,

$$H \cap S = R.$$

Hence, there exists $c \in C_G(A)$ such that $S^c = P$ (8.2.3 (b)), and by our hypothesis $H^c = H$. It follows that

$$H \cap P = H \cap S^c \in \mathrm{Syl}_p H. \qquad\qquad\qquad\qquad □$$

In Chapter 11 we will need variations of 8.2.3, 8.2.4, and 8.2.5 for *solvable* groups G. Note that in the previous proofs we only used—apart from the theorem of Schur-Zassenhaus—Sylow's Theorem for a prime $p \in \pi(G)$.

If we replace p be a nonempty set $\pi \subseteq \pi(G)$ for which the π-Sylow Theorem holds (see 6.4.7 on page 137), then the above arguments yield results with the term Sylow p-*subgroup* replaced by the term Hall π-*subgroup*.

Since the π-Sylow Theorem holds in solvable groups (6.4.7 on page 137) we get:

8.2.6 *Suppose that the action of A on the solvable group G is coprime.*

(a)   *There exist A-invariant Hall π-subgroups of G.*

(b)   *The A-invariant Hall π-subgroups of G are conjugate under $C_G(A)$.*

(c)   *Every A-invariant π-subgroup is contained in an A-invariant Hall π-subgroup of G.*

(d)   The intersection of all A-invariant Hall π-subgroups of G is the largest
       A-invariant π-subgroup of G that is normalized by $C_G(A)$.

(e)   If P is an A-invariant Hall π-subgroup of G and H an A-invariant
       subgroup of G normalized by $C_G(A)$, then $P \cap H$ is a Hall π-subgroup
       of H.   □

The correspondence between the fixed-point group of A in G and in factor
groups of G described in 8.2.2 (a) has some interesting consequences.

8.2.7   *Suppose that the action of A on G is coprime.*

(a) $G = [G, A] \, C_G(A)$,

(b) $[G, A] = [G, A, A]$.

Proof. (a) follows from 8.2.2 (a) with $N := [G, A]$; note 8.1.1 (a). The
commutator formula 1.5.4 on page 25 shows that (a) implies (b).   □

8.2.8   Thompson's $P \times Q$ -Lemma. *Let $A = P \times Q$ be the direct
product of a p-group P and a p'-group Q. Suppose that G is a p-group such
that*
$$C_G(P) \leq C_G(Q).$$
Then Q acts trivially on G.

Proof. $C_U(P) \leq C_U(Q)$ for all A-invariant subgroups $U \leq G$. Thus, we
may assume be induction on $|G|$ that $[U, Q] = 1$ for all proper A-invariant
subgroups of G. As by 8.1.4 (b) $[G, P]$ is a proper subgroup, we get
$$[G, P, Q] = 1 \quad \text{and} \quad [P, Q, G] = 1,$$
the second equality holds since $[P, Q] = 1$. The Three-Subgroups Lemma
gives $[Q, G, P] = 1$, i.e.,
$$[Q, G] \leq C_G(P) \leq C_G(Q),$$
and $[G, Q, Q] = 1$. Now $[G, Q] = 1$ follows from 8.2.7 (b) (with Q in place
of A).   □

8.2.9   *Suppose that A acts trivially on $G/\Phi(G)$.*

(a)   *If the action of A on $\Phi(G)$ is coprime, then A acts trivially on G.*

(b)   *If $\Phi(G)$ is a p-group, then also $A/C_A(G)$ is a p-group.*

Proof. (a) 8.2.2 (a) gives $G = \Phi(G)\,C_G(A)$ and thus $G = C_G(A)$ (5.2.3).

(b) By (a) every p'-subgroup of A acts trivially on G.   □

8.2.10   *Let G be a p-group and $\mathcal{K}$ the set of all A-composition factors of G. Suppose that the action of A on G is coprime. Then*

$$\bigcap_{K \in \mathcal{K}} C_A(K)/C_A(G) = O_p(A/C_A(G)).$$

Proof. We may assume that A acts faithfully on G. By 8.1.5 $O_p(A)$ acts trivially on each A-composition factor $K \in \mathcal{K}$. On the other hand, by 8.2.2 (b) every p'-subgroup $B \le A$ acts trivially on G, if B acts trivially on each A-composition factor $K \in \mathcal{K}$. This shows the assertion.   □

The next result will be used in Chapter 11.

8.2.11   *Suppose that the action of A on G is coprime. Let G be the product of two A-invariant subgroups X and Y. Then $C_G(A) = C_X(A)\,C_Y(A)$.*

Proof. Let $g = xy \in C_G(A)$, $x \in X$, $y \in Y$. Then $xy = (xy)^a = x^a y^a$ and thus

$$x^{-1}x^a = y\,y^{-a} \in X \cap Y =: U$$

for all $a \in A$. This implies $(xU)^A = xU$ and $(Uy)^A = Uy$. By 8.2.1 there exist elements $c \in C_X(A)$, $d \in C_Y(A)$ and $u, w \in U$ such that

$$x = cu \quad \text{and} \quad y = wd.[11]$$

---

[11]See footnote 7.

Since $cuwd = xy \in C_G(A)$ also

$$uw \in C_G(A) \cap X \cap Y,$$

and $xy \in C_X(A)\,C_Y(A)$ follows.   □

We conclude this section with a particularly important application of the $P \times Q$-Lemma:

8.2.12 Let $p \in \pi(G)$ and $\overline{G} := G/O_{p'}(G)$. Suppose that

$(*)$ $$C_{\overline{G}}(O_p(\overline{G})) \le O_p(\overline{G}).$$

Then for every p-subgroup $P \le G$

$$O_{p'}(N_G(P)) = O_{p'}(G) \cap N_G(P).$$

Proof. $C_G(P) \trianglelefteq N_G(P)$ implies

$$O_{p'}(N_G(P)) = O_{p'}(C_G(P)).$$

Hence, it suffices to show that

$$O_{p'}(G) \cap C_G(P) = O_{p'}(C_G(P)).$$

The inclusion $O_{p'}(G) \cap C_G(P) \le O_{p'}(C_G(P))$ is trivial.

For the proof of the other inclusion we may assume that $O_{p'}(G) = 1$ (3.2.8 on page 66). Set

$$G_1 := O_p(G) \quad \text{and} \quad Q := O_{p'}(C_G(P)).$$

By our assumption $C_G(G_1) \le G_1$, and $PQ = P \times Q$ acts on the p-group G_1. As $C_{G_1}(P)$ is a normal p-subgroup of $C_G(P)$, the group Q acts trivially on $C_{G_1}(P)$. Hence the $P \times Q$-Lemma (8.2.8) gives $Q \le C_G(G_1) \le G_1$, so $Q = 1$.   □

Observe that solvable groups—more generally p-separable groups—satisfy hypothesis $(*)$ in 8.2.12; see 6.4.3 and 6.4.1.

As a corollary of 8.2.12 we get:

8.2.13   *Let P be a p-subgroup of G and $U \leq O_{p'}(N_G(P))$. Suppose that U and P are contained in the solvable subgroup $L \leq G$ Then $U \leq O_{p'}(L)$.*

Proof. We have $U \leq O_{p'}(N_L(P))$. Hence, the assertion follows from 8.2.12 (with L in place of G) since L is solvable. $\square$

Exercises

Let A be a group that acts on G.

1. Suppose that the action of A on G is nilpotent and faithful (see page 183). Then $\pi(A) \subseteq \pi(G)$.

2. Let $U \leq C_G(A)$ and $x \in G$ such that $U^x \leq C_G(A)$. If the action of A on G is coprime, then there exists a $y \in C_G(A)$ such that $U^x = U^y$.

3. (Zassenhaus [102]) Let $|A| = 2 = |C_G(A)|$. Then there exists an Abelian normal subgroup N of G such that

    (a) $x^a = x^{-1}$ for $a \in A^{\#}$.

    (b)   If $|G/N| \neq 2$, then $N = Z(G)$ and $G/N \cong A_4$.

4. Let G be π-separable. Then the $\pi \cup \{p\}$-Sylow Theorem holds in G for every $p \in \pi'$. (Use the fact that every π- or π'-section of G is solvable.)

8.3   Action on Abelian Groups

In the next two sections we investigate the action of groups on Abelian groups. So in the following let A be a group that acts on the Abelian group V. Here the choice of notation should remind the reader that in many applications V is an elementary Abelian p-group and thus also a vector space over $\mathbb{F}_p$.

The action of A on V is **irreducible** if 1 and V are the only A-invariant subgroups of V and $V \neq 1$. For the semidirect product AV this means that V is a minimal normal subgroup and A a maximal subgroup.

The following remark is fundamental for this section:

8.3.1   *Let $\mathcal{A}$ be a nonempty set of proper subgroups of A such that*

$$(\#) \qquad\qquad A^{\#} = \dot{\bigcup_{B \in \mathcal{A}}} B^{\#}, \ ^{12}$$

and $k := |\mathcal{A}| - 1$. If

$$(k, |V|) = 1$$

and $V \neq 1$, then there exists $B \in \mathcal{A}$ such that $C_V(B) \neq 1$.

Proof. For $v \in V$ and $B \leq A$ set

$$v_B := \prod_{a \in B} v^a = v \prod_{a \in B^{\#}} v^a.$$

Then

$$(v_B)^b = \prod_{a \in B} v^{ab} = v_B$$

for every $b \in B$, so $v_B \in C_V(B)$. Since $\mathcal{A}$ is a partition of A we get

$$v_A = \left(\prod_{B \in \mathcal{A}} v_B \right) v^{-k}.$$

Assume that $1 = C_V(B) \ (\geq C_V(A))$ for all $B \in \mathcal{A}$. Then $v_B = v_A = 1$ and thus $v^{-k} = 1$ for every $v \in V$. But now $(k, |V|) = 1$ implies $V = 1$, a contradiction (compare with 2.2.1 on page 49). $\qquad\qquad\square$

8.3.2   Theorem. *Let $V \neq 1$ and*

$$(+) \qquad\qquad C_V(a) = 1 \ \text{for all } a \in A^{\#}.$$

Then A is cyclic provided one of the following conditions holds:

(a)   *A is Abelian.*

(b)   *A is a p-group for $p \neq 2$.*

(c)   *A is a 2-group but not a quaternion group.*[13]

---

[12]$\mathcal{A}$ is called a **partition** of A. A beautiful treatment of groups possessing a partition can be found in [16].

[13]See 8.6 for *such an action of the quaternion group.*

(d) $|A| = pq$, where p, q are (not necessarily different) primes.

Proof. Note first that hypothesis $(+)$ also holds for every subgroup $A_1 \leq A$ in place of A.

Let $p \in \pi(A)$ and $A_1 \in \mathrm{Syl}_p A$. Then A_1 acts on the Sylow p-subgroup V_p of V (see 2.1.6 on page 46). If $V_p \neq 1$, then by 8.1.4 A_1 has a nontrivial fixed point in V_p. Hence $V_p = 1$, and the action of A on V is coprime.

We now assume that A is not cyclic. In the cases (a), (b), (c) we apply 2.1.7 and 5.3.8 to get an elementary Abelian subgroup $A_1 \leq A$ of order p^2. By induction on $|A|$ we may assume that $A = A_1$. In case (d) A is—again by 2.1.7—non-Abelian of order pq, $p \neq q$, or elementary Abelian of order p^2 and $p = q$.

Let $\mathcal{A}$ be the set of all subgroups of prime order of A. Since $|A| = pq$ the set $\mathcal{A}$ is a partition of A as in 8.3.1.

If A is elementary Abelian, then

$$|\mathcal{A}| = \frac{p^2 - 1}{p - 1} = p + 1.$$

Let A be non-Abelian of order pq and $q < p$. Then $\mathcal{A}$ is the set of nontrivial Sylow subgroups of A, and Sylow's Theorem shows that G possesses exactly p Sylow q-subgroups and exactly one Sylow p-subgroup. As before $|\mathcal{A}| = p + 1$.

The coprime action of A on G gives $(p, |V|) = 1$. Hence 8.3.1 contradicts hypothesis $(+)$.   □

Compare the next result with 8.6.1 on page 211.

8.3.3   *Let A be Abelian. Suppose that the action of A on V is irreducible. Then $A/C_A(V)$ is cyclic.*

Proof. We may assume that $C_A(V) = 1$. Then $C_V(a) \neq V$ for all $a \in A^\#$. Moreover for all $x \in A$

$$C_V(a)^x = C_V(a^x) = C_V(a)$$

since A is Abelian. The irreducible action of A on V gives $C_V(a) = 1$ for all $a \in A^{\#}$, and the assertion follows from 8.3.2. $\qquad\square$

As a corollary of 8.3.3 one gets that the multiplicative group of a finite field is cyclic:

Let F be a finite field with additive group V and multiplicative group A. Then the distributive law shows that A acts by right multiplication on V. Moreover, this action is faithful and transitive on the $A^{\sharp}$, thus also irreducible. Now 8.3.3 shows that A is cyclic.

Another important corollary of 8.3.3 is:

8.3.4 *Let A be an Abelian group acting on the group G. Suppose that the action of A on G is coprime.*

(a) $\quad G = \langle C_G(B) \mid B \leq A \ \ and \ \ r(A/B) \leq 1 \rangle.$[14]

(b) $\quad$ *If A is not cyclic, then* $G = \langle C_G(a) \mid a \in A^{\#} \rangle.$

(c) $\quad [G, A] = \langle [C_G(B), A] \mid B \leq A \ \ and \ \ r(A/B) \leq 1 \rangle.$

Proof. Let $\mathcal{B}$ be the set of all subgroups $B \leq A$ such that A/B is cyclic.

(a) We first treat two particular cases and then show that the general case can be reduced to these cases.

First assume that G is Abelian. If A acts irreducibly on G, then $B := C_A(G) \in \mathcal{B}$ by 8.3.3, and we get $G = C_G(B)$. Hence, we may assume that A is not irreducible on G. Let W be an A-invariant subgroup of G such that $1 \neq W \neq G$. Induction on $|G|$, applied to the pairs (W, A) and $(G/W, A)$, shows that

$$W = \langle C_W(B) \mid B \in \mathcal{B} \rangle \quad \text{and} \quad G/W = \langle C_{G/W}(B) \mid B \in \mathcal{B} \rangle.$$

Since the action of A on G is coprime 8.2.2 implies that

$$C_{G/W}(B) = C_G(B)W/W,$$

and thus

$$G = \langle C_G(B)W \mid B \in \mathcal{B} \rangle = \langle C_G(B) \mid B \in \mathcal{B} \rangle.$$

[14] $r(A/B) \leq 1$ means that A/B is cyclic; see page 48.

Assume next that G is a p-group. Then A acts on the Abelian factor group $G/\Phi(G)$. Hence 8.2.2 (a) and the case already proved above give

$$G = \langle\, C_G(B) \mid B \in \mathcal{B}\,\rangle\, \Phi(G)$$

and then together with 5.2.3 the assertion.

In the general case 8.2.3 (a) shows that for every $p \in \pi(G)$ there exists an A-invariant Sylow p-subgroup G_p of G. As seen above the assertion holds for the pair (G_p, A) and thus also for (G, A) since $G = \langle\, G_p \mid p \in \pi(G)\,\rangle$.

(b) follows from (a).

(c) For $B \in \mathcal{B}$ the subgroup $G_B := C_G(B)$ is A-invariant since A is Abelian. Hence by 8.2.7

$$G_B = [G_B, A]\, C_{G_B}(A) = [G_B, A]\, C_G(A).$$

For $G_1 := \langle\, [G_B, A] \mid B \in \mathcal{B}\,\rangle$ this implies

$$G_1\, C_G(A) = \langle\, [G_B, A]\, C_G(A) \mid B \in \mathcal{B}\,\rangle = \langle\, G_B \mid B \in \mathcal{B}\,\rangle \overset{\text{(a)}}{=} G.$$

In particular $[G, A] \le G_1$. The other inclusion is trivial. $\square$

We conclude this section with a further look on Frobenius groups. This is done using Frobenius's Theorem that Frobenius kernels are subgroups (4.1.6 on page 80).

8.3.5   *Let G be a Frobenius group with Frobenius complement H and Frobenius kernel K. Suppose that G acts on a nontrivial Abelian group V such that*

$$(|V|, |K|) = 1 \quad \text{and} \quad C_V(K) = 1.$$

Then $C_V(H) \neq 1$.

Proof. Set $A := G$ and

$$\mathcal{A} := \{K\} \cup \{H^a \mid a \in A\}.$$

Then $\mathcal{A}$ is a partition of G as in (#) of 8.3.1. Moreover by 4.1.5 on page 80

$$|\mathcal{A}| - 1 = |\{H^a \mid a \in A\}| = |K|.$$

According to 8.3.1 there exists $B \in \mathcal{A}$ such that $C_V(B) \neq 1$. Moreover, $B \in \{H^a \mid a \in A\}$ since by our hypothesis $C_V(K) = 1$. Thus $C_V(H) \neq 1$.

□

We now use 8.3.5 to answer the question about the uniqueness of Frobenius complements raised in Section 4.1 and to discuss their structure.

8.3.6 *Let G be a Frobenius group with Frobenius complement H and Frobenius kernel K, and let G_1 be a subgroup of G. Suppose $G = G_1 K$. Then G_1 contains a conjugate of H.*

Proof. Since $G_1 \not\leq K$ we may assume after suitable conjugation that $H \cap G_1 \neq 1$. If $G_1 \leq H$, then the Dedekind identity (1.1.11) shows that $H = G_1(H \cap K)$ and thus $H = G_1$ since $H \cap K = 1$.

If $H \not\leq G_1$, then G_1 is a Frobenius group with Frobenius complement $H \cap G_1$ and Frobenius kernel $K \cap G_1$ (4.1.8 (a)). In particular $|G_1| = |K \cap G_1| \, |H \cap G_1|$. On the other hand, the Homomorphism Theorem gives $G/K \cong G_1/G_1 \cap K$ and thus

$$|H| = |G/K| = |G_1/G_1 \cap K| = |H \cap G_1|,$$

which contradicts $H \not\leq G_1$.

□

8.3.7 *Let G be a Frobenius group. Then all Frobenius complements of G are conjugate.*

Proof. Let H and H_0 be two Frobenius complements of G. By 4.1.8 (b) on page 81 we may assume that $H_0 \leq H$, so it suffices to show that $H_0 = H$.

Assume that $H_0 < H$. By 4.1.8 (a) H is a Frobenius group with Frobenius complement H_0. Let K be the Frobenius kernel of G with respect to H and K_0 the Frobenius kernel of H with respect to H_0. Note that KK_0 is the Frobenius kernel of G with respect to H_0.

Let

$$p \in \pi(K), \ P \in \mathrm{Syl}_p K, \ V := Z(P), \ \text{and} \ G_1 := N_G(P).$$

The Frattini argument gives $G = KG_1$. Thus, by 8.3.6 we may assume that $H \leq G_1$. The Frobenius group H acts on V such that $C_V(K_0) = 1$ and

$(|V|, |K_0|) = 1$; see 8.1.12 (applied to the Frobenius complement H) and 4.1.5. Hence 8.3.5 (with $A := H$) yields $C_V(H_0) \neq 1$. But this contradicts 8.1.12, this time applied to the Frobenius complement H_0. □

The argument used in the proof of 8.3.7 also reveals the structure of the Sylow subgroups of Frobenius complements:

8.3.8 *Let G be a Frobenius group with Frobenius complement H. Then the Sylow subgroups of H are cyclic or quaternion groups.*

Proof. Let K be the Frobenius kernel of G, $p \in \pi(K)$, and $P \in \mathrm{Syl}_p K$. As in the proof of 8.3.7 we may assume that H normalizes $V := Z(P)$. Moreover, as there 8.1.12 implies that

$$C_V(h) = 1 \text{ for } 1 \neq h \in H.$$

Now 8.3.2 gives the assertion. □

Exercises

Let G be a group and A a group acting on G.

1. Let G be Abelian and $A_1, \ldots, A_{n+1}$ be a partition of A. Let

    $$G_0 := \langle C_G(A_i) \mid i = 1, \ldots, n+1 \rangle.$$

    Then G/G_0 has exponent $\leq n$.

2. Let G be nilpotent and $(|A|, |G|) = 1$. If A is Abelian and $r(A) \geq 2$, then

    $$G = \prod_{a \in A^\#} C_G(a).$$

3. Let G be a solvable Frobenius group. Then the Frobenius kernel K of G is nilpotent and $F(G) = K$.[15]

4. Let G be p-separable ($p \in \mathbb{P}$), $A = \langle a \rangle \cong C_p$ and $H := AG$. Suppose that for all $x \in G$

    $$x x^a \cdots x^{a^{p-1}} = 1.$$

    (a) The elements $y \in H \setminus G$ act fixed-point-freely on every H-invariant p'-section of G.

---

[15] Use Frobenius's Theorem 4.1.6 on p. 80.

    (b) G is p-closed.

    (c) $o(y) = p$ for all $y \in H \setminus G$.

8.4   The Decomposition of an Action

As in Section 8.3 let A be a group that acts on the *Abelian* group V. We now use the fact that the set of endomorphisms together with the composition of mappings and the addition

$$v^{\alpha+\beta} := v^{\alpha} v^{\beta} \quad (\alpha, \beta \text{ endomorphisms}, \ v \in V)$$

is a ring, the *endomorphism ring* $\operatorname{End} V$ of V.

Since for every $a \in A$ the mapping

$$v \mapsto v^{a} \quad (v \in V)$$

is an endomorphism of V we can compose the endomorphisms of V and the elements of A in their action on V:

$$v^{a+\beta} := v^{a} v^{\beta}, \quad v^{\beta a} := (v^{\beta})^{a} \quad \text{and} \quad v^{a\beta} := (v^{a})^{\beta} \quad (a \in A, \ \beta \in \operatorname{End} V).$$

In other words, we identify the elements of $a \in A$ with the element of $\operatorname{End} V$ induce by a.

For example, for $a \in A$ the **commutator mapping**

$$\kappa \colon v \mapsto [v, a] = [a, v] = v^{a} v^{-1} = v^{a - \mathrm{id}} \quad (v \in V)$$

is the endomorphism $a - \mathrm{id}$ with $\operatorname{Ker} \kappa = C_V(a)$ and

$$\operatorname{Im} \kappa = \{ [v, a] \mid v \in V \} = [V, a].^{16}$$

Since $[V, a]$ is invariant under $\langle a \rangle$ and the factor group $V/[V, a]$ is centralized by $\langle a \rangle$, we get

$$\operatorname{Im} \kappa = [V, \langle a \rangle].$$

The Homomorphism Theorem gives:

8.4.1 $V/C_V(a) \cong [V, \langle a \rangle]$.   □

---

[16]Recall: $[V, a] = \langle [v, a] \mid v \in V \rangle$.

The next result and 8.4.5 below are corollaries of Gaschütz's Theorem (3.3.2 on page 73). But they have been known long before this theorem and both have short and elementary proofs that we will give here.

8.4.2 *Suppose that the action of A on V is coprime. Then*

$$V = C_V(A) \times [V, A].$$

Proof. By 8.2.7 (a) it suffices to show $C_V(A) \cap [V, A] = 1$. To do this we investigate the endomorphism

$$\varphi \colon V \to V \quad \text{with} \quad v \mapsto \prod_{x \in A} v^x.$$

For a commutator $v = [w, a] \in [V, A]$

$$v^\varphi = (w^a)^\varphi w^{-\varphi} = \left(\prod_{x \in A} w^{ax} \right) \left(\prod_{x \in A} w^{-x} \right) = 1,$$

and thus $[V, A] \leq \operatorname{Ker} \varphi$. On the other hand, for $v \in C_V(A)$ we get $v^\varphi = v^{|A|}$ and

$$v^{|A|} = 1 \iff v = 1$$

since $(|A|, |V|) = 1$. Hence $v = 1$ for $v \in C_V(A) \cap [V, A]$. $\square$

As a corollary we get:

8.4.3 *Suppose that the action of A on V is coprime and that A acts trivially on $\Omega(V)$. Then A acts trivially on V.*

Proof. The decomposition in 8.4.2 gives

$$\Omega([V, A]) \leq C_V(A) \cap [V, A] = 1,$$

and this implies $[V, A] = 1$. $\square$

Here is another consequence of 8.4.2 that will be needed in Chapter 10:

8.4.4 *Suppose that the action of the group A on the group G is coprime and $|G : C_G(A)| = p$ ($p \in \mathbb{P}$). Then $[G, A]$ has order p and $A/C_A(G)$ is cyclic.*

Proof. Let $G_1 := [G, A]$. From 8.2.7 we get

$$G = G_1 C_G(A) \quad \text{and} \quad G_1 = [G_1, A],$$

so $|G_1 : C_{G_1}(A)| = p$ and $C_A(G_1) = C_A(G)$. If $G_1 < G$, then the assertion follows by induction on $|G|$. Hence, we may assume that

$$(*) \quad G = [G, A].$$

Let G_p be an A-invariant Sylow p-subgroup of G (8.2.3 (a)). By our hypothesis $G = G_p C_G(A)$, and thus $G = [G, A] = [G_p, A]$. In particular G is a p-group.

For $\overline{G} := G/\Phi(G)$ we get from $(*)$

$$[\overline{G}, A] = \overline{G}.$$

But $\overline{G}$ is Abelian (5.2.7 on page 106), and 8.4.2 gives

$$\overline{G} = [\overline{G}, A] \times C_{\overline{G}}(A).$$

Hence $C_{\overline{G}}(A) = 1$, and the hypothesis $|G : C_G(A)| = p$ implies $|\overline{G}| = p$. Now 5.2.7 (b) on page 106 shows that G is cyclic, and by 8.4.3 $|G| = p$. Thus, 2.2.4 on page 50 yields the assertion. $\square$

A similar consideration as in 8.4.2 gives:

8.4.5 *Suppose that the action of A on V is coprime. Let U be an A-invariant subgroup of V. If U has a complement in V, then U also has an A-invariant complement in V.*

Proof. Let W be a complement of U in V; i.e.,

$$V = U \times W.$$

If $V = U$, then clearly W is A-invariant. Thus, we may assume that $V \neq U$. The projection

$$\eta : V \to U \quad \text{with} \quad uw \mapsto u \quad (u \in U, w \in W)$$

is an endomorphism of V, hence also

$$\eta_A : V \to V \quad \text{with} \quad v \mapsto \prod_{x \in A} v^{x^{-1} \eta x}.$$

Now $U^A = U$ implies

$$u^{\eta_A} = \prod_{x \in A} u^{x^{-1}x} = u^{|A|} \quad \text{for} \quad u \in U,$$

and as in 8.4.2 the coprime action of A gives

$$\text{Ker } \eta_A \cap U = 1.$$

On the other hand $\text{Im } \eta_A \leq U$ and by 1.2.5 on page 13

$$|\text{Ker } \eta_A| \, |\text{Im } \eta_A| = |V|.$$

Hence $\text{Ker } \eta_A$ is a complement of U in V that is invariant under A, as

$$\eta_A \, a = \sum_{x \in A} (x^{-1}\eta x)a = \sum_{x \in A} aa^{-1}x^{-1}\eta xa = a \sum_{x \in A} (xa)^{-1}\eta(xa) = a\, \eta_A$$

for $a \in A$.[17]   □

The action of A on V is **semisimple** if every A-invariant subgroup of V has an A-invariant complement in V. Evidently, every irreducible action is also semisimple.

Suppose that V is an Abelian p-group. If $V \neq \Omega(V)$, then the action of A on V is not semisimple since $\Omega(V)$ has no complement in V. On the other hand if $V = \Omega(V)$, then every subgroup of V has a complement in V (see 2.1.2 on page 44).[18] Hence 8.4.5 gives:

8.4.6   Maschke's Theorem.[19] *Suppose that the action of A on V is coprime and V is an elementary Abelian p-group. Then the action of A on V is semisimple.*   □

The minimal A-invariant subgroups of V are the minimal normal subgroups of the semidirect product AV that are contained in V. Hence 1.7.2 on page 37 applies to this situation:

8.4.7   *Let $\mathcal{M}$ be the set of all minimal A-invariant subgroups of V. Then the following statements are equivalent:*

---

[17] The calculation carried out in the endomorphism ring.

[18] According to 2.1.8 on page 46 this is the well-known fact that every subspace of a finite-dimensional vector space has a complement.

[19] Compare with [76].

(i)   The action of A on V is semisimple.

(ii)   There exist $U_1, \ldots, U_n \in \mathcal{M}$ such that $V = U_1 \times \cdots \times U_n$.

(iii) $V = \displaystyle\prod_{U \in \mathcal{M}} U$.

Proof. The implication (ii) $\Rightarrow$ (iii) is trivial, and the implication (i) $\Rightarrow$ (ii) can be shown by a trivial induction on $|V|$.

(iii) $\Rightarrow$ (i): Let U_1 be an A-invariant subgroup of V. By 1.7.2 (a) on page 37 there exist $U_2, \ldots, U_n \in \mathcal{M}$ such that $V = U_1 \times U_2 \times \cdots \times U_n$. Hence $U_2 \times \cdots \times U_n$ is an A-invariant complement of U_1 in V. $\qquad\square$

The semisimple action of A on V induces a semisimple action of A on each A-invariant subgroup of V (Exercise 2). But if we restrict this action to subgroups of A the situation gets more complicated. For normal subgroups the resulting action is again semisimple, but not for subgroups in general (Exercise 1).

We discuss this elementary fact and suggest that the reader compare the following with the notion of an A-composition series (of V) introduced in 1.8 on page 39.

From now on we assume that $V \neq 1$. As above, $\mathcal{M}$ is the set of all minimal A-invariant subgroups of V. For $U, W \in \mathcal{M}$, define

$$U \sim W \iff U \text{ is } A\text{-isomorphic to } W.$$

Then $\sim$ is an equivalence relation on $\mathcal{M}$. The equivalence classes we denote by $\mathcal{M}_1, \ldots, \mathcal{M}_n$. For $i = 1, \ldots, n$

$$V_i := \prod_{U \in \mathcal{M}_i} U \text{ and } V_0 := \prod_{i=1}^{n} V_i.$$

The subgroups V_i, $i = 1, \ldots, n$, are the **homogeneous A-components** of V. From 1.7.2 on page 37 and 8.4.7 we get:

8.4.8   (a) V_i is the direct product of subgroups from $\mathcal{M}_i$ $(i = 1, \ldots, n)$.

(b) $V_0 = \displaystyle\mathop{\times}_{i=1}^{n} V_i$.

(c)   The action of A on V_0 is semisimple. $\square$

By 8.4.7 the action of A on V is semisimple if and only if $V = V_0$.

8.4.9   Clifford's Theorem.[20] *Let H be a group that acts on V and A be a normal subgroup of H. Suppose that the action of H on V is semisimple. Then also the action of A on V is semisimple. Moreover, if H acts irreducibly on V, then H acts transitively on the homogeneous A-components of V, and $AC_H(A)$ is contained in the kernel of this action.*

Proof. With respect to the action of A on V we use the notation introduced earlier. Then H acts on $\mathcal{M}$ since A is normal in H. We show that this action preserves the equivalence relation $\sim$ on $\mathcal{M}$:

Let $U, W \in \mathcal{M}$ such that $U \sim W$ and $h \in H$. By definition there exists an A-isomorphism $\varphi \colon U \to W$. Now

$$\varphi^h := h^{-1}\varphi h \quad {}^{21}$$

is an isomorphism from U^h in W^h, and $A^h = A$ implies

$$\varphi^h a = h^{-1}\varphi h a = h^{-1}\varphi a^{h^{-1}} h = h^{-1}a^{h^{-1}}\varphi h = a h^{-1}\varphi h = a\,\varphi^h.$$

Hence φ^h is an A-isomorphism and $U^h \sim W^h$.

We have shown that H acts on the equivalence classes of $\sim$ and thus on the set of homogeneous A-components $V_1, \ldots, V_n$. The kernel of this action contains $AC_H(A)$.

The subgroup $V_0 = \prod_{i=1}^{n} V_i$ is H-invariant, and every H-invariant subgroup of V contains some element of $\mathcal{M}$. Now the semisimple action of H on V gives $V = V_0$, and the action of A on V is semisimple (8.4.7).

Suppose that H acts irreducibly on V. Then $V = \langle V_1{}^H \rangle$, and 8.4.8 (b) yields

$$V_1{}^H = \{V_1, \ldots, V_n\}. \qquad \square$$

Exercises

Let A be a group, which acts on the elementary Abelian p-group V.

1.   Suppose that for every $U \leq A$ the action of U on V is semisimple. Then $p \notin \pi(A/C_A(V))$.

---

[20]Compare with [39].
[21]The product is taken in End V.

2. Suppose that the action of A on V is semisimple and W is an A-invariant subgroup of V. Then the action of A on W is also semisimple.

3. Let $S \in \mathrm{Syl}_p A$ and $L \trianglelefteq\trianglelefteq A$. If $[C_V(S), A] = 1$, then also

$$[C_V(S \cap L), L] = 1.$$

4. Let $V = \langle v_1, \dots, v_n \rangle$ be an n-dimensional vector space over $\mathbb{F}_2$. The symmetric group S_n acts on V according to

$$v_i{}^g := v_{i^g} \quad (g \in S_n, \ i \in \{1, \dots, n\}).$$

For which n is this action semisimple?

8.5 Minimal Nontrivial Action

In this section we investigate a situation that frequently occurs in proofs by induction: A group A acts nontrivially on a group G but trivially on each proper A-invariant subgroup of G. If in addition the action of A on G is coprime, the structure of G can be described fairly well. Clearly in this case G is a p-group since A normalizes a Sylow p-subgroup for every $p \in \pi(G)$ (8.2.3 on page 185). The analysis of this situation—in the literature called the Hall-Higman-reduction—is given in 8.5.1.[22]

The second result of this section, (8.5.3), is a generalization of the $P \times Q$-Lemma, which as the $P \times Q$-Lemma itself is due to Thompson. Our proof follows Bender, who uses a nice idea that goes back to Baer.

Apart from these two theorems some more special results are proved that will be used later.

For practical reasons we formulate the Hall-Higman-reduction slightly more generally:

8.5.1 *Let B be a group that acts on the p-group P. Suppose that B contains a normal p'-subgroup A that acts nontrivially on P but trivially on every proper B-invariant subgroup of P. Then $P = [P, A]$, and P is either elementary Abelian or a special p-group. Moreover, B acts irreducibly on $P/\Phi(P)$. If in addition $p \neq 2$, then $x^p = 1$ for all $x \in P$.*

---

[22]See [67].

Proof. Since $A \trianglelefteq B$ the subgroup $[P, A]$ is B-invariant, and by 8.2.7 $[P, A, A] = [P, A]$. This gives

(1) $$P = [P, A].$$

Every characteristic subgroup C of P is B-invariant. Hence either $[C, A] = 1$ or $C = P$, in particular

$$[P', A] = 1 = [\Phi(P), A].$$

Let $\overline{P} := P/P'$. The coprime action of A on P and 8.4.2 imply

$$\overline{P} = [\overline{P}, A] \times C_{\overline{P}}(A) = \overline{[P, A]} \times C_{\overline{P}}(A)$$

and thus $C_{\overline{P}}(A) = 1$ by (1). Now the trivial action of A on every proper B-invariant subgroup shows that B acts irreducibly on $\overline{P}$. In particular, $\overline{C} = 1$ for every proper characteristic subgroup C of P. Hence $P' \leq \Phi(P)$ implies

$$P' = \Phi(P).$$

If $P' = 1$, then P is elementary Abelian and we are done. Thus, we may assume now that $P' \neq 1$. Then $Z(P) \neq P$ and $Z(P) \leq P'$ since $Z(P)$ is characteristic in P. The inclusion $P' \leq Z(P)$ follows from (1) and the Three-Subgroups Lemma:

$$[P, P', A] = 1 = [P', A, P]$$

and thus

$$[P, P'] = [P, A, P'] = 1.$$

This gives $Z(P) = P'$.

As $P/Z(P)$ is elementary Abelian, for every $x, y \in P$

$$1 = [x^p, y] \overset{1.5.4}{=} [x, y]^p.$$

Hence $Z(P) = P' = \Omega(Z(P))$, and P is a special p-group.

Let $p \neq 2$. Then 5.3.4 (a) on page 111 yields

$$[x, a]^p = (x^{-1}x^a)^p = x^{-p}(x^p)^a = x^{-p}x^p = 1$$

for $x \in P$ and $a \in A$, so $P = [P, A] = \Omega(P)$. Hence 5.3.5 on page 112 applies, and we are done. $\square$

The following result deals with a situation that will occur in Chapter 11:

8.5.2 *Let A be an Abelian p-group that acts on the p'-group G, and set $A_0 := C_A(G)$. Suppose that $[G, A] \neq 1$, but $[U, A] = 1$ for every A-invariant subgroup $U \neq G$.*

(a) *$r(A/A_0) = 1$.*

(b) *If the semidirect product AG acts on the elementary Abelian p-group V such that $C_G(V) = 1$, then AG/A_0 acts faithfully on $C_V(A_0)$.*

Proof. (a) As already mentioned in the introduction of this section, G is a q-group ($q \in \mathbb{P}$). By 8.5.1 and 8.2.9 (a) the factor group A/A_0 acts irreducibly and faithfully on $G/\Phi(G)$, so (a) follows from 8.3.3.

(b) Set $K := C_{AG}(C_V(A_0))$. The $P \times Q$-Lemma (8.2.8, with $P = A_0$) shows that K is a p-group. As $A_0 \leq K$ and G is a p'-group, we get that $K = A_0$. □

An important generalization of the $P \times Q$-Lemma for $p \neq 2$ is the following result:

8.5.3   Theorem (Thompson [93]). *Let $p \neq 2$ and A be the semidirect product of a p-subgroup P with a normal p'-subgroup Q. Suppose that A acts on a p-group G such that*

$$(') \qquad\qquad C_G(P) \leq C_G(Q).$$

Then Q acts trivially on G.

Proof (Bender [27]). We may assume that $[G, Q] \neq 1$. As every proper A-invariant subgroup of G also satisfies (') (in place of G), we may further assume by induction on $|G|$ that Q acts trivially on every such proper subgroup. Hence 8.5.1 yields

$$G = [G, Q] \quad \text{und} \quad G' \leq Z(G).$$

We first treat the case $G' = 1$: Then 8.4.2 gives $C_G(Q) = 1$ and thus also $C_G(P) = 1$ by hypothesis ('). Now 8.1.4 implies $G = 1$, which contradicts $[G, Q] \neq 1$.

We now use an idea of Baer[23] to show that the case $G' \neq 1$ already follows from the Abelian case. This is done by defining an addition on the set G

---
[23]Bender applied this idea to this situation ([24], [27]).

that turns it into an Abelian group and that is compatible with the action of A. Then the result follows from the case just treated.

As G has odd order, we have $\langle x^2 \rangle = \langle x \rangle$ for every $x \in G$. In particular, for $x, y \in G$

$$x^2 = y^2 \Rightarrow x^2 y^{-2} = (xy^{-1})^2 = 1 \Rightarrow xy^{-1} = 1$$

and thus

$$x^2 = y^2 \iff x = y.$$

Hence, for every $g \in G$ there exists a unique $x \in G$ such that $x^2 = g$. Set

$$\sqrt{g} := x.$$

Then the following hold:

$$g, h \in Z(G) \quad \Rightarrow \quad \sqrt{g}, \sqrt{h} \in Z(G) \text{ and } \sqrt{gh} = \sqrt{g}\sqrt{h};$$

$$g \in G, \ a \in A \quad \Rightarrow \quad \sqrt{g}^a = \sqrt{g^a};$$

$$g \in G \quad \Rightarrow \quad g^{-1}\sqrt{g} = \sqrt{g^{-1}}.$$

We now define an addition on the set G by

$$(+) \qquad\qquad\qquad g + h := gh\sqrt{[h, g]}.$$

The commutator identity $[g, h]^{-1} = [h, g]$ implies

$$\begin{aligned} g + h &= gh\sqrt{[h, g]} = hg[g, h]\sqrt{[h, g]} = hg[g, h]\sqrt{[g, h]^{-1}} = hg\sqrt{[g, h]} \\ &= h + g, \end{aligned}$$

so this addition is commutative.

The proof of the associativity of $+$ uses the fact that $G' \leq Z(G)$: For $g, h, f \in G$ (1.5.4 on page 25) yields

$$\begin{aligned} [f, g + h] &= [f, gh] = [f, g][f, h] \quad \text{and} \\ [h + f, g] &= [hf, g] = [h, g][f, g]. \end{aligned}$$

This shows that

$$(g + h) + f = ghf\sqrt{[h, g]}\sqrt{[f, g]}\sqrt{[f, h]} = g + (h + f).$$

Evidently, 1 is the identity of $G(+)$ and $-g := g^{-1}$ is the inverse of g (with respect to $+$). Hence $G(+)$ is an Abelian group.

The group A acts on the Abelian group $G(+)$ since

$$(g+h)^a = g^a h^a \sqrt{[h^a, g^a]} = g^a + h^a,$$

where the action of A on the set G is as before. Now the above-treated Abelian case shows that Q acts trivially on G. □

The following two results are used in the proof of 8.5.6, and this result is used in Section 11 to prove Glauberman's Signalizer Functor Theorem.

8.5.4 *Let P be a special 2-group such that $\Omega(P) = Z(P)$ and $|Z(P)| = 4$. Then $2^3 \leq |P/Z(P)| \leq 2^4$.*

Proof. Set $Z := Z(P)$ and $2^n := |P/Z|$. Since P is special we have $P' = Z = \Phi(P)$, and by our hypothesis $P' = \Omega(P)$ and $|P'| = 4$. In particular, every element in $P \setminus Z$ has order 4. If $n = 2$, then there exist $x, y \in P$ such that $P = \langle x, y, Z(P) \rangle$. Hence $P' = \langle [x, y] \rangle \cong C_2$ which contradicts $|P'| = 4$. Thus we have $n \geq 3$.

Let $a \in P \setminus Z$ and set

$$C := C_P(a), \quad \overline{C} := C/\langle a \rangle.$$

Note that $|\overline{Z}| = 2$ since $1 \neq a^2 \in Z$. Pick $x \in C$ such that $\overline{x}$ is an involution. In the case $\overline{x} \notin \overline{Z}$ the element x has order 4 and $x^2 \in \langle a \rangle$, i.e., $x^2 = a^2$. Hence $o(xa) = 2$, which contradicts $\Omega(P) = Z$ and $xa \notin Z$.

We have shown that $\overline{Z}$ is the unique subgroup of order 2 in $\overline{C}$. Thus, by 5.3.7 on page 114 either $\overline{C}$ is cyclic (of order ≤ 4) or a quaternion group of order 8. This gives

$$|C| \in \{8, 16, 32\}.$$

On the other hand, the conjugacy class a^P is contained in aZ since $\langle a \rangle Z$ is normal in P. It follows that

$$|P/C| = |a^P| \leq |aZ| = 4$$

and thus $|P| \leq 2^7$, i.e., $|P/Z| \leq 2^5$.

In the case $|P/C| \leq 2$ we get $|P| \leq 2^6$, so $|P/Z| \leq 2^4$ and we are done. Hence it suffices to show that the assumption

$$(+) \qquad |P : C_P(a)| = 4 \quad \text{for all } a \in P \setminus Z \text{ and } |P/Z| = 2^5$$

leads to a contradiction. Let $z \in Z^{\#}$ and $\widetilde{P} := P/\langle z \rangle$.[24] Then

$$\Phi(\widetilde{P}) = \widetilde{P}' = \widetilde{Z} \cong C_2.$$

If $\widetilde{Z} = Z(\widetilde{P})$, then $\widetilde{P}$ is extraspecial and $|\widetilde{P}/\widetilde{Z}|$ a square (5.2.9), which contradicts (+).

Assume that $\widetilde{Z} \neq Z(\widetilde{P})$. Then there exists $a \in P \setminus Z$ such that $\widetilde{a} \in Z(\widetilde{P})$, so

$$\{x^{-1}x^a \mid x \in P\} = \{1, z\}.$$

Now 8.1.7 yields $|P : C_P(a)| = 2$, which again contradicts (+). $\qquad\square$

8.5.5 *Let $\langle d \rangle$ be a cyclic 3-group that acts faithfully and fixed-point-freely on the 2-group P. Suppose that*

(+) $\qquad\qquad r(V) \leq 2$ *for every Abelian subgroup V of P.*

Then $o(d) = 3$.

Proof. Set $o(d) = 3^n$. By induction on $|P|$ we may assume that $\langle d^3 \rangle$ acts trivially on every proper $\langle d \rangle$-invariant subgroup of P. Hence 8.5.1 shows that $\langle d \rangle$ acts irreducibly on

$$\overline{P} := P/\Phi(P),$$

and either P is elementary Abelian or a special p-group. Moreover, by 8.2.9 (a) $\langle d \rangle$ also acts faithfully on $\overline{P}$. Hence $C_{\overline{P}}(x) = 1$ for all $1 \neq x \in \langle d \rangle$. In particular, every orbit of $\langle d \rangle$ on $\overline{P}^{\#}$ has length $o(d) = 3^n$. This gives

(1) $\qquad\qquad\qquad |\overline{P}| \equiv 1 \pmod{3^n}.$

As $Z(P)$ is elementary Abelian, the fixed-point-free action of $\langle d \rangle$ gives $|Z(P)| \geq 4$. Now (+) implies

(2) $\qquad\qquad\qquad |Z(P)| = 4 \quad \text{and} \quad \Omega(P) = Z(P).$

The case $P' = 1$ gives $|P| = 4$, and $o(d) = 3$ since $\langle d \rangle$ acts faithfully on P.

In the remaining case P is special and satisfies the hypothesis of 8.5.4 (see (2)). It follows that $|\overline{P}| = 2^3$ or $|\overline{P}| = 2^4$. Hence (1) gives $|\overline{P}| = 2^4$ and $n = 1$. $\qquad\square$

---

[24]Tilde instead of bar convention.

8.5.6   *Let G be a group and $1 \neq d \in G$ a 3-element. Suppose that the following hold:*

(1) $C_G(O_2(G)) \leq O_2(G)$.

(2)   *There exists an element of order 6 in G.*

(3)   *There exists a $\langle d \rangle$-invariant elementary Abelian 2-subgroup W in G such that*
$$C_W(d) = 1 \neq W.$$

Then G contains an elementary Abelian subgroup of order 8.

Proof. Set
$$P := O_2(G), \quad Z := \Omega(Z(P)) \quad \text{and} \quad C := C_G(Z).$$

We assume that G is a counterexample, so

(+) $\qquad r(V) \leq 2$ for every Abelian 2-subgroup V of G.

This shows that $|Z| \leq 4$ and $|WC_Z(W)| \leq 4$. As d normalizes W and Z and acts fixed-point-freely on W, we get
$$W = Z \cong C_2 \times C_2 \quad \text{and} \quad C_Z(d) = 1.$$

Let $D \in \mathrm{Syl}_3 G$ such that $d \in D$ and pick $x \in D$. If $[Z, x] \neq 1$, then $C_Z(x) = 1$ since $|Z| = 4$, and (+) yields $C_P(x) = 1$. Hence P and x satisfy the hypothesis of 8.5.5. We get

(∗) $C_P(x) = 1$ and $o(x) = 3$ for all $x \in D$ such that $[Z, x] \neq 1$.

Assume first that D is cyclic. Then by (∗) $D = \langle d \rangle \cong C_3$, and hypothesis (2) together with Sylow's Theorem shows that there exists an involution $t \in G$ such that $[t, d] = 1$. Since $Z\langle t \rangle$ is nilpotent we get $|[Z, t]| \leq 2$ (5.1.6 on page 101). Now $[Z, t] = 1$ follows since $[Z, t]$ is $\langle d \rangle$-invariant. Moreover, t is not contained in Z, so $Z\langle t \rangle$ is elementary Abelian of order 8. This contradicts (+).

Assume now that D is not cyclic. Then $C_D(Z)$ is a nontrivial normal subgroup of D. Let b be an element of order 3 in $C_D(Z) \cap Z(D)$ and

$E := \langle b, d \rangle$. By $(*)$ E is an elementary Abelian subgroup of order 9. Hence 8.3.4 gives

$$P = \langle C_P(x) \mid x \in E^{\#} \rangle.$$

In addition, again by $(*)$, $C_P(x) = 1$ for all $x \in E \setminus \langle b \rangle$. Now $P = C_P(b)$ follows, which contradicts hypothesis (1). $\square$

Exercises

1.   Let G be a group. Suppose that every proper subgroup of G is nilpotent, but not G. Then there exist different primes p, r and a p-element $a \in G$ such that

$$G = F(G)\langle a \rangle \quad \text{and} \quad F(G) = \langle a^p \rangle \times R,$$

where R is an elementary Abelian or special r-group.

8.6   Linear Action and the Two-Dimensional Linear Groups

In this section we introduce the action of a group on a vector space. As examples of such an action we present the groups $\mathrm{GL}_2(q)$ and $\mathrm{SL}_2(q)$, but we also need an important property of $\mathrm{SL}_2(q)$ in the next chapter.

Let p be a prime. It is well known that (up to isomorphism) for every power

$$q = p^m \quad (m \in \mathbb{N})$$

there exists exactly one finite field $\mathbb{F}_q$ with $|\mathbb{F}_q| = q$. We set

$$K := \mathbb{F}_q.$$

The additive group $K(+)$ is an elementary Abelian p-group and the multiplicative group K^* a cyclic group; see the remark after 8.3.3.

Let V be an n-dimensional vector space over K. Then the additive group $V(+)$ is an elementary Abelian p-group of order q^n.

Let $\mathrm{GL}(V)$ be the group of automorphisms of the vector space V; i.e.,

$$\mathrm{GL}(V) = \{x \in \mathrm{Aut}\, V(+) \mid \lambda v^x = (\lambda v)^x \text{ for all } v \in V \text{ and } \lambda \in K\}.$$

A group G is said to **act on the vector space** V if G acts on the Abelian group $V(+)$, so $\mathcal{O}_1 - \mathcal{O}_3$ hold, and in addition:

$\mathcal{O}_4$ $(\lambda v)^g = \lambda v^g$ $(\lambda \in K, v \in V,\ g \in G)$.

Any such action gives rise to a homomorphism from G into $\mathrm{GL}(V)$, which describes the action of G on V.

The action of G on V is **irreducible** if $V \neq 0$ and 0 and V are the only G-invariant *subspaces* of V.[25]

8.6.1   Schur's Lemma. *Suppose that G acts irreducibly and faithfully on the K-vector space V. Then the following hold:*

(a) $Z(G)$ *is a cyclic p'-group.*

(b)   *If $|Z(G)| = n$ such that $n|(q-1)$, then there exists a monomorphism $\varphi\colon Z(G) \to K^*$ such that*

$$v^z = z^\varphi v \quad \text{for all } z \in Z(G),\quad v \in V.[26]$$

Proof. Let $Z := Z(G)$ and $z \in Z^{\#}$. Then $C_V(z)$ is a proper G-invariant subspace of V. The irreducible action of G gives $C_V(z) = 0$ for all $z \in Z^{\#}$. Now (a) follows from 8.3.2 and 8.1.4.

(b) For $\lambda \in K^*$ the mapping

$$z_\lambda\colon V \to V \quad \text{with} \quad v \mapsto \lambda v$$

is in $Z(\mathrm{GL}(V))$, and $M := \{z_\lambda \mid \lambda \in K^*\}$ is a subgroup of $Z(\mathrm{GL}(V))$ that is isomorphic to K^*.

As G acts faithfully on V, this action is described by a monomorphism of G into $GL(V)$. After identifying the elements of G with their images in $GL(V)$ we may regard G and thus also Z as subgroup of $GL(V)$. Then $H := MZ$ is an Abelian subgroup of $GL(V)$ which is centralized by G. The irreducible action of G on V yields:

$$C_V(h) = 0 \text{ for all } h \in H^{\#}.$$

As seen above, this shows that H is cyclic. In particular, H contains exactly one subgroup of order $|Z|$ (1.4.3 on page 22). Moreover, since n divides $|K^*|$, this subgroup is in M ($\cong K^*$), and $Z \leq M$ follows. $\square$

---

[25] In general this does not imply an irreducible action of G on $V(+)$.

[26] z acts by *scalar multiplication*.

It should be pointed out that decomposition theorems similar to those of
Section 8.4, in particular Maschke's Theorem and Clifford's Theorem, can
be proved for groups acting on vector spaces using the same arguments as
in Section 8.4. Of course, now irreducibility and semisimplicity refers to
subspaces rather than subgroups.

It is well known that the mapping

$$\det\colon \mathrm{GL}(V) \to K^*$$

that maps every element $x \in \mathrm{GL}(V)$ on its determinant $\det x$ is an epimorphism. Hence

$$\mathrm{SL}(V) := \{x \in \mathrm{GL}(V) \mid \det x = 1\}$$

is a normal subgroup of $\mathrm{GL}(V)$ such that

$$\mathrm{GL}(V)/\mathrm{SL}(V) \cong K^*.$$

Since $(q-1,p) = 1$ all p-elements of $\mathrm{GL}(V)$ are contained in $\mathrm{SL}(V)$; we
will use this fact frequently.

With respect to a fixed basis $v_1, \ldots, v_n$ of V, every $x \in \mathrm{GL}(V)$ corresponds
to an invertible matrix

$$A(x) = \begin{pmatrix} \lambda_{11} & \cdots & \lambda_{1n} \\ \vdots & & \vdots \\ \lambda_{n1} & \cdots & \lambda_{nn} \end{pmatrix},$$

where the $\lambda_{ij} \in K$ are determined by the equations

$$v_i{}^x = \sum_{j=1}^{n} \lambda_{ij} v_j, \quad i = 1, \ldots, n.$$

The mapping $x \mapsto A(x)$ is an isomorphism of $\mathrm{GL}(V)$ into the group
$\mathrm{GL}_n(q)$ of all invertible $n \times n$-matrices over K. In particular $\mathrm{SL}(V)$ is
mapped onto the group $\mathrm{SL}_n(q)$ of all matrices with determinant 1.

Frequently we will describe the elements $x \in \mathrm{GL}(V)$ by the matrices $A(x)$
(for a fixed basis of V) writing

$$x \equiv A(x).$$

Despite the fact that most of the following results can easily be generalized
to n-dimensional vector spaces, we will from now on assume that V is a
2-dimensional vector space over K.

The order of $GL(V)$ is equal to the number of ordered pairs (v, w), where v, w is a basis of V. Hence

$$| GL_2(q)| = | GL(V)| = (q^2 - 1)(q^2 - q)$$

and

$$| SL_2(q)| = | SL(V)| = (q - 1)q(q + 1).$$

For $\lambda \in K^*$ the *scalar multiplication* by λ

$$z_\lambda \colon V \to V \quad \text{with} \quad v \mapsto \lambda v$$

is an element of $Z(GL(V))$, and

$$z_\lambda \equiv \begin{pmatrix} \lambda & 0 \\ 0 & \lambda \end{pmatrix}$$

with respect to every basis of V. We set

$$Z := \{z_\lambda \mid \lambda \in K^*\} \quad (\leq Z(GL(V))),$$

clearly Z is isomorphic to K^*. Moreover $z_\lambda \in SL(V)$ only if $\lambda = \pm 1$. Let

$$z := z_{-1}.$$

If $p = 2$ then $z = 1$, and if $p \neq 2$ then z is an involution in $SL(V)$.

8.6.2 *Let $p \neq 2$. Then z is the unique involution in $SL(V)$.*

Proof. Let $t \in SL(V)$ be an involution. Then

$$v + v^t \in C_V(t)$$

for all $v \in V$.

If $C_V(t) = 0$, then $v^t = -v$ and thus $t = z$. Assume that $C_V(t) \neq 0$. Then there exists a basis v, w of V with $v \in C_V(t)$, and with respect to this basis

$$t \equiv \begin{pmatrix} 1 & 0 \\ * & -1 \end{pmatrix}.$$

But now $\det t = -1$, which contradicts $p \neq 2$ and $t \in SL(V)$. $\qquad \square$

Let $V^\# := \{v \in V \mid v \neq 0\}$ and $v \in V^\#$. By

$$Kv := \{\lambda v \mid \lambda \in K\}$$

we denote the subspace of V generated by v. Let v, w be a basis of V; i.e., $Kv \neq Kw$. We set:[27]

$$
\begin{aligned}
\widehat{S}(v) \quad &:= \quad \{x \in \mathrm{GL}(V) \mid (Kv)^x = Kv\} \\
&\equiv \quad \left\{ \begin{pmatrix} \delta_1 & 0 \\ \lambda & \delta_2 \end{pmatrix} \, \middle| \, \delta_1, \delta_2 \in K^*, \ \lambda \in K \right\}, \\[2mm]
\widehat{P}(v) \quad &:= \quad \{x \in \mathrm{GL}(V) \mid v^x = v\} \\
&\equiv \quad \left\{ \begin{pmatrix} 1 & 0 \\ \lambda & \delta_2 \end{pmatrix} \, \middle| \, \delta_2 \in K^*, \ \lambda \in K \right\}, \\[2mm]
\widehat{D}(v,w) \quad &:= \quad \widehat{S}(v) \cap \widehat{S}(w) \equiv \left\{ \begin{pmatrix} \delta_1 & 0 \\ 0 & \delta_2 \end{pmatrix} \, \middle| \, \delta_1, \delta_2 \in K^* \right\}, \\[2mm]
S(v) \quad &:= \quad \widehat{S}(v) \cap \mathrm{SL}(V) \equiv \left\{ \begin{pmatrix} \delta & 0 \\ \lambda & \delta^{-1} \end{pmatrix} \, \middle| \, \delta \in K^*, \ \lambda \in K \right\}, \\[2mm]
P(v) \quad &:= \quad \widehat{P}(v) \cap \mathrm{SL}(V) \equiv \left\{ \begin{pmatrix} 1 & 0 \\ \lambda & 1 \end{pmatrix} \, \middle| \, \lambda \in K \right\}, \\[2mm]
D(v,w) \quad &:= \quad \widehat{D}(v,w) \cap \mathrm{SL}(V) \equiv \left\{ \begin{pmatrix} \delta & 0 \\ 0 & \delta^{-1} \end{pmatrix} \, \middle| \, \delta \in K^* \right\}.
\end{aligned}
$$

It is easy to see that all the above sets are subgroups of $\mathrm{GL}(V)$. We now discuss their structure:

8.6.3 (a) $\widehat{D}(v,w) \cong K^* \times K^*$ and $D(v,w) \cong K^*$. In particular $D(v,w)$ is a cyclic group of order $q - 1$.

(b) $P(v) \cong K(+)$.

(c) $P(v)$ is a normal subgroup of $\widehat{S}(v)$, and $\widehat{S}(v)$ (resp. $S(v)$) is a semidirect product of $P(v)$ with $\widehat{D}(v,w)$ (resp. $D(v,w)$). In particular, $\widehat{S}(v)$ and $S(v)$ are solvable groups.

(d) $C_{P(v)}(x) = 1$ and $[P(v), x] = P(v)$ for $x \in D(v,w) \setminus \langle z \rangle$. [28]

---

[27] With respect to the basis v, w.
[28] Hence $S(v)/\langle z \rangle$ is a Frobenius group; see 8.1.12 on p. 182.

Proof. Most of the assertions follow directly from the matrix representation given in the definition of the subgroups. For (c) note that $\widehat{P}(v)$ is the kernel of the action of $\widehat{S}(v)$ on Kv. Hence $\widehat{P}(v) \trianglelefteq \widehat{S}(v)$ and $P(v) \trianglelefteq S(v)$.

Claim (d) can be verified using

$$\begin{pmatrix} \delta^{-1} & 0 \\ 0 & \delta \end{pmatrix}^{-1} \begin{pmatrix} 1 & 0 \\ \lambda & 1 \end{pmatrix} \begin{pmatrix} \delta^{-1} & 0 \\ 0 & \delta \end{pmatrix} = \begin{pmatrix} 1 & 0 \\ \delta^{-2}\lambda & 1 \end{pmatrix};$$

note here: $\delta^{-2} = 1 \iff \delta = \pm 1$. $\qquad\qquad\qquad\qquad\qquad\qquad\square$

8.6.4 $[V, P(v)] = Kv$ *and* $[V, P(v), P(v)] = 0$ *for* $v \in V^{\#}$.

Proof. The matrix representation of $P(v)$ shows that $P(v)$ acts trivially on V/Kv. Hence

$$0 \neq [V, P(v)] \subseteq Kv,$$

and $[V, P(v), P(v)] = 0$ follows. Since $[V, P(v)]$ is a subspace of V we also get $[V, P(v)] = Kv$. $\qquad\qquad\qquad\qquad\qquad\qquad\qquad\qquad\qquad\qquad\square$

The next result describes the Sylow p-structure of $GL(V)$ and thus also of $SL(V)$.

8.6.5   (a) $\mathrm{Syl}_p\, GL(V) = \{P(v) \mid v \in V^{\#}\}$ *and*

$$P(v) = P(u) \iff Kv = Ku.$$

In particular $|\mathrm{Syl}_p\, GL(V)| = q + 1$.

(b) $N_{GL(V)}(P(v)) = \widehat{S}(v)$, $v \in V^{\#}$.

(c) $P(v) \cap P(u) = 1$ *if* $Kv \neq Ku$ $(v, u \in V^{\#})$.

(d) $\mathrm{Syl}_p\, GL(V) = \{P(u)\} \cup \{P(v)^x \mid x \in P(u)\}$ *for* $v, u \in V^{\#}$ *such that* $Kv \neq Ku$.

Proof. Clearly $P(v)$ $(v \in V^{\#})$ is a Sylow p-subgroup of $GL(V)$ since $|P(v)| = q = p^m$. Conversely, a Sylow p-subgroup P of $GL(V)$ acts on the set V. Since $0^P = 0$, 3.1.7 on page 59 shows that there also exists $v \in V^\#$ such that $v^P = v$. This implies $P \leq P(v)$ and thus $P = P(v)$.

Let $v, u \in V^\#$. Since v, u is a basis of V if $Kv \neq Ku$, we get

$$P(v) \neq P(u) \iff Kv \neq Ku \iff P(v) \cap P(u) = 1.$$

This implies (c) and also (a); note here that $\frac{q^2-1}{q-1} = q+1$ is the number of subspaces Kv, $v \in V^\#$. Since $P(v)^x = P(v^x)$, $x \in \mathrm{GL}(V)$, we also get (b).

(d) $P(u)$ acts on $\mathrm{Syl}_p \mathrm{GL}(V)$ by conjugation. If $P(v)^x = P(v)$, $x \in P(u)$, then $x = 1$ follows from (b) and (c). Hence there exist $q = |P(u)|$ pairwise distinct Sylow p-subgroups $P(v)^x$, $x \in P(u)$. Together with $P(u)$ there are $q+1$ Sylow p-subgroups, and the assertion follows from (a). □

Next we collect some properties about the relations between the above introduced subgroups. As before v, w is a basis of V and with respect to this basis $t \in \mathrm{SL}(V)$ such that

$$t \equiv \begin{pmatrix} 0 & -1 \\ 1 & 0 \end{pmatrix}.$$

8.6.6   (a)   For $u \in V \setminus (Kv \cup Kw)$

$$\widehat{S}(v) \cap \widehat{S}(w) \cap \widehat{S}(u) = Z.$$

In particular $Z(\mathrm{GL}(V)) = Z$ and $Z(\mathrm{SL}(V)) = \langle z \rangle$.

(b) $N_{\mathrm{GL}(V)}(\widehat{D}(v,w)) = \langle t \rangle \widehat{D}(v,w)$ or $q = 2$ and $\widehat{D}(v,w) = 1$.

(c) $N_{\mathrm{SL}(V)}(D(v,w)) = \langle t \rangle D(v,w)$ or $q \leq 3$ and $D(v,w) = \langle z \rangle$.

(d) $[\widehat{D}(v,w), x] = D(v,w)$ for all $x \in \widehat{D}(v,w)t$.

(e) $C_{\mathrm{GL}(V)}(a) = P(v)Z(\mathrm{GL}(V))$ for all $a \in P(v)^\#$.

Proof. (a) Since

$$Z \leq \widehat{S}(v) \cap \widehat{S}(w) \cap \widehat{S}(u) =: H$$

it suffices to show that $H \leq Z$. Let $\lambda_1, \lambda_2 \in K^*$ such that

$$u = \lambda_1 v + \lambda_2 w,$$

and let $h \in H$. Then there exist $\mu_1, \mu_2, \mu_3 \in K^*$ such that $v^h = \mu_1 v$, $w^h = \mu_2 w$ and $u^h = \mu_3 u$. It follows that

$$\mu_3 \lambda_1 v + \mu_3 \lambda_2 w = \mu_3 u = u^h = \lambda_1 v^h + \lambda_2 w^h = \lambda_1 \mu_1 v + \lambda_2 \mu_2 w$$

and thus, since v, w is a basis,

$$\mu_3 \lambda_1 = \lambda_1 \mu_1 \quad \text{and} \quad \mu_3 \lambda_2 = \lambda_2 \mu_2.$$

Hence $\mu_3 = \mu_1 = \mu_2$ and $h = z_{\mu_1} \in Z$.

(b) and (c) For $x \in N_{\mathrm{GL}(V)}(\widehat{D}(v, w))$

$$\widehat{D}(v, w) = (\widehat{D}(v, w))^x = \widehat{D}(v^x, w^x).$$

Now (a) implies $\{Kv, Kw\} = \{Kv^x, Kw^x\}$ or $\widehat{D}(v, w) = Z(\mathrm{GL}(V))$. In the first case $x \notin \widehat{D}(v, w)$ yields

$$(Kv)^x = Kw, \quad (Kw)^x = Kv,$$

and $tx \in \widehat{D}(v, w)$. In the second case

$$\widehat{D}(v, w) \cong K^* \times K^* \quad \text{and} \quad Z(\mathrm{GL}(V)) \cong K^*$$

implies $\widehat{D}(v, w) = 1$ and $q = 2$.

With $D(v, w)$ in place of $\widehat{D}(v, w)$ the case $D(v, w) = Z(\mathrm{SL}(V))$ gives

$$q - 1 = |D(v, w)| = |\langle z \rangle| \leq 2$$

and thus $q \leq 3$.

(d) Since $\widehat{D}(v, w)$ is Abelian we may assume that $x = t$. For the element $d \equiv \begin{pmatrix} \delta_1 & 0 \\ 0 & \delta_2 \end{pmatrix}$

$$d^{-1} t^{-1} d t \equiv \begin{pmatrix} \delta_1^{-1} & 0 \\ 0 & \delta_2^{-1} \end{pmatrix} \begin{pmatrix} 0 & 1 \\ -1 & 0 \end{pmatrix} \begin{pmatrix} \delta_1 & 0 \\ 0 & \delta_2 \end{pmatrix} \begin{pmatrix} 0 & -1 \\ 1 & 0 \end{pmatrix}$$

$$= \begin{pmatrix} \delta_1^{-1} \delta_2 & 0 \\ 0 & \delta_2^{-1} \delta_1 \end{pmatrix};$$

and (d) follows.

(e) Clearly $Z(\mathrm{GL}(V))P(v) \leq C_{\mathrm{GL}(V)}(a)$ since $P(v)$ is Abelian. On the other hand $C_V(a) = Kv$, and thus $C_{\mathrm{GL}(V)}(a) \leq \widehat{S}(v)$. Now (e) follows from 8.6.3 (c) and an elementary calculation. $\square$

8.6.7 *Let P_1, P_2 be two different Sylow p-subgroups of $GL(V)$. Then*

$$SL(V) = \langle P_1, P_2 \rangle.$$

Proof. According to 8.6.5 (a) there exists a basis v, w of V such that $P_1 = P(v)$ and $P_2 = P(w)$. By 8.6.3 (c) and 8.6.5 (b) $N_{SL(V)}(P(v)) = P(v)D(v, w)$, and by 8.6.5 (d)

$$G := \langle P_1, P_2 \rangle \quad (\leq SL(V))$$

is the group generated by all Sylow p-subgroups of $SL(V)$. The Frattini argument implies

$$SL(V) = G D(v, w).$$

Since $D(v, w)$ is Abelian, t acts trivially on $SL(V)/G$. Hence 8.6.6 (d) shows that $D(v, w) \leq G$ and $G = SL(V)$.   □

If $x \in SL(V)$ and $v \in V^{\#}$ such that $v^x = v$, then x is a p-element (8.6.3 (b)). This shows:

8.6.8 *Let R be a p'-subgroup of $SL(V)$. Then $C_V(x) = 0$ for all $x \in R^{\#}$.*   □

8.6.9 *Let $r \in \mathbb{P}$, $r \neq p$, and R a Sylow r-subgroup of $SL(V)$. If $r \neq 2$, then R is cyclic, and if $r = 2$, then R is a quaternion group.*

Proof. Because of 8.6.8 we can apply 8.3.2. Then R is cyclic or a quaternion group. It remains to exclude the case where $r = 2$ and R is cyclic. In this case $SL(V)$ has a normal 2-complement (7.2.2), and this 2-complement contains all the Sylow p-subgroups of $SL(V)$. Now 8.6.7 and $R \neq 1$ contradict each other.   □

$SL_2(2)$ is a non-Abelian group of order 6, and thus isomorphic to S_3.

$SL_2(3)$ is a group of order 24 that is not 3-closed (8.6.5). A Sylow 2-subgroup Q is a quaternion group (8.6.9), and its center has order 2 (8.6.6 (a)). Hence 4.3.4 on page 90 shows that Q is a normal subgroup of $SL_2(3)$.

We sum up:

8.6.10 $SL_2(2)$ *and* $SL_2(3)$ *are solvable groups. The group* $SL_2(2)$ *is isomorphic to* S_3, *and the group* $SL_2(3)$ *is a semidirect product of a cyclic group of order 3 with a quaternion group of order 8 that is not direct. Conversely, every such semidirect product is isomorphic to* $SL_2(3)$.

Proof. Only the last statement requires a proof. Let A, B be two groups of order 3, which act nontrivially on a quaternion group Q_8. It suffices to show

$$A \ltimes Q_8 \cong B \ltimes Q_8.$$

Both groups act faithfully on Q_8 and thus can be regarded as subgroups of $\operatorname{Aut} Q_8$. Now 5.3.3 on page 110 shows that A and B are conjugate in $\operatorname{Aut} Q_8$, and the corresponding semidirect products are isomorphic.   □

Assume now that $q \geq 4$. Then $D(v, w) \neq \langle z \rangle$, and for $x \in D(v, w) \setminus \langle z \rangle$ we get from 8.6.3 (d)

$$P(v) = [P(v), x] \leq \operatorname{SL}(V)'.$$

Hence, every Sylow p-subgroup of $\operatorname{SL}(V)$ is contained in $\operatorname{SL}(V)'$. Now 8.6.7 gives:

8.6.11 $\operatorname{SL}(V)$ *is perfect for* $q \geq 4$.   □

Since $GL(V)/\operatorname{SL}(V)$ $(\cong K^*)$ is Abelian, the commutator group of $GL(V)$ is contained in $\operatorname{SL}(V)$. Hence, For $q \geq 4$ 8.6.11 implies

$$GL(V)' = \operatorname{SL}(V).^{29}$$

The next statement about the structure of $\operatorname{SL}(V)$ will be of relevance in the next chapter.

8.6.12   Let $p \neq 2$. *Suppose that a is a p-element and R an $\langle a \rangle$-invariant p'-subgroup of $\operatorname{SL}(V)$ such that $1 \neq [R, a]$. Then $p = 3$, R is a quaternion group of order 8, and $R\langle a \rangle \cong SL_2(3)$.*

---

[29] This is also true for $q = 3$.

Proof. By 8.2.7 on page 187 we have

$$R = [R, a] \, C_R(a) \quad \text{and} \quad [R, a, a] = [R, a].$$

In addition, 8.6.6 (e) gives $C_R(a) \le \langle z \rangle$.

We show:

(1) $[R, a]$ is a quaternion group.

(1) implies that $\langle z \rangle = Z([R, a])$ (8.6.2) and thus $R = [R, a]$. Hence R is a quaternion group Q_{2^n}. For $n \ge 4$ R possesses a cyclic characteristic subgroup of index 2; and a acts trivially on R (8.2.2 (b) on page 184). Hence $R = Q_8$, and the assertion follows from 8.6.10.

We now prove (1) by induction on $|R|$. We may assume that $R = [R, a]$ and, because of 8.2.3 on page 185, also that R is an r-group, r a prime different from p.

If R is not cyclic, then (1) follows from 8.6.9 (b). We will show that the other case leads to a contradiction:

Since $R = [R, a]$ the element a acts as an automorphism of order p on the cyclic group R. By 2.2.5 on page 51

(2) p divides $(r - 1)$.

Hence $r \ne 2$, and

(3)   either r divides $(q - 1)$ or r divides $(q + 1)$

since $|\,\mathrm{SL}(V)| = (q - 1)q(q + 1)$. If $q = p$, then (2) and (3) contradict each other.

The case $r | (q - 1)$ is also elementary. By Sylow's Theorem we may assume that $R \le D(v, w)$, v, w a basis of V (8.6.3 (c)). Then Kv and Kw are the only R- invariant 1-dimensional subspaces of V (8.6.6 (a)). Hence $R^a = R$ shows that $\langle a \rangle$ is either trivial or transitive on the set $\{Kv, Kw\}$. This contradicts $a \ne 1$ (8.6.5 (c)) and $p \ne 2$, respectively.

We now show that the case $r | (q + 1)$ follows from the case just treated. Note first that

$$r \text{ divides } (q^2 - 1).$$

Next we use from the theory of finite fields the well-known fact that there exists a field extention L of $K = \mathbb{F}_q$ such that $L \cong \mathbb{F}_{q^2}$. This shows that

$$SL(V) \cong SL_2(q) \le SL_2(q^2) \cong SL(\widetilde{V}),$$

where $\widetilde{V}$ is a 2-dimensional vector space over L. Hence $R\langle a \rangle$ is isomorphic to a subgroup of $\mathrm{SL}(\widetilde{V})$. Since $r|(q^2 - 1)$ we are back in the previous case, and the contradiction follows as there. $\square$

8.6.13   *Let $p \neq 2$ and G a subgroup of $\mathrm{SL}(V)$ that is not p-closed. Then the Sylow 2-subgroups of G are quaternion groups.*

Proof. Assume that $G \leq \mathrm{SL}(V)$ is a minimal counterexample. As $O^{p'}(G)$ is p-closed if and only if G is p-closed, we get

$$(') \qquad\qquad O^{p'}(G) = G.$$

Using 8.6.9 we may assume that for all prime divisors $r \neq p$ of $|G|$ the Sylow r-subgroups of G are cyclic. Let r be the smallest such divisor. Since G has no normal p-complement by $(')$, we get from 7.2.1 for $R \in \mathrm{Syl}_r\, G$

$$C_G(R) \neq N_G(R).$$

Hence, there exists a prime divisor s of $|N_G(R)|$ (and thus of $|G|$) and an s-element $a \in N_G(R) \setminus C_G(R)$. Since R is cyclic 2.2.5 implies

$$s \text{ divides } r - 1.$$

The minimality of r yields $s = p$. But this contradicts 8.6.12 since R is cyclic. $\square$

A well-known theorem of Dickson shows that a group G as in 8.6.13 possesses a subgroup isomorphic to $\mathrm{SL}_2(p)$.[30]

Let Ω be the set of 1-dimensional subspaces of V, i.e.,

$$\Omega := \{Kv \mid v \in V^{\#}\}.\text{ [31]}$$

Then $GL(V)$ acts on Ω:

$$Kv \mapsto Kv^x, \quad x \in \mathrm{GL}(V),$$

$\widehat{S}(v)$ is the stabilizer of the point Kv,

$$\widehat{D}(v,w) = \widehat{S}(v) \cap \widehat{S}(w) \quad (Kv \neq Kw)$$

---

[30]See [6], chap. 8, or more modern [12], p. 44.
[31]Ω is the 1-dimensional **projective space**, the **projective line** over K.

is the stabilizer of two points, and

$$Z = Z(\mathrm{GL}(V)) \overset{8.6.6(a)}{=} \{z_\lambda \mid \lambda \in K^*\}$$

is the kernel of this action. Hence

$$\langle z \rangle = Z \cap \mathrm{SL}(V) \overset{8.6.6(a)}{=} Z(\mathrm{SL}(V))$$

is the kernel of the action of $\mathrm{SL}(V)$ on Ω.

Moreover $P(v)$ is a normal subgroup of $\widehat{S}(v)$ that acts regularly on $\Omega \backslash \{Kv\}$ (use $|P(v)| = q = |\Omega \backslash \{Kv\}|$ and 8.6.5 (c)). As $P(v) \le \mathrm{SL}(V)$, both groups $GL(V)$ and $\mathrm{SL}(V)$ are 2-transitive on Ω. The subgroup $\widehat{D}(v,w)/Z$ has order $q-1$ and acts regularly on $\Omega \backslash \{Kv, Kw\}$ (8.6.6 (a)). Thus, the action of $\mathrm{GL}(V)$ on Ω is 3-transitive.

If $p = 2$, then $SL(V)$ acts faithfully on Ω since $z = 1$. Hence, $D(v,w)$ is regular on $\Omega \backslash \{Kv, Kw\}$, and in this case also $\mathrm{SL}(V)$ is 3-transitive on Ω.

If $p \ne 2$, then $z \ne 1$ and $D(v,w)/\langle z \rangle$ cannot be transitive on $\Omega \backslash \{Kv, Kw\}$. Consequently, in this case $\mathrm{SL}(V)$ is not 3-transitive on Ω.

Our discussion also shows that the group $\widehat{S}(v)/Z$ (resp. $S(v)/\langle z \rangle$) acts as a Frobenius group on $\Omega \backslash \{Kv\}$, where $P(v)$ ($\cong K(+)$) is the Frobenius kernel and $\widehat{D}(v,w)/Z$ ($\cong K^*$) (resp. $D(v,w)/\langle z \rangle$) a Frobenius complement.

It should be pointed out that by 8.6.5 (a) the mapping

$$\rho\colon \Omega \to \mathrm{Syl}_p\, \mathrm{GL}(V) \quad \text{such that} \quad Kv \mapsto P(v)$$

is a bijection with

$$((Kv)^x)^\rho = ((Kv)^\rho)^x.$$

Hence the action of G on Ω and on $\mathrm{Syl}_p\, \mathrm{GL}(V)$ (by conjugation) are equivalent.

The factor groups

$$\mathrm{PGL}(V) := \mathrm{GL}(V)/Z(\mathrm{GL}(V)) \quad \text{and} \quad \mathrm{PSL}(V) := \mathrm{SL}(V)/Z(\mathrm{SL}(V))$$

are the 2-dimensional **projective linear group** and the **special projective linear group**, respectively.

Similarly one defines

$$\mathrm{PGL}_2(q) := \mathrm{GL}_2(q)/Z(\mathrm{GL}_2(q)) \quad \text{and} \quad \mathrm{PSL}_2(q) := \mathrm{SL}_2(q)/Z(\mathrm{SL}_2(q)).[32]$$

---

[32] A list of all subgroups of $\mathrm{PSL}_2(q)$ was first given by Dickson ([6]); see [13], II.2.8.

8.6.14   Theorem. PSL(V) *is a non-Abelian simple group for every* $q \geq 4$.

Proof. Let $G := SL(V)$, and let N be a normal subgroup of G such that

$$\langle z \rangle < N \leq G.$$

It suffices to show that $N = G$. As G is 2-transitive on Ω, by 4.2.2 on page 85 and 4.2.4 on page 86 N is transitive on Ω. Hence the Frattini argument gives $G = N\, S(v)$, in particular

$$G/N \cong S(v)/S(v) \cap N.$$

The solvability of $S(v)$ (8.6.3 (c)), and thus of G/N yields either $G/N = 1$ or $(G/N)' \neq G/N$. In the first case we are done, in the second case $G' \neq G$ (1.5.1 on page 24), which contradicts 8.6.11.   □

Exercises

1. PSL$_2(3) \cong A_4$ and PSL$_2(4) \cong$ PSL$_2(5) \cong A_5$.

2. A_6 and PSL$_2(9)$ are isomorphic.

3. PSL$_2(9)$ has a unique conjugacy class of involutions.

4. Every nonsolvable group of order 120 is either isomorphic to S_5, SL$_2(5)$ or $A_5 \times C_2$.

5. For $q \equiv 3 \pmod 8$ and $q \equiv 5 \pmod 8$ the Sylow 2-subgroups of PSL$_2(q)$ are isomorphic to $C_2 \times C_2$.[33]

---

[33]By a theorem of Gorenstein-Walter these are the only simple groups having a Sylow 2-subgroup isomorphic to $C_2 \times C_2$; see the Appendix, p. 369.

Chapter 9

Quadratic Action

9.1 Quadratic Action

In the following p is a prime and G a group that acts on the elementary Abelian p-group V. An element $a \in G$ acts **quadratically** on V if

$$[V, a, a] = 1.$$

This means that a and thus also $\langle a \rangle$ acts trivially on $[V, a]$ and $V/[V, a]$. In particular $\langle a \rangle C_G(V)/C_G(V)$ is a p-group (see 8.2.2 (b) on page 184).

In the endomorphism ring of $End\, V$ the quadratic action of a gives

$$v^{(a-1)(a-1)} = 1 \text{ for all } v \in V,$$

i.e., $(a-1)^2 = 0.$[1] Hence either a acts trivially on V or possesses a *quadratic* minimal polynomial.

The group G acts **quadratically** on V if

$$[V, G, G] = 1.$$

In the following examples the action of G on V is quadratic:

(a) G acts trivially on V.

(b) $|G| = 2 = p$: Then for $a \in G$ and $v \in V$

$$[v, a]^a = [v, a]^{-1} = [v, a].$$

---

[1] Since we write V multiplicatively the zero of $End\, V$ maps every element of V to 1_V.

(c) G is p-group such that $|V/C_V(G)| = p$; see 8.1.4 on page 177.

(d) V is the additive group of a 2-dimensional vector space W over $\mathbb{F}_{p^m}$, and G is a Sylow p-subgroup of $SL(W)$; see 8.6.4 on page 215.

(e) V and G are normal subgroups of the group H, and G is Abelian;

$$[V, G, G] = [[V, G], G] \le [V \cap G, G] = 1.$$

The action of G on V is **p-stable** if for all $a \in G$:

$$[V, a, a] = 1 \quad \Rightarrow \quad a\, C_G(V) \in O_p(G/C_G(V)).$$

For $p = 2$ every involution acts quadratically on V (Example (b)). Thus, p-stability is only interesting for $p \ne 2$.

We first collect some elementary properties:

9.1.1   *Suppose that G acts quadratically on V.*

(a) $[v, a^n] = [v^n, a] = [v, a]^n$ *for all* $v \in V$, $a \in G$, $n \in \mathbb{N}$.

(b) $|V| \le |C_V(a)|^2$ *for all* $a \in G$.

(c) $G/C_G(V)$ *is an elementary Abelian p-group.*

Proof. (a) follows from 1.5.4 on page 25 since $[v, a]^a = [v, a]$ for $a \in G$. Now $v^p = 1$ and (a) give $[v, a^p] = 1$; in particular

$$a^p \in C_G(V) \quad \text{for all } a \in G.$$

Moreover, the quadratic action of G and the Three-Subgroups Lemma imply $[V, G'] = 1$. Hence $G/C_G(V)$ is Abelian, and (c) holds.

Finally by 8.4.1 on page 197

$$V/C_V(a) \cong [V, a] \le C_V(a),$$

and (b) follows.   □

The next example shows that for $p > 2$ elements of order p need not act quadratically.

Let $G = S_4$ and V be a 4-dimensional vector space over $\mathbb{F}_3$ with basis $v_1, \ldots, v_4$. Then

$$v_i^g := v_{i^g}, \quad i = 1, \ldots, 4 \quad (g \in G)$$

defines a faithful action of G on V. By 9.1.1 (c) only 3-elements of G can act quadratically on V.

Let $g := (123)$. Then

$$[v_1, g, g] = [-v_1 + v_2, g] = v_1 + v_2 + v_3 \neq 0,$$

so no nontrivial 3-element of G acts quadratically on V. In particular the action of G on V is 3-stable.

In most of the following it would suffice to regard V as a vector space over the prime field $\mathbb{F}_p = \mathbb{Z}/p\mathbb{Z}$ (2.1.8 on page 46). But in the proof of 9.1.4 it is more appropriate to investigate the action of G on an $\mathbb{F}_q$-vector space W, where $\mathbb{F}_q$ has characteristic p,[2] we take a slightly more general point of view and say that a group or element acts quadratically on the $\mathbb{F}_q$-vector space W, if it acts quadratically on the additive group of W.

9.1.2 *Let G act on the $\mathbb{F}_q$-vector space $W \neq 0$, $q = p^m$. Suppose that*

(1) $G = \langle a, b \rangle$, *where a and b act quadratically on W,*

(2) $G/C_G(W)$ *is not a p-group, and*

(3) $o(ab) = p^e k$ *and $k|(q-1)$.*

Then there exists a homomorphism

$$\varphi \colon G \to \mathrm{SL}_2(q),$$

such that G^φ is not a p-group.

Proof. We proceed by induction on $|G| + \dim W$. If the action of G is not faithful, then $|G/C_G(W)| < |G|$ and the assertion follows by induction.

Let W_1 be a maximal G-invariant subspace of W. If $G/C_G(W_1)$ is not a p-group, then by induction the claim follows for the pair (G, W_1) and thus also for the pair (G, W). Finally, if $G/C_G(W_1)$ is a p-group, then by

---

[2] All vector spaces under consideration are finite-dimensional.

hypothesis (2) and 8.2.2 on page 184 $G/C_G(W/W_1)$ is not a p-group. The case $W_1 \neq 0$ gives dim $W/W_1 <$ dim W, and again the assertion follows by induction.

We are left with the case that G acts faithfully and irreducibly on W. In particular a and b are p-elements since they act quadratically. Hypothesis (2) shows that G is not Abelian.

By hypothesis (3) and Schur's Lemma (8.6.1 on page 211) the cyclic group $\langle ab \rangle$ acts as scalar multiplication on a minimal $\langle ab \rangle$-invariant subspace of W. In particular, there exists a vector $w \neq 0$ in W and $\lambda \in \mathbb{F}_q^*$ such that

$$w^{ab} = \lambda w, \quad \text{and thus} \quad w^a = \lambda w^{b^{-1}}.$$

If $w^a \in \mathbb{F}_q w$ then $\mathbb{F}_q w$ is G-invariant and thus $W = \mathbb{F}_q w$ by the irreducibility of W. But then G is Abelian, a contradiction.

We have shown that $W_1 := \mathbb{F}_q w + \mathbb{F}_q w^a$ is 2-dimensional. The quadratic action of a and b gives

$$w^a - w = [w,a] \in C_{W_1}(a), \quad \text{and}$$
$$w^{b^{-1}} - w = \lambda^{-1} w^a - w \in C_{W_1}(b).$$

Hence $(w^a)^a \in W_1$, $w^b \in W_1$ and of course also $(w^a)^b \in W_1$. This shows that W_1 is G-invariant and thus $W_1 = W$. In particular $G \leq \text{SL}(W) \cong \text{SL}_2(q)$ since G is generated by the p-elements a,b. □

9.1.3   *Let $p \neq 2$ and G be faithful on V. Suppose that*

(1)   *$G = \langle a,b \rangle$, where a and b act quadratically on V, and*

(2)   *G is not a p-group.*

Then the following hold:

(a)   *The Sylow 2-subgroups of G are not Abelian.*

(b)   *If Q is a normal p'-subgroup of G and $[Q,a] \neq 1$, then $p = 3$, and there exists a section of G isomorphic to $\text{SL}_2(3)$.*

Proof. Let $o(ab) = p^e k$ such that $(p, k) = 1$, and let q be a power of p such that $k | (q - 1)$.[3] We write V additively as a vector space over $\mathbb{F}_p$ ($\leq \mathbb{F}_q$) and choose a basis $v_1, \ldots, v_n$ of V. Let W be an $\mathbb{F}_q$-vector space with basis $v_1, \ldots, v_n$, i.e., $V \subseteq W$. The action of G on V is uniquely determined by the images of the basis $v_1, \ldots, v_n$ under G. Hence this action of G on V can be extended in a unique way to an action of G on the vector space W.[4] Now G and W satisfy the hypothesis of 9.1.2, so there exists a homomorphism $\varphi \colon G \to \mathrm{SL}_2(q)$ such that

$$G^\varphi = \langle a^\varphi, b^\varphi \rangle$$

is not a p-group. In particular, G^φ is not p-closed since a^φ and b^φ are p-elements. Thus, claim (a) follows from 8.6.13 on page 221.

For the proof of claim (b) assume that Q^φ is an a^φ-invariant p'-subgroup such that $[Q^\varphi, a^\varphi] \neq 1$. Then (b) follows from 8.6.12 on page 219. $\qquad\square$

9.1.4 Theorem. *Let $p \neq 2$. Suppose that the action of G on V is faithful and not p-stable. Then the following hold:*

(a) *The Sylow 2-subgroups of G are non-Abelian.*

(b) *If in addition G is p-separable, then $p = 3$, and there exists a section of G isomorphic to $\mathrm{SL}_2(3)$.*

Proof. Since G is not p-stable on V there exists $a \in G \setminus O_p(G)$ such that $[V, a, a] = 1$. Let $\mathcal{K}$ be the set of G-composition factors of V. By 8.2.10 on page 188

$$O_p(G) = \bigcap_{W \in \mathcal{K}} C_G(W).$$

Hence there exists $W \in \mathcal{K}$ such that $a \notin C_G(W)$ and thus

$$a C_G(W) \notin O_p(G / C_G(W)) = 1.$$

Now $G / C_G(W)$ and W satisfy the hypotheses, and by induction on $|G| + |V|$ we may assume that $W = V$ and $O_p(G) = 1$. By Baer's Theorem (6.7.6) there exists $b \in a^G$ such that

$$G_1 := \langle a, b \rangle$$

---

[3]As $\mathbb{Z}/k\mathbb{Z}$ is finite, there exist positive integers $i < j$ such that $p^j \equiv p^i \pmod{k}$, so $p^{j-i} \equiv 1 \pmod{k}$.

[4]In matrices: $G \leq \mathrm{GL}_n(p) \leq \mathrm{GL}_n(q)$.

is not a p-group. Now claim (a) follows from 9.1.3 (a) (with G_1 in place of G); note that b also acts quadratically on W.

Assume that G is p-separable and set $Q := O_{p'}(G)$. Then $[Q, a] \neq 1$ (6.4.3 on page 134), and the element b can be chosen in a^Q. Hence claim (b) follows from 9.1.3 (b). $\square$

With the Theorem of Dickson mentioned on page 221 the proof of 9.1.4 (a) provides a stronger statement: There exists a section of G isomorphic to $\mathrm{SL}_2(p)$.

Conversely, let V be a 2-dimensional vector space over $\mathbb{F}_p$. Then the isomorphism $\mathrm{SL}_2(p) \cong \mathrm{SL}(V)$ gives rise to an action of $\mathrm{SL}_2(p)$ on V, and this action is not p-stable since $O_p(\mathrm{SL}(V)) = 1$ (see Example (d)).

We close this section with a lemma that will be needed later. The proof refers to a situation where quadratic action occurs in a natural way.

9.1.5 *Suppose that G acts faithfully on V. Let E_1, E_2 be two subnormal subgroups of G such that $[V, E_1, E_2] = 1$. Then $[E_1, E_2] \leq O_p(G)$.*

Proof. By our hypothesis $V_1 := [V, E_1]$ is invariant under $E := \langle E_1, E_2 \rangle$, so

$$E_0 := C_E(V_1) \quad \text{and} \quad E^0 := C_E(V/V_1)$$

are normal subgroups of E. As $E_0 \cap E^0$ acts quadratically on V, $E_0 \cap E^0$ is a p-group (9.1.1 (c)), so $E_0 \cap E^0 \leq O_p(E)$. Now $E_1 \leq E^0$ and $E_2 \leq E_0$ imply that

$$[E_1, E_2] \leq [E^0, E_0] \leq E_0 \cap E^0 \leq O_p(E).$$

By 6.7.1 on page 156 E and thus also $O_p(E)$ is subnormal in G, so $O_p(E) \leq O_p(G)$ (6.3.1 on page 130). $\square$

9.2 The Thompson Subgroup

As in the first section of this chapter let G be a group that acts on the elementary Abelian p-group V. In this section we discuss the question how to find subgroups of G that act quadratically *and* nontrivially on V.

Obviously, for every subgroup $A \leq G$ the subgroup

$$A^* := C_A([V, A])$$

acts quadratically on V. Thus, we are interested in conditions that guarantee $C_A(V) < A^*$. The next lemma gives a first hint. The crucial argument in the proof is due to Thompson.

9.2.1 *Suppose that the group A acts on the elementary Abelian p-group V such that $A/C_A(V)$ is Abelian. Let U be a subgroup of V. Then there exists a subgroup $A^* \leq A$ such that one of the following holds:*

(a) $|A| \, |C_V(A)| < |A^*| \, |C_V(A^*)|$ *or*

(b) $A^* = C_A([U, A])$, $C_V(A^*) = [U, A]C_V(A)$ *and*

$$|A| \, |C_V(A)| = |A^*| \, |C_V(A^*)|.$$

Proof. We may assume that for all subgroups $B \leq A$

(∗) $|A| \, |C_V(A)| \geq |B| \, |C_V(B)|,$

and we verify the equations in (b) for

$$A^* := C_A([U, A]).$$

Clearly $[U, A, A^*] = 1$, and since $A/C_A(V)$ is Abelian also $[A, A^*, U] = 1$. Hence the Three-Subgroups Lemma gives $[U, A^*, A] = 1$, i.e.,

(1) $[U, A^*] \leq C_V(A).$

First we show that the inequality

(2) $|A| \, |C_V(A)| \leq |A^*| \, |[U, A]C_V(A)|$

implies (b). We have

$$|A^*| \, |C_V(A^*)| \overset{(∗)}{\leq} |A| \, |C_V(A)| \overset{(2)}{\leq} |A^*| \, |[U, A]C_V(A)|$$
$$\leq |A^*| \, |C_V(A^*)|$$

and thus $|A^*| \, |C_V(A^*)| = |A| \, |C_V(A)|$ and $|[U, A]C_V(A)| = |C_V(A^*)|$. The last equality implies $C_V(A^*) = [U, A]C_V(A)$ since $[U, A] \leq C_V(A^*)$.

It remains to prove (2): We may assume that $U \neq 1$. Let

$$Y := C_V(A) \quad \text{and} \quad X := [U, A].$$

We first treat the case $|U| = p$, the general case then will be reduced to this one by induction.

Assume that $U = \langle u \rangle$. Then 1.5.4 on page 25 shows that $[u, A] = [U, A]$ since V is Abelian. The mapping

$$\varphi \colon A/A^* \to XY/Y \quad \text{such that} \quad aA^* \mapsto [u, a]Y$$

is well-defined since for all $a^* \in A^*$ by (1)

$$[u, a^*a] = [u, a]\,[u, a^*]^a \in [u, a]Y.$$

If φ is injective, then
$$|A/A^*| \leq |XY/Y|,$$

and (2) follows. Let $a_1, a_2 \in A$ such that $[u, a_1]Y = [u, a_2]Y$, i.e., $[u, a_1][u, a_2]^{-1} = u^{a_1} u^{-a_2} \in Y$. Then

$$[u, a_1 a_2^{-1}] = u^{-1} u^{a_1 a_2^{-1}} = (u^{-a_2} u^{a_1})^{a_2^{-1}} = u^{a_1} u^{-a_2} \in Y$$

and thus
$$[u, a_1 a_2^{-1}, A] = 1 = [a_1 a_2^{-1}, A, u],$$

so $[u, A, a_1 a_2^{-1}] = 1$ and $a_1 a_2^{-1} \in C_A([U, A]) = A^*$. This shows that φ is injective, and (2) follows in the case $U = \langle u \rangle$.

Assume now that $|U| > p$. Let U_1 be a subgroup of index p in U. Then $U = U_1 \langle u \rangle$ for a suitable $u \in U$. Let

$$\begin{aligned}
X_1 &:= [U_1, A], \quad A_1 := C_A(X_1) \quad \text{and} \\
X_2 &:= [\langle u \rangle, A], \quad A_2 := C_A(X_2).
\end{aligned}$$

Note that
$$X_1 X_2 \, C_V(A) = X \, C_V(A), \quad A^* = A_1 \cap A_2,$$

and
$$X_1 \, C_V(A) \cap X_2 \, C_V(A) \leq C_V(A_1 A_2).$$

By induction on $|U|$ we may assume for $i = 1, 2$

$$|A||C_V(A)| = |A_i||X_i \, C_V(A)|.$$

Hence

$$|A||C_V(A)| \overset{(*)}{\geq} |A_1A_2||C_V(A_1A_2)|$$
$$\geq |A_1A_2||X_1C_V(A) \cap X_2C_V(A)|$$
$$\overset{1.1.6}{=} \frac{|A_1||A_2|}{|A_1 \cap A_2|} \frac{|X_1C_V(A)||X_2C_V(A)|}{|X_1C_V(A)\,X_2C_V(A)|}$$
$$= \frac{|A|^2|C_V(A)|^2}{|A^*||XC_V(A)|},$$

and $(*)$ implies (2). □

In view of 9.2.1 candidates for quadratic action are subgroups A of G that satisfy:

$\mathcal{Q}_1$ $|A|\,|C_V(A)| \geq |A^*|\,|C_V(A^*)|$, for all subgroups A^* of A, and

$\mathcal{Q}_2$ $A/C_A(V)$ is an elementary Abelian p-group.

Note that by 9.1.1 (c) every quadratically acting subgroup A satisfies $\mathcal{Q}_2$. Hence $\mathcal{Q}_2$ is a necessary condition.

Let $\mathcal{A}_V(G)$ be the set of subgroups A of G that satisfy $\mathcal{Q}_1$ and $\mathcal{Q}_2$. For every such A we obtain from 9.2.1 (b) (with $U = V$):

9.2.2   Let $A \in \mathcal{A}_V(G)$ and $A^* := C_A([V,A])$. Then

$$|A/A^*| = |C_V(A^*)/C_V(A)| \quad \text{and} \quad C_V(A^*) = [V,A]\,C_V(A). \quad □$$

9.2.3   **Timmesfeld Replacement Theorem [95].**[5] Let $A \in \mathcal{A}_V(G)$ and U be a subgroup of V. Then

$$C_A([U,A]) \in \mathcal{A}_V(G) \quad \text{and} \quad C_V(C_A([U,A])) = [U,A]C_V(A).$$

Moreover $[V, C_A([U,A])] \neq 1$ if $[V,A] \neq 1$.

Proof. Let $A^* := C_A([U,A])$. Since $A \in \mathcal{A}_V(G)$ 9.2.1 (b) applies, so

$(')$ $|A^*||C_V(A^*)| = |A||C_V(A)| \quad \text{and} \quad C_V(A^*) = [U,A]C_V(A).$

[5] See also [38].

In addition, for every $A_0 \leq A^*$ $\mathcal{Q}_1$ gives

$$|A_0||C_V(A_0)| \leq |A^*||C_V(A^*)|.$$

Hence $A^* \in \mathcal{A}_V(G)$.

For the proof of the additional claim we may assume that $[V, A^*] = 1$. Then
($'$) implies that

$$V = [U, A]C_V(A) = [V, A]C_V(A).$$

In particular $[V, A, A] = [V, A]$. But then $[V, A] = 1$ since $A/C_A(V)$ is a
p-group (8.1.4 b on page 177). $\qquad\qquad\square$

By $\mathcal{A}_V(G)_{\min}$ we denote the set of minimal elements of the set

$$\{A \in \mathcal{A}_V(G) \,|\, [V, A] \neq 1\}.$$

9.2.4   *Every element of $\mathcal{A}_V(G)_{\min}$ acts quadratically and nontrivially on
V.*

Proof. Let $A \in \mathcal{A}_V(G)_{\min}$. By 9.2.3 $A^* := C_A([V, A])$ is also in $\mathcal{A}_V(G)$ and
$[V, A^*] \neq 1$. The minimality of A implies $A^* = A$ and thus $[V, A, A] = 1$. $\qquad\square$

Up to now we have discussed candidates for quadratic action but never could
exclude the possibility that these subgroups act trivially on V (Example on
page 226). In this context 9.2.4 yields:

9.2.5   *Suppose that G is p-stable on V and $O_p(G/C_G(V)) = 1$. Then
every element of $\mathcal{A}_V(G)$ acts trivially on V.* $\qquad\square$

The property

$$(*) \qquad\qquad\qquad O_p(G/C_G(V)) = 1$$

is not only useful for p-stability. We will meet it later in other situations.
For example $(*)$ is satisfied if G acts irreducibly on V (8.1.5 on page 178).
A more general condition that implies $(*)$ is the following:

9.2.6   *Let $V = \langle C_V(S) \,|\, S \in \mathrm{Syl}_p\, G \rangle$. Then $O_p(G/C_G(V)) = 1$.*

Proof. Let S be a Sylow p-subgroup of G. Set

$$Z := C_V(S) \quad \text{and} \quad C := C_G(V).$$

Since all Sylow p-subgroups of G are conjugate we get

$$V = \langle Z^G \rangle.$$

Let $C \leq D \leq G$ such that $D/C = O_p(G/C)$. Then $D \cap S \in \text{Syl}_p D$, $D = C(D \cap S)$ (3.2.5), and

$$G = CN_G(D \cap S) \quad \text{(Frattini argument)}.$$

This gives

$$V = \langle Z^{N_G(D \cap S)} \rangle$$

and thus $[V, D \cap S] = 1$. Hence $D \cap S \leq C$ and $D = C$. $\square$

The following application of 9.2.6 can be used frequently:

9.2.7 *Let G be a group and $C_G(O_p(G)) \leq O_p(G)$. Then*

$$V := \langle \Omega(Z(S)) \mid S \in \text{Syl}_p G \rangle$$

is an elementary Abelian normal subgroup of G and $O_p(G/C_G(V)) = 1$.

Proof. Let $S \in \text{Syl}_p G$. Then $\Omega(Z(S)) \leq C_G(O_p(G)) \leq O_p(G) \leq S$, so V is contained in $\Omega(Z(O_p(G)))$ and $\Omega(Z(S)) = C_V(S)$. Now the assertion follows from 9.2.6. $\square$

In the following we are interested in finding conditions for the existence of elements in $\mathcal{A}_V(G)$ that act nontrivially on V. As in 9.2.7 we investigate a situation that holds in most of the applications. From now on we assume:

- V is an elementary Abelian normal subgroup of G, and

- G acts on V by conjugation.

Fundamental for the following investigations is a subgroup of G that was introduced by Thompson and carries his name. For the definition of this subgroup let p be a prime and $\mathcal{E}(G)$ be the set of elementary Abelian subgroups of G. Let

$$
\begin{aligned}
m &:= \max\{|A| \mid A \in \mathcal{E}(G)\}, \\
\mathcal{A}(G) &:= \{A \in \mathcal{E}(G) \mid |A| = m\}, \text{ and} \\
J(G) &:= \langle A \mid A \in \mathcal{A}(G)\rangle.
\end{aligned}
$$

$J(G)$ is the **Thompson subgroup** of G with respect to p. It will be always clear from the context which p is meant in the definition of the Thompson subgroup.

Before we go back to quadratic action, we first list some elementary properties of the Thompson subgroup that are easy consequences of the definition.

9.2.8 (a) $J(G)$ *is a characteristic subgroup of* G, *which is nontrivial if* $p \in \pi(G)$.

(b) *If* $J(G) \leq U \leq G$, *then* $J(G) = J(U)$.

(c) $J(G) = \langle J(S) \mid S \in \mathrm{Syl}_p\, G\rangle$.

(d) *If* $x \in C_G(J(G))$ *and* $o(x) = p$, *then* $x \in Z(J(G))$.

(e) *If* $\mathcal{B} \subseteq \mathcal{A}(G)$, *then* $J(\langle \mathcal{B}\rangle) = \langle \mathcal{B}\rangle$.   □

The following result gives a connection between $\mathcal{A}(G)$ and $\mathcal{A}_V(G)$:

9.2.9 (a) $\mathcal{A}(G) \subseteq \mathcal{A}_V(G)$.

(b) *If* $V \nleq Z(J(G))$, *then there exists* $A \in \mathcal{A}(G)$ *such that* $[V, A] \neq 1$.

Proof. (a) Let A^* be a subgroup of $A \in \mathcal{A}(G)$. Then $A^* C_V(A^*)$ is in $\mathcal{E}(G)$. It follows that

$$
|A| \geq |A^* C_V(A^*)| = \frac{|A^*||C_V(A^*)|}{|A^* \cap V|} \geq \frac{|A^*||C_V(A^*)|}{|C_V(A)|},
$$

and A satisfies $\mathcal{Q}_1$.

(b) is obvious.   □

According to 9.2.9 (a) we can apply our earlier results to $\mathcal{A}(G)$ and get:

9.2.10   Theorem. *Let* $A \in \mathcal{A}(G)$ *and* $A_0 := [V, A]C_A([V, A])$.

(a) A_0 *is in* $\mathcal{A}(G)$ *and acts quadratically on* V.

(b)   *If* $[V, A] \neq 1$, *then also* $[V, A_0] \neq 1$.

Proof. Let $X := [V, A]$ and $A^* := C_A(X)$; i.e., $A_0 = A^*X$. Then A_0 is an elementary Abelian p-group. By its definition

$$[V, A_0, A_0] \leq [V, A, A_0] = 1,$$

so A_0 acts quadratically on V. For the proof of $A_0 \in \mathcal{A}(G)$ it suffices to show that $|A| = |A_0|$. The maximality of A gives

$$C_V(A) = V \cap A = V \cap A^*,$$

and by the definition of A^*

$$X \cap A = X \cap A^*.$$

It follows

$$|A|\,|A \cap V| = |A|\,|C_V(A)| \overset{9.2.2}{=} |A^*|\,|XC_V(A)|$$

and by 1.1.6

$$|A| = \frac{|A^*|\,|XC_V(A)|}{|C_V(A)|} = \frac{|A^*|\,|X|}{|X \cap C_V(A)|} = \frac{|A^*|\,|X|}{|X \cap A^*|} = |A^*X| = |A_0|.$$

This yields (a), and 9.2.3 implies (b). $\qquad\qquad\qquad\qquad\qquad\qquad\square$

A further property that implies nontrivial quadratic action, can be derived from the observation that for $A^* := C_A(V)$ condition $\mathcal{Q}_1$ implies

$$\mathcal{Q}_1' \qquad\qquad |A/C_A(V)| \geq |V/C_V(A)|.$$

9.2.11   *Let* $\mathcal{B}$ *be the set of subgroups* $A \leq G$ *satisfying* $\mathcal{Q}_1'$ *and* $\mathcal{Q}_2$. *Let* $A \in \mathcal{B}$ *and suppose that*

(m) $\qquad\qquad |A^*/C_{A^*}(V)|\,|C_V(A^*)| \leq |A/C_A(V)|\,|C_V(A)|.$

for all subgroups $A^* \leq A$ *that are in* $\mathcal{B}$. *Then* $A \in \mathcal{A}_V(G)$.

Proof. We have to verify $\mathcal{Q}_1$ for A. Let $A^* \leq A$. If A^* does not satisfy $\mathcal{Q}_1'$, then A^* is not in $\mathcal{B}$ and

$$|A^*/C_{A^*}(V)|\,|C_V(A^*)| \overset{\mathcal{Q}_1'}{<} |V| \leq |A/C_A(V)|\,|C_V(A)|.$$

It follows

$$
\begin{aligned}
|A/C_A(V)|\,|C_V(A)| &\geq |A^*/C_{A^*}(V)|\,|C_V(A^*)| \\
&\overset{1.2.6}{=} |A^*C_A(V)/C_A(V)|\,|C_V(A^*)|.
\end{aligned}
$$

This inequality is also true for $A^* \in \mathcal{B}$ since then (m) holds. Thus, we have for all $A^* \leq A$

$$|A^*|\,|C_V(A^*)| \leq |A^*C_A(V)|\,|C_V(A^*)| \leq |A|\,|C_V(A)|,$$

and A satisfies $\mathcal{Q}_1$. $\square$

Assume that there exists an $A \in \mathcal{B}$ that acts nontrivially on V. Among all such A we choose A with the additional property that

$$|A/C_A(V)|\,|C_V(A)|$$

is maximal. Then (m) in 9.2.11 holds for A, i.e., $A \in \mathcal{A}_V(G)$ and $\mathcal{A}_V(G)_{\min}$ $\neq \varnothing$. Now 9.2.4 gives the existence of subgroups that act quadratically and nontrivially on V.

Let S be a Sylow p-subgroup of G. Then G is **Thompson factorizable** with respect to p if

$$G = O_{p'}(G)\,C_G(\Omega(Z(S)))\,N_G(J(S)).$$

Note that $\overline{S}$ is a Sylow p-subgroup of $\overline{G} := G/O_{p'}(G)$. The Frattini argument implies (see 3.2.8 on page 66)

$$N_{\overline{G}}(J(\overline{S})) = \overline{N_G(J(S))} \quad \text{and} \quad C_{\overline{G}}(\Omega(Z(\overline{S}))) = \overline{C_G(\Omega(Z(S)))}.$$

Hence G is Thompson factorizable if and only if $\overline{G}$ is.

9.2.12   *Let $O_{p'}(G) = 1$ and $V := \langle \Omega(Z(S)) \mid S \in \mathrm{Syl}_p\, G \rangle$. Then G is Thompson factorizable if and only if $J(G) \leq C_G(V)$.*

Proof. Let $S \in \mathrm{Syl}_p G$ and $C := C_G(V)$. Assume that G is Thompson factorizable. Then Sylow's Theorem gives

$$V = \langle \Omega(Z(S))^g \mid g \in G \rangle = \langle \Omega(Z(S))^g \mid g \in N_G(J(S)) \rangle.$$

Since $\Omega(Z(S)) \leq Z(J(S))$ this implies $V \leq Z(J(S))$ and thus $J(G) \leq C$ (9.2.8 (c)).

Assume now that $J(G) \leq C$. Then $J(S) \leq C \cap S$. Since

$$J(S) \text{ char } C \cap S \in \mathrm{Syl}_p C$$

and $\Omega(Z(S)) \leq Z(J(S))$ the Frattini argument yields the factorization

$$G = CN_G(C \cap S) = C_G(\Omega(Z(S)))N_G(J(S)). \qquad \square$$

9.3 Quadratic Action in p-Separable Groups

In this section we consider p-separable groups that are not Thompson factorizable with respect to p. The following observation provides us with conditions for a suitable set-up:

9.3.1 *Suppose that G is a p-separable group that is not Thompson factorizable with respect to p. Let $O_{p'}(G) = 1$ and set*

$$V := \langle \Omega(Z(S)) \mid S \in \mathrm{Syl}_p G \rangle \quad \text{and} \quad H := J(G)C_G(V)/C_G(V).$$

Then the following hold:

$\mathcal{S}_1$ $C_H(O_{p'}(H)) \leq O_{p'}(H)$.

$\mathcal{S}_2$ V *is an elementary p-group, and H acts faithfully on V.*

$\mathcal{S}_3$ $H = \langle A \mid A \in \mathcal{A}_V(H) \rangle \neq 1$.

Proof. By 6.4.3 $C_G(O_p(G)) \leq O_p(G)$. Hence $\mathcal{S}_2$ and $O_p(G/C_G(V)) = 1$ follow from 9.2.7, in particular $O_p(H) = 1$. As H is p-separable and thus also p'-separable, 6.4.3 on page 134 yields $\mathcal{S}_1$. Moreover, 9.2.9 and 9.2.12 imply $H \neq 1$ and $\mathcal{S}_3$. $\qquad \square$

In the following we will use these properties $\mathcal{S}_1 - \mathcal{S}_3$ rather than the p-separability of G.

We begin with an example that later in 9.3.7 will turn out to be typical. Let $V_1, \ldots, V_r$ be elementary Abelian p-groups of order p^2. We regard these groups as 2-dimensional $\mathbb{F}_p$-vector spaces. Then

$$E_i := \mathrm{SL}(V_i), \quad i = 1, \ldots, r,$$

acts its natural way on V_i, and

$$\mathcal{A}_{V_i}(E_i) = \{A \mid A \in \mathrm{Syl}_p\, E_i\}$$

(compare with 8.6.4 on page 215). Hence, the pair (E_i, V_i) satisfies $\mathcal{S}_2$ and $\mathcal{S}_3$. For $p = 2$ and $p = 3$ the groups E_i are solvable. Thus, in these cases also $\mathcal{S}_1$ holds, see 8.6.10 on page 219. Let

$$H := E_1 \times \cdots \times E_r \quad \text{and} \quad V := V_1 \times \cdots \times V_r.$$

Then the action of the components E_i induce an action of H on V, i.e., E_i acts as $\mathrm{SL}(V_i)$ on V_i and $[V_j, E_i] = 1$ for $i \neq j$. It follows that

$$\mathcal{A}_{V_i}(E_i) = \mathcal{A}_V(E_i) \subseteq \mathcal{A}_V(H),$$

so for $p \in \{2, 3\}$ the pair (H, V) satisfies $\mathcal{S}_1 - \mathcal{S}_3$.

From now on let (V, H) be a pair satisfying $\mathcal{S}_1$, $\mathcal{S}_2$ and $\mathcal{S}_3$.

9.3.2 Let $1 \neq A \in \mathcal{A}_V(H)$.

(a) $|A| = |V/C_V(A)|$.

(b) There exist $A_1, \ldots, A_k \in \mathcal{A}_V(H)_{\min}$ such that $A = A_1 \times \cdots \times A_k$.

(c) $|B| = |[V, B]| = |V/C_V(B)| = p$ for all $B \in \mathcal{A}_V(H)_{\min}$.

Proof. The hypothesis $A \in \mathcal{A}_V(H)$ gives

$$(*) \qquad\qquad |A_i||C_V(A_i)| \leq |A||C_V(A)|$$

for all subgroups $A_i \leq A$. Let $\mathcal{B}$ be the set of maximal subgroups of A, i.e.,

$$|A/A_i| = p \quad \text{for } A_i \in \mathcal{B},$$

and set $Q := O_{p'}(H)$. We apply 8.3.4 (c) on page 193 to A and Q. Then

$$(') \qquad 1 \neq [Q, A] = \langle [C_Q(A_i), A] \mid A_i \in \mathcal{B} \rangle.$$

Hence there exists $A_0 \in \mathcal{B}$ such that

$$Q_0 := [C_Q(A_0), A] \neq 1,$$

and $Q_0 = [Q_0, A]$ (8.2.7 on page 187). If $C_V(A_0) = C_V(A)$, then Q_0 acts trivially on $C_V(A_0)$, and the $P \times Q$-Lemma (applied to $Q_0 \times A_0$ and V) yields $[V, Q_0] = 1$ and thus $Q_0 = 1$, a contradiction.

We have shown that $C_V(A_0) \neq C_V(A)$, in particular $|C_V(A_0)/C_V(A)| \geq p$ and

$$|A||C_V(A)| \leq |A_0||C_V(A_0)|.$$

Since by $(*)$ also the opposite inequality holds we get

$$(**) \qquad |A||C_V(A)| = |A_0||C_V(A_0)|,$$

and $A_0 \in \mathcal{A}_V(H)$. Moreover, if $A \in \mathcal{A}_V(H)_{min}$, then $A_0 = 1$ and

$$|A| = |V/C_V(A)| = p.$$

This, together with $(**)$, gives (a) and (c).

It remains to prove (b). We may assume that $|A| > p$. According to $(')$ there exists a second subgroup $A_1 \in \mathcal{B}$ such that $[C_Q(A_1), A] \neq 1$, and as we have seen above $A_1 \in \mathcal{A}_V(H)$.

By induction on $|A|$ we may assume that (b) holds for A_0 and A_1 in place of A. But then (b) also holds for A since $A = A_0 A_1$. $\qquad \square$

9.3.3 Let $A \in \mathcal{A}_V(H)_{min}$ and $x \in O_{p'}(H) \setminus C_{O_{p'}(H)}(A)$, and set

$$E_x := \langle A, A^x \rangle, \quad Q_x := E_x \cap O_{p'}(H) \quad \text{and} \quad V_x := [V, E_x].$$

Then the following hold:

(a) $p \in \{2, 3\}$.

(b) $|V_x| = p^2$.

(c) $E_x = \mathrm{SL}(V_x) \cong \mathrm{SL}_2(p)$.

In particular, Q_x acts irreducibly on V_x and

$$Q_x \cong \begin{cases} C_3 & \\ Q_8 \end{cases} \text{if} \quad \begin{matrix} p = 2 \\ p = 3 \end{matrix}$$

Proof. The subgroups A and A^x are are two different Sylow p-subgroups of E_x since

$$1 \neq [x, A] \leq E_x \cap O_{p'}(H) = Q_x \text{ and } AQ_x = E_x.$$

Thus they are conjugate under Q_x, and

$(')$ $\qquad\qquad\qquad\qquad [Q_x, A] \neq 1.$

Since E_x acts trivially on $V/[V, A][V, A^x]$ we get

$$V_x = [V, A][V, A^x].$$

By 9.3.2 $|A| = p$ and $|[V, A]| = p$, so $|V_x| \leq p^2$. Moreover, by 8.2.2 on page 184 the p'-group Q_x acts faithfully on V_x. Hence, the quadratic action of A gives (b). In addition, 8.6.12 on page 219 and $(')$ imply (a). Now the structure of the groups $SL_2(2)$ and $SL_2(3)$ given in 8.6.10 on page 219 yields (c) and the additional claim. $\qquad\qquad\qquad\qquad\qquad\qquad\qquad\square$

9.3.4 *Let $A \in \mathcal{A}_V(H)_{\min}$. Then $[O_{p'}(H), A]$ is a normal 3-subgroup of $O_{p'}(H)$ and $p = 2$, or $[O_{p'}(H), A]$ is a non-Abelian normal 2-subgroup and $p = 3$. In particular, the subgroups Q_x defined in 9.3.3 are subnormal in $O_{p'}(H)$.*

Proof. The subgroup $[O_{p'}(H), A]$ is normal in $O_{p'}(H)$. Let r be a prime divisor of $|O_{p'}(H)|$ and R an A-invariant Sylow r-subgroup of $O_{p'}(H)$ (8.2.3 on page 185). If $[R, A] \neq 1$, then 9.3.3 for $x \in R$ and $[x, A] \neq 1$ shows that $p = 3$ and Q_x is a quaternion group, or $p = 2$ and Q_x is a nontrivial 3-group. In particular, this shows that $O_{p'}(H) = RC_{O_{p'}(H)}(A)$ and $[O_{p'}(H), A] \leq R$ (8.1.1 on page 177). $\qquad\qquad\qquad\qquad\square$

In the situation of 9.3.4 the following lemma is crucial:

9.3.5   *Let E be a group that acts faithfully on the elementary Abelian p-group V, and let E_1, E_2 be two subnormal subgroups of E and*

$$V_i := [V, E_i], \quad i = 1, 2.$$

Suppose that the following hold:

(1) $E = \langle E_1, E_2 \rangle$ *and* $O_p(E) = 1$.

(2) E_i *acts irreducibly on* V_i, $i = 1, 2$.

(3) $V_1 \not\leq V_2$ *and* $V_2 \not\leq V_1$.

(4) $|E_1| > 2$ *and* $|E_2| > 2$.

Then $E = E_1 \times E_2$ and $[V, E] = V_1 \times V_2$.

Proof. Clearly $[V, E] = V_1 V_2$, and E acts trivially on $V/V_1 V_2$. By 8.2.2 (b) $C_E(V_1 V_2)$ is a p-group and thus by (1)

$$C_E(V_1 V_2) \leq O_p(E) = 1.$$

Similarly $C_{E_i}(V_i) \leq O_p(E_i) = 1$ since E_i is subnormal in E (6.3.1). The irreducibility of V_i gives $V_i = [V_i, E_i]$. Hence, $V_1 V_2$ and E satisfy the hypotheses, and we may assume that

$$V = V_1 V_2.$$

Assume first that $V_1^E = V_1$. Then $V_1 \cap V_2$ is invariant under E_2 and thus $V_1 \cap V_2 = 1$ by (2) and (3). It follows that $V = V_1 \times V_2$ and

$$[V, E_1, E_2] = [V_1, E_2] \leq V_1 \cap V_2 = 1.$$

Now 9.1.5 implies $[E_1, E_2] \leq O_p(E) = 1$. Hence also $V_2^E = V_2$. Now a symmetric argument shows that also $[V_2, E_1] = 1$, and V is centralized by $E_1 \cap E_2$. The faithful action of E on V gives $E_1 \cap E_2 = 1$, i.e., $E = E_1 \times E_2$.

We may assume now that neither V_1 nor V_2 is normalized by E. In particular,

$$K := N_{E_2}(V_1) < E_2.$$

We choose our notation such that

$$|V_1| \geq |V_2|.$$

Let $x \in E_2 \setminus K$ and $E^* := \langle E_1, E_1^x \rangle$. Then E^* is subnormal in E (6.7.1 on page 156) and thus $O_p(E^*) \leq O_p(E) = 1$. Hence (E_1, E_1^x, V_1, V_1^x) satisfies the hypotheses in place of (E_1, E_2, V_1, V_2). Moreover $E^* < E$ since $E \neq E_1$ and $E_1 \trianglelefteq\trianglelefteq E$. By induction on $|E|$ we may assume that

$$E^* = E_1 \times E_1^x \quad \text{and} \quad V_1 \cap V_1^x = 1.$$

In particular

$$[V_1^x, E_1] = 1 \text{ and } V_1 \times V_1^x \leq V = V_1 V_2.$$

Thus $|V_1| \geq |V_2|$ implies

$$|V_1| = |V_2| \quad \text{and} \quad V = V_1 \times V_2 = V_1 \times V_1^x.$$

In the case $|E_2 : K| > 2$ there exists $y \in E_2 \setminus K$ such that $V_1^x \neq V_1^y$. The same argument as above—this time applied to $(E_1^x, E_1^y, V_1^x, V_1^y)$—gives

$$V = V_1 \times V_1^y = V_1^x \times V_1^y$$

and $[V_1^y, E_1] = 1$. But this implies $[V, E_1] = 1$, a contradiction.

We have shown that $|E_2 : K| = 2$; in particular $K \trianglelefteq E_2$ and $V_1^{xK} = V_1^x$. It follows that

$$[V_1, K] \leq V_1 \cap V_2 = 1 \quad \text{and} \quad [V_1^x, K] \leq V_1^x \cap V_2 = 1.$$

But this shows that $[V, K] = 1$ and thus $K = 1$ and $|E_2| = 2$, which contradicts (4).   □

9.3.6   Let $A \in \mathcal{A}_V(H)_{\min}$ and set

$$E := [O_{p'}(H), A]A \quad \text{and} \quad F := C_H([V, E]).$$

(a) $E \cong SL_2(p)$ and $p \in \{2, 3\}$.

(b) $V = [V, E] \times C_V(E)$ and $|[V, E]| = p^2$.

(c) $H = E \times F$ and $\mathcal{A}_V(H)_{\min} = \mathcal{A}_V(E) \cup \mathcal{A}_V(F)_{\min}$.

(d) $[V, F] \leq C_V(E)$ and $\mathcal{A}_V(F) = \mathcal{A}_{C_V(E)}(F)$.

Proof. From 9.3.2 (c) we get

$$(') \qquad\qquad |A| = |[V, A]| = p.$$

Let $Q := [O_{p'}(H), A]$, and for $x \in Q \setminus C_Q(A)$ let E_x, Q_x, V_x be defined as in 9.3.3. We will use the properties of E_x given there without reference.

The coprime action of A on $O_{p'}(H)$ gives

$$O_{p'}(H) = C_{O_{p'}(H)}(A)Q,$$

so

$$E = \langle E_x \mid x \in Q \setminus C_Q(A) \rangle \text{ and } O_p(E) = C_A(O_{p'}(H)) = 1.$$

Pick $x, y \in Q \setminus C_Q(A)$. Then Q_x and Q_y are subnormal in $O_{p'}(H)$ (9.3.4). If $V_x \neq V_y$, then $E_1 := Q_x$ and $E_2 := Q_y$ satisfy the hypotheses of 9.3.5. Hence $V_x \cap V_y = 1$, which contradicts $[V, A] \leq V_x \cap V_y$.

We have shown that $V_x = V_y$ for all $x, y \in Q \setminus C_Q(A)$. In particular

$$[V, Q] = [V, E] = V_x \cong C_p \times C_p.$$

Now 8.4.2 on page 198 yields

$$\mathcal{Z}, \qquad\qquad V = [V, Q] \times C_V(Q).$$

and $(')$ implies $C_V(Q) = C_V(E)$. It follows that

$$E = \text{SL}(V_x) = E_x = E \text{ and } V = [V, E] \times C_V(E),$$

and (a) and (b) hold.

The decomposition $\mathcal{Z}$ is invariant under $O_{p'}(H)$ since Q is a normal subgroup of H. Let $B \in \mathcal{A}_V(H)_{\min}$ and

$$\widetilde{Q} = [O_{p'}(H), B].$$

As above for A and Q we get

$$|[V, \widetilde{Q}]| = p^2 \quad \text{and} \quad C_V(\widetilde{Q}) = C_V(B\widetilde{Q}).$$

Moreover, by 9.3.3 $\widetilde{Q}$ is irreducible on $[V, \widetilde{Q}]$. The invariance of the decomposition $\mathcal{Z}$ under $\widetilde{Q}$ gives

$$[V, Q] = [V, \widetilde{Q}] \quad \text{and} \quad C_V(Q) = C_V(\widetilde{Q})$$

or

$$[V, Q] \leq C_V(\widetilde{Q}) \quad \text{and} \quad [V, \widetilde{Q}] \leq C_V(Q).$$

In both cases the decomposition $\mathcal{Z}$ is invariant under B. Thus, by 9.3.2 (b) this decomposition is also invariant under H. As $C_H(C_V(Q))$ acts faithfully on $[V, Q]$, we also get that $C_H(C_V(Q)) = E$ and $H = E \times F$. Moreover, 9.3.2 (b) implies

$$\mathcal{A}_V(H)_{\min} \subseteq \mathcal{A}_V(E) \cup \mathcal{A}_V(F).$$

This is (c), and the H-invariance of $\mathcal{Z}$ gives (d). $\qquad\qquad\square$

9.3.7   Theorem (Glauberman [51]). *Let $E_1, \ldots, E_r$ be the different subgroups of the form $[O_{p'}(H), A]A$, $A \in \mathcal{A}_V(H)_{\min}$. Then the following hold:*

(a) $p \in \{2, 3\}$.

(b) $H = E_1 \times \cdots \times E_r$ *and* $V = C_V(H) \times [V, E_1] \times \cdots \times [V, E_r]$. *In particular, E_i acts faithfully on $[V, E_i]$ and trivially on $[V, E_j]$ for $j \neq i$.*

(c) $|[V, E_i]| = p^2$ *and* $E_i = \mathrm{SL}([V, E_i]) \cong \mathrm{SL}_2(p)$ *for* $i = 1, \ldots, r$.

(d) $A = \underset{i=1}{\overset{r}{\times}} (A \cap E_i)$ *and* $|A||C_V(A)| = |V|$ *for all* $A \in \mathcal{A}_V(H)$.

Proof. By 9.3.6 $H = E_1 \times H_1$ for $H_1 := C_H([V, E_1])$, and $(H_1, C_V(E_1))$ satisfies $\mathcal{S}_1 - \mathcal{S}_3$. Now (a)–(c) follow form 9.3.6 by an elementary induction. Claim (d) is 9.3.2 (b). $\qquad\qquad\square$

According to 9.3.1 we now can apply 9.3.7 to p-separable groups:

9.3.8   *Let G be a p-separable group with $O_{p'}(G) = 1$ that is not Thompson factorizable with respect to p, and let*

$$V = \langle \Omega(Z(S)) \mid S \in \mathrm{Syl}_p G \rangle.$$

Then the statements (a) – (d) of 9.3.7 hold for $H := J(G)C_G(V)/C_G(V)$. $\qquad\square$

We give two corollaries that will be needed in Chapter 12, resp. 11.

9.3.9 *Let G be a p-separable group and V an elementary Abelian normal p-subgroup of of G such that $O_p(G/C_G(V)) = 1$. Then for $C := C_G(V)$*

$$[\Omega(Z(J(C))), J(G)] \leq V.$$

Proof.[6]   Let $H = J(G)C/C$. If $H = 1$, then $J(C) = J(G)$ by 9.2.8 (b) and thus

$$[\Omega(Z(J(C))), J(G)] = 1.$$

Hence, we may assume that $H \neq 1$. The pair (V, H) satisfies condition $\mathcal{S}_2$ and by 6.4.3 on page 134 also $\mathcal{S}_1$. Moreover 9.2.9 (a) implies $\mathcal{S}_3$. We apply 9.3.2 (a). Then for $A \in \mathcal{A}(G)$

$$|A/C_A(V)| = |AC/C| = |V/C_V(A)|.$$

The maximality of A yields $C_V(A) = A \cap V$ and

$$|A| \geq |VC_A(V)| = |V/V \cap A||C_V(A)| = |A|.$$

It follows that

$$(') \qquad\qquad VC_A(V) \in \mathcal{A}(C) \subseteq \mathcal{A}(G).$$

Note that V is contained in $V_0 := \Omega(Z(J(C)))$, so by $(')$ $C_A(V) = C_A(V_0)$. Because of 9.2.9 (a) we are allowed to use $\mathcal{Q}'_1$ on page 237 for A and V_0:

$$|A/C_A(V)| = |V/C_V(A)| = |VC_{V_0}(A)/C_{V_0}(A)| \leq |V_0/C_{V_0}(A)|$$
$$\leq |A/C_A(V_0)| = |A/C_A(V)|.$$

This gives $V_0 = VC_{V_0}(A)$ and thus $[V_0, A] \leq V$ for $A \in \mathcal{A}(G)$. Now $[V_0, J(G)] \leq V$ follows. $\qquad\square$

9.3.10 *Let X be an elementary Abelian q-group (q a prime) that acts on the p-separable q'-group G. Suppose that $O_{p'}(G) = 1$.*

(a) $G = \langle N_G(J(S)), C_G(\Omega(Z(S))), C_G(X)\rangle$ *for $S \in \mathrm{Syl}_p G$.*

---

[6]The proof uses an argument of B. Baumann; see [26].

(b) If G is not Thompson factorizable with respect to p, then $p = 2$ or $p = 3$, and there exist subgroups W and D of $C_G(X)$ such that

$$W \cong C_p \times C_p, \quad W^D = W \quad \text{and} \quad D/C_D(W) \cong \mathrm{SL}_2(p).$$

Proof. If G is Thompson factorizable, then clearly (a) holds. Thus, we may assume that G is not Thompson factorizable. Let S be an X-invariant Sylow p-subgroup of G (see 8.2.3). As in 9.3.8 we set

$$V := \langle \Omega(Z(S)))^G \rangle, \quad C := C_G(V) \quad \text{and} \quad H := J(G)C/C.$$

Note that the semidirect product XG acts on V. According to 9.3.1 we can apply 9.3.7 to H. Let E_i and V_i, $i = 1, \ldots, r$, be defined as there. Then X acts on $\mathcal{A}(G)$ (by conjugation) and thus also on $\{E_1, \ldots, E_r\}$. We choose notation such that $\{E_1, \ldots, E_k\}$ is an orbit under X. Since $\pi(E_i) = \{2, 3\}$ and $(q, |G|) = 1$ we get $q \geq 5$. Hence $N_X(E_1)$ acts trivially on E_1, as $|E_1| = 6$ resp. $|E_1| = 24$.

Set

$$N := J(G)C, T := N \cap S \, (\in \mathrm{Syl}_p N) \text{ and } P := (TC/C) \cap E_1 \, (\in \mathrm{Syl}_p E_1).$$

Since $E_1 = \langle P^{E_1} \rangle$ we can apply 8.1.6 on page 178 and get

$$E_1 \times \cdots \times E_k \leq \langle C_H(X), P^X \rangle.$$

The corresponding statement holds for every other X-orbit of $\{E_1, \ldots, E_r\}$. It follows that

$$H = \langle C_H(X), TC/C \rangle \text{ and thus } N = \langle C_N(X), T \rangle C$$

since P^X is in TC/C. Now the Frattini argument implies

$$G = N_G(T)N \overset{9.2.8(\mathrm{b})}{=} N_G(J(S))N = \langle N_G(J(S)), C, C_G(X) \rangle.$$

This is (a).

For the proof of (b) we note that $N_X(V_i) = N_X(E_i)$ and investigate

$$\langle V_1{}^X \rangle = V_1 \times \cdots \times V_k, \quad \langle E_1{}^X \rangle = E_1 \times \cdots \times E_k.$$

Let $\mathcal{S}$ be a transversal of $N_X(E_1)$ in X. Then

$$W := \Big\{ \prod_{s \in \mathcal{S}} v^s \,\Big|\, v \in V_1 \Big\} \leq C_G(X) \quad (8.1.6 \text{ (a) on page 178}).$$

For the corresponding *diagonal* in $\langle E_1{}^X \rangle$ we get

$$\overline{D} := \left\{ \prod_{s \in S} e^s \mid e \in E_1 \right\} \leq C_H(X).$$

Here the action of $\overline{D}$ on W is equivalent to that of E_1 on V_1; thus $W \cong V_1$ and $\overline{D} = \mathrm{SL}(W)$. The coprime action of X on $\overline{D}$ gives a subgroup D of $C_N(X)$ such that $DC/C = \overline{D}$. This implies (b).   □

9.4   A Characteristic Subgroup

Let p be a prime, G a group such that $O_{p'}(G) = 1$, and $S \in \mathrm{Syl}_p\, G$. By definition G is Thompson factorizable with respect to p if

$$G = N_G(J(S))\, C_G(\Omega(Z(S))).$$

In this section we investigate the question under which additional hypotheses one can find a nontrivial characteristic subgroup of S that is normal in G.

The most important and best-known answer to this question is Glauberman's ZJ-Theorem [50]. It states that

$$G = N_G(Z(J(S)))$$

whenever G is a group such that $C_G(O_p(G)) \leq O_p(G)$ and the action of G on the chief factors of G in $O_p(G)$ is p-stable. Note here that the definition of $Z(J(S))$ only depends on S but not on G.

In this section we prove an analogue of Glauberman's ZJ-Theorem using a different approach. Instead of showing that a given characteristic subgroup of S has the desired property, we will *approximate* such a subgroup using suitable subgroups of $Z(J(S))$.

In the following let S be a p-group. By $\mathcal{C}_J(S)$ we denote the class of all pairs (τ, H) satisfying the following four conditions:[7]

$\mathcal{C}_1$ H is a group with $C_H(O_p(H)) \leq O_p(H)$, and τ is a monomorphism from S into H.

$\mathcal{C}_2$ S^τ is a Sylow p-subgroup of H.

---

[7]Thus $(\tau, H) \in \mathcal{C}_J(S)$ means that $\mathcal{C}_1$–$\mathcal{C}_4$ hold for (τ, H).

$\mathcal{C}_3$ $J(S^\tau)$ is a normal subgroup of H.

$\mathcal{C}_4$ H is p-stable on every normal subgroup of H that is contained in $\Omega(Z(J(S^\tau)))$.

It is evident that (id, S) is in $\mathcal{C}_J(S)$.

For a p-group P we set

$$A(P) := \Omega(Z(P)) \quad \text{and} \quad B(P) := \Omega(Z(J(P))).$$

Then $A(P^\eta) = A(P)^\eta$ and $B(P^\eta) = B(P)^\eta$ for every isomorphism η of P. We now define recursively a subgroup $W(S) \leq B(S)$. We start with

$$W_0 := A(S) \leq B(S).$$

Assume that for $i \geq 1$ the subgroups $W_0, W_1, \ldots, W_{i-1}$ with

$$A(S) = W_0 < W_1 < \cdots < W_{i-1} \leq B(S)$$

are already defined. If $W_{i-1}^\tau \trianglelefteq H$ for all $(\tau, H) \in \mathcal{C}_J(S)$, then we define $W(S) := W_{i-1}$. In the other case we choose $(\tau_i, H_i) \in \mathcal{C}_J(S)$ such that $W_{i-1}^{\tau_i}$ is not normal in H_i and define

$$W_i := \langle (W_{i-1}^{\tau_i})^{H_i} \rangle^{\tau_i^{-1}}.$$

Note that in this case

$$A(S^{\tau_i}) \leq W_{i-1}^{\tau_i} < W_i^{\tau_i} \leq B(S^{\tau_i}) \overset{\mathcal{C}_3}{\trianglelefteq} H_i$$

and thus

$$A(S) \leq W_{i-1} < W_i \leq B(S).$$

Since $B(S)$ is finite there exists an integer m where this recursive definition terminates, i.e., where we have

$$W(S) := W_m.$$

Then

$\mathcal{R}$ $A(S) = W_0 < \cdots < W_i < \cdots < W_m = W(S) \leq B(S)$

and

$(')$ $W(S)^\tau \trianglelefteq H$ for all $(\tau, H) \in \mathcal{C}_J(S)$.

At first sight this definition of $W(S)$ seems to depend on the choice of the pairs (τ_i, H_i). But if one defines in an analogous way

$$W_0 = \widetilde{W}_0 < \cdots < \widetilde{W}_{\widetilde{m}} =: \widetilde{W}(S)$$

for suitable pairs $(\widetilde{\tau}_i, \widetilde{H}_i)$, $i = 0, \ldots, \widetilde{m}$, then by ($'$) $\widetilde{W}(S) \leq W(S)$. The symmetric argument shows that also $W(S) \leq \widetilde{W}(S)$ and thus $W(S) = \widetilde{W}(S)$.

Let η be an isomorphism of S. Then

$$(\tau, H) \mapsto (\eta^{-1}\tau, H)$$

defines a bijection from $\mathcal{C}_J(S)$ to $\mathcal{C}_J(S^\eta)$, and the series $\mathcal{R}$ corresponds to

$$A(S^\eta) = A(S)^\eta = W_0{}^\eta < \cdots < W_m{}^\eta = W(S)^\eta \leq B(S)^\eta = B(S^\eta).$$

It follows:

9.4.1 *Let η be an isomorphism of S. Then $W(S^\eta) = W(S)^\eta$. In particular, $W(S)$ is a characteristic subgroup of S satisfying*

$$W(S) \neq 1 \iff S \neq 1. \qquad \square$$

The additional statement follows from the fact that $\Omega(Z(S)) \leq W(S)$ and $Z(S) \neq 1$ if $S \neq 1$.

9.4.2 *Let $x \in S$ such that $[W(S), x, x] = 1$. Then $[W(S), x] = 1$.*

Proof. For W_0 in $\mathcal{R}$ we have $[W_0, x] = 1$, for all $x \in S$. Assume now that S is a counterexample. Then there exists $i \in \{1, \ldots, m\}$ such that the implication

$$[W_i, x, x] = 1, \quad x \in S \quad \Rightarrow \quad [W_i, x] = 1$$

does not hold. We choose i minimal with that property. Then

$$(+) \qquad [W_{i-1}, x] \neq 1, \quad x \in S \quad \Rightarrow \quad [W_{i-1}, x, x] \neq 1.$$

Let $y \in S$ such that $[W_i, y, y] = 1$ but $[W_i, y] \neq 1$. For $a := y^{\tau_i}$ this yields

$$[W_i{}^{\tau_i}, a, a] = 1 \quad \text{and} \quad [W_i{}^{\tau_i}, a] \neq 1,$$

where (τ_i, H_i) is the pair used in the construction of W_i.

Now let $C := C_{H_i}(W_i^{\tau_i})$ and $C \leq L \leq H_i$ such that $L/C = O_p(H_i/C)$. Then C_4 implies $aC \in O_p(H_i/C)$, and $P := S^{\tau_i} \cap L$ is a Sylow p-subgroup of L, so $L = CP$. The Frattini argument gives $H_i = N_{H_i}(P)L = N_{H_i}(P)C$ and thus

$$W_i^{\tau_i} = \langle (W_{i-1}^{\tau_i})^{N_{H_i}(P)} \rangle.$$

Hence, there exists $h \in N_{H_i}(P)$ such that

$$[(W_{i-1}^{\tau_i})^h, a] \neq 1.$$

For $x := (a^{h^{-1}})^{\tau_i^{-1}}$ we get $[W_{i-1}, x] \neq 1$. This contradicts $(+)$ since

$$[W_{i-1}, x, x] = [(W_{i-1}^{\tau_i})^h, a, a]^{h^{-1}\tau_i^{-1}} \leq [W_i^{\tau_i}, a, a]^{h^{-1}\tau_i^{-1}} = 1. \qquad \square$$

For technical reasons the proof of the main theorem of this section requires us to investigate—besides $\mathcal{C}_J(S)$—the class $\mathcal{C}_0(S)$ of all pairs (τ, H) that satisfy:

$\mathcal{C}_0 1$ H is a group with $C_H(O_p(H)) \leq O_p(H)$, and $\tau \colon S \to H$ is a monomorphism.

$\mathcal{C}_0 2$ S^τ is a Sylow p-subgroup of H.

$\mathcal{C}_0 3$ $J(S^\tau)$ is not normal in H and $(\tau, N_H(J(S^\tau))) \in \mathcal{C}_J(S)$.

$\mathcal{C}_0 4$ H is p-stable on every elementary Abelian normal p-subgroup of H and on $O_p(H)/\Phi(O_p(H))$.

9.4.3 $W(S)^\tau$ *is normal in* H *for every* $(\tau, H) \in \mathcal{C}_0(S)$.

Proof. Let $(\tau, H) \in \mathcal{C}_0(S)$ and

$$W := W(S)^\tau.$$

Since $O_p(H)^{\tau^{-1}} \leq S$ and $W(S) \trianglelefteq S$ we get $[O_p(H), W] \leq W \cap O_p(H)$, i.e.,

(1) $[O_p(H), W, W] = 1.$

Because of $\mathcal{C}_0 4$ this implies for $V := O_p(H)/\Phi(O_p(H))$

$$W C_H(V)/C_H(V) \leq O_p(H/C_H(V))$$

and thus $W \leq O_p(H)$ by $\mathcal{C}_0 1$ and 8.2.9 (b) on page 188. Moreover, $W^{\tau^{-1}} \leq O_p(H)^{\tau^{-1}}$ and $W^{\tau^{-1}} \trianglelefteq S$ give $W \trianglelefteq O_p(H)$ and thus $W^h \trianglelefteq O_p(H)$ for all $h \in H$. It follows that

$$[W, W^h, W^h] = 1 \quad \text{and} \quad [W(S), (W^h)^{\tau^{-1}}, (W^h)^{\tau^{-1}}] = 1.$$

Now 9.4.2 yields

$$[W(S), (W^h)^{\tau^{-1}}] = 1 \quad \text{and} \quad [W, W^h] = 1.$$

Hence

$$W^* := \langle W^H \rangle$$

is elementary Abelian.

Assume first that $[W^*, J(S^\tau)] = 1$. Then also $[W^*, J(S^\tau)^h] = 1$ for $h \in H$, so $[W^*, J(H)] = 1$. Since $J(S^\tau) \leq J(H)$ there exists $T \in \mathrm{Syl}_p J(H)$ such that $J(S^\tau) = J(T)$. The Frattini argument implies

$$H = J(H)N_H(T) = J(H)N_H(J(S^\tau)) = C_H(W^*)N_H(J(S^\tau))$$

and $W^* = \langle W^{N_H(J(S^\tau))} \rangle$. By $\mathcal{C}_0 3$ $(\tau, N_H(J(S^\tau))) \in \mathcal{C}_J(S)$ and thus $W = W(S)^\tau \trianglelefteq N_H(J(S^\tau))$, so $W^* = W$ follows, and W is normal in H.

We now assume that $[W^*, J(S^\tau)] \neq 1$ and show that this leads to a contradiction. Let $C_H(W^*) \leq L \leq H$ such that

$$L/C_H(W^*) = O_p(H/C_H(W^*))$$

and $P := S^\tau \cap L$. The Frattini argument gives

$$H = LN_H(P) = C_H(W^*)N_H(P)$$

and thus

$$(') \qquad\qquad W^* = \langle W^{N_H(P)} \rangle.$$

Because of $\mathcal{C}_0 4$ and 9.2.10 there exists $A^* \in \mathcal{A}(S^\tau)$ such that $[W^*, A^*] \neq 1$ and $A^* \leq P$. This implies $A^* \leq J(P) \leq J(S^\tau)$, so $[W, J(P)] = 1$. With $(')$ we get

$$[W^*, A^*] \leq [W^*, J(P)] = 1,$$

which contradicts $[W^*, A^*] \neq 1$. $\qquad\square$

We say that a group G (with $S \in \mathrm{Syl}_p G$) is p-**stable** if the following two conditions hold:

- G is p-stable on every elementary Abelian normal p-subgroup of G and on $O_p(G)/\Phi(O_p(G))$.

- $N_G(J(S))$ is p-stable on every normal subgroup V of $N_G(J(S))$, which is contained in $\Omega(Z(J(S)))$.

9.4.4   Theorem [85]. *Let S be a p-group. Then there exists a characteristic subgroup $W(S)$ of S satisfying:*

(a) $\Omega(Z(S)) \leq W(S) \leq \Omega(Z(J(S)))$.

(b)   *If G is a p-stable group such that $C_G(O_p(G)) \leq O_p(G)$ and S is a Sylow p-subgroup of G, then $W(S)$ is a normal subgroup of G.*

(c) $W(S^\eta) = W(S)^\eta$ *for every isomorphism η of S..*

Proof. Let $W(S)$ be defined as above. Because of 9.4.1 we only have to prove (b). Let G be as in (b). Then (id, G) is in $\mathcal{C}_J(S)$, if $J(S)$ is normal in G. In this case the construction of $W(S)$ shows that it is normal in G. If $J(S)$ is not normal in G, then (id, G) is in $\mathcal{C}_0(S)$, and (b) follows from 9.4.3.   □

Here we want to emphasize again that the subgroup $W(S)$ only depends on S but not on the group G in (b). Thus, given a group Y with $S \in \mathrm{Syl}_p Y$, all p-stable subgroups $M \leq Y$ satisfying

$$S \leq M \quad \text{and} \quad C_M(O_p(M)) \leq O_p(M)$$

are contained in $N_Y(W(S))$.

Concerning the notion of p-stability we collect (see 8.6.12 on page 219):

9.4.5   *Let $p \neq 2$. A group G is p-stable, if G satisfies one of the following conditions:*

(1)   *G is p-separable and $p \geq 5$.*

(2)   *G is of odd order.*

(3)   *G has Abelian Sylow 2-subgroups.*   □

If one uses Dickson's Theorem quoted after 8.6.13 on page 221, then condition (3) in 9.4.5 can be substituted by

(3′) No section of G is isomorphic to $\mathrm{SL}_2(p)$.

As we have already noted before 9.1.1 on page 226 p-stability is only interesting for $p \neq 2$. But if one replaces p-stability by condition (3′) (for example in 9.4.4 (b)), one also gets nontrivial results for $p = 2$; see [53] and [88].[8]

We will use the following corollary later:

9.4.6   *Let G be a p-separable group, $p \geq 5$ and $S \in \mathrm{Syl}_p G$. Then $G = O_{p'}(G)N_G(W(S))$.*

Proof. Let $\overline{G} := G/O_{p'}(G)$. Then $\overline{S} \in \mathrm{Syl}_p \overline{G}$ and by 9.4.4 (c)

$$W(\overline{S}) = \overline{W(S)}.$$

From 9.4.4 (b) together with 9.4.5 (1) and 6.4.4 (a) on page 134 we get that $\overline{W(S)}$ is a normal subgroup of $\overline{G}$, i.e.,

$$O_{p'}(G)W(S) \trianglelefteq G.$$

Now the assertion follows with the Frattini argument.   □

We conclude this section with a theorem of Thompson. With some right the proof of this theorem can be regarded as the beginning of modern group theory. It already contains the nucleus of the ideas described in this chapter. We recommend that the reader read [91] and [92]. The original version of this theorem differs from the one given here since at that time the ZJ-Theorem (resp. 9.4.4) was not yet available.

9.4.7   Normal p-Complement Theorem of Thompson. *Let G be a group, p an odd prime, and $S \in \mathrm{Syl}_p G$. Then G has a normal p-complement provided $N_G(W(S))$ has such a complement.*

---

[8]This paper uses the same approach as in the proof of 9.4.4.

Proof. Note first that subgroups and factor groups of groups having a normal p-complement, also have one. Now let G be a minimal counterexample, and let $\mathcal{K}$ be the set of all subgroups of G that have a normal p-complement. The minimality of G implies

(1) $S \leq G_1 < G \quad \Rightarrow \quad G_1 \in \mathcal{K}$

Set $N := O_{p'}(G)$ and $\overline{G} := G/N$. Assume that $N \neq 1$. Then $\overline{S}$ is a Sylow p-subgroup of $\overline{G}$ isomorphic to S. In particular $W(\overline{S}) = \overline{W(S)}$ by 9.4.1. Now 3.2.8 on page 66 shows that

$$N_{\overline{G}}(W(\overline{S})) = \overline{N_G(W(S))}.$$

Hence, by the above remark $N_{\overline{G}}(W(\overline{S}))$ has a normal p-complement; so $\overline{G}$ satisfies the hypothesis. If $|\overline{G}| < |G|$, then by induction also $\overline{G}$ has a normal p-complement and thus also G, since N is a normal p'-subgroup of G. As G is a counterexample, we have

(2) $O_{p'}(G) = 1.$

By the Normal p-Complement Theorem of Frobenius (7.2.4 on page 170) the set $\mathcal{W}$ of nontrivial p-subgroups W such that $N_G(W) \notin \mathcal{K}$ is not empty. We choose $P \in \mathcal{W}$ such that $|N_G(P)|_p$ is maximal and show:

(3) $P \trianglelefteq G$, in particular $O_p(G) \neq 1.$

In a counterexample to (3) $G_1 := N_G(P) \neq G$. After conjugating P by a suitable element of G we may assume that $T := N_G(P) \cap S$ is a Sylow p-subgroup of G_1. Then $T \neq S$ by (1) and thus $T < N_S(T)$ (3.1.10 on page 61). For every characteristic subgroup U of T, in particular for $U = W(T)$, we get $T < N_S(T) \leq N_G(U)$. Now the maximal choice of $|N_G(P)|_p$ implies that $N_G(W(T)) \in \mathcal{K}$ and thus also $N_{G_1}(W(T)) \in \mathcal{K}$. Hence, by induction G_1 has a normal p-complement, since $|G_1| < |G|$ and G is a minimal counterexample. But then also the subgroup $P \leq G_1$ has such a complement, which contradicts $P \in \mathcal{W}$. This contradiction proves claim (3).

Let

$$\overline{G} := G/O_p(G)$$

and N be the inverse image of $N_{\overline{G}}(W(\overline{S}))$ in G. By (3) $|\overline{G}| < |G|$. On the other hand, $\overline{N} < \overline{G}$ since $O_p(\overline{G}) = 1$ and $W(\overline{S}) \neq 1$, so also $N < G$. Now (1) implies that N has a normal p-complement. As G is a minimal counterexample we conclude that $\overline{G}$ has a normal p-complement. This yields together with (2):

(4) $C_G(O_p(G)) \le O_p(G)$, and $\overline{G}$ has a normal p-complement $\overline{K}$. In particular G is p-separable,

If G is p-stable, then 9.4.4 implies $G = N_G(W(S))$, which contradicts $G \not\subseteq \mathcal{K}$. Thus, G is not p-stable, and $\overline{K}$ has non-Abelian Sylow 2-subgroups (9.4.5). Since $\overline{S}$ and $\overline{K}$ are coprime there exists an $\overline{S}$-invariant Sylow 2-subgroup $\overline{T}$ of $\overline{K}$ (8.2.3). But then also $Z(\overline{T})$ is $\overline{S}$-invariant.

Let U be the inverse image of $Z(\overline{T})\overline{S}$ in G. Then $U \ne G$ since T is non-Abelian, and (1) shows that U has a normal p-complement $U_0 \ne 1$. From

$$[U_0, O_p(G)] \le U_0 \cap O_p(G) = 1$$

we get $U_0 \le C_G(O_p(G)) \not\le O_p(G)$, which contradicts (4). This final contradiction shows that G is not a counterexample.   □

9.5   Fixed-Point-Free Action

As promised in Section 8.1 we show in this section—using 9.4.7—that a group admitting a fixed-point-free automorphism of prime order is nilpotent. We then take this as an opportunity to discuss fixed-point-free action in a more general context proving a *post-classification theorem*[9] that states that in general every group admitting a fixed-point-free automorphism is solvable.

It should be pointed out that this section is independent from the other sections of this chapter, if one takes 9.4.7 for granted.

9.5.1   Theorem (Thompson [90]). *Every group admitting a fixed-point-free automorphism of prime order is nilpotent.*

Proof. Let G be a group and α a fixed-point-free automorphism of prime order. Then G is a p'-group (8.1.4 on page 177). Now let G be a counterexample of minimal order. Then we obtain:

(1)   If $N < G$ such that $N^\alpha = N$, then N is nilpotent.

(2)   If N is a nontrivial proper α-invariant normal subgroup of G, then G/N is nilpotent and G is solvable (6.1.2 on page 122).

---

[9]That is, a theorem whose proof uses the classification of the finite simple groups.

For (2) note that $\langle \alpha \rangle$ acts fixed-point-freely on G/N (8.2.2 or 8.1.11 (c)).

We first treat the case that G is solvable. Then G contains a minimal α-invariant normal subgroup V that is an elementary Abelian q-group.[10] As by (2) G/V is nilpotent but G is not, we get $C_G(V) \neq G$ (5.1.2 on page 100). Again by (2)

$$\overline{G} := G/C_G(V)$$

is nilpotent, and there exists a prime r such that

$$\overline{G}_1 := O_r(\overline{G}) \neq 1.$$

Now 8.1.5 on page 178 implies $r \neq q$. Moreover $C_V(\overline{G}_1) = 1$ since $C_V(\overline{G}_1)$ is an α-invariant normal subgroup of G and $C_G(V) \neq G$. Every nontrivial power of α is also fixed-point-free on $\overline{G}_1$. Hence, the semidirect product $\langle \alpha \rangle \overline{G}_1$ is a Frobenius group with Frobenius complement $\langle \alpha \rangle$ (8.1.12 on page 182). But now 8.3.5 shows that $1 \neq C_V(\alpha)$, a contradiction.

We have shown that the minimal counterexample G is not solvable. According to (1) and (2) this implies that G does not contain any nontrivial proper α-invariant normal subgroup. Thus, by induction we have:

(3)   If $1 \neq U < G$ such that $U^\alpha = U$, then $N_G(U)$ is nilpotent.

Since G is not solvable there exists an odd prime divisor q of $|G|$ and an α-invariant Sylow q-subgroup Q of G (8.2.3). But then α leaves invariant every characteristic subgroup W of Q and thus also $N_G(W)$. According to (3) $N_G(W)$ is nilpotent provided $W \neq 1$, and nilpotent groups possess normal q-complements. Hence, the Normal p-Complement Theorem of Thompson 9.4.7 (here for q in place of p) shows that G has a normal q-complement. Since this complement is characteristic in G it is invariant under α. But this contradicts (3). □

From 8.1.12 on page 182 we obtain as a corollary (using Frobenius's Theorem 4.1.6):

9.5.2   *The Frobenius kernel of a Frobenius group is nilpotent.*   □

It is not too difficult to construct solvable groups that are not nilpotent but admit a fixed-point-free automorphism of composite order. The conjecture

---

[10]V is a minimal normal subgroup of the semidirect product $\langle \alpha \rangle G$.

that every group admitting a fixed-point-free automorphism is *solvable* could only be verified by means of the classification of the finite simple groups. In the following we discuss this result as a typical *post-classification theorem*.

Recall that 8.1.11 shows that the fixed-point-free action of an automorphism and coprime action have some basic properties in common.

Let $\mathcal{E}$ be the class of simple groups E satisfying:

There exists $p \in \pi(E)$ such that E has a cyclic Sylow p-subgroup.

Let $\mathcal{K}$ be the class of groups all of whose composition factors are in $\mathcal{E}$.

From the classification of the finite simple groups one can conclude that *every* simple group is in $\mathcal{E}$, so $\mathcal{K}$ is the class of all (finite) groups. With this in mind the following theorem proves the above mentioned conjecture.

9.5.3   Let $G \in \mathcal{K}$ and A be a group that acts fixed-pont-freely on G, i.e., $C_G(A) = 1$. Suppose that the action of A on G is coprime if A is noncyclic. Then G is solvable.[11]

Proof. Let G be a minimal counterexample. If G contains an A-invariant normal subgroup N such that $1 \neq N \neq G$, then 8.1.11 on page 181 resp. 8.2.2 allows to apply induction to N and G/N. Thus N and G/N are solvable, and G is not a counterexample (see 6.1.2).

Thus, G is a nonsolvable minimal normal subgroup of the semidirect product AG. By 1.7.3 on page 38 there exists a nonsolvable simple subgroup E of G such that

$$G = E_1 \times \cdots \times E_n \quad \text{and} \quad E^A = \{E_1, \ldots, E_n\}.$$

The fixed-point-free action of A on G implies a fixed-point-free action of $N_A(E_1)$ on E_1 (8.1.6 on page 178). If $E_1 \neq G$, then by induction E_1 is solvable, a contradiction. Hence, $G = E_1$ is a simple group from $\mathcal{E}$. Let $p \in \pi(G)$ such that $P \in \mathrm{Syl}_p G$ is cyclic. Because of 8.1.11 (b) resp. 8.2.3 we may assume that $P^A = P$. Then A acts on $N_G(P)/C_G(P)$, and this action is trivial since the automorphism group of a cyclic group is Abelian. On the other hand, A acts fixed-point-freely on $N_G(P)/C_G(P)$ (8.1.11 (c)), and 8.2.3 on page 185) yields $N_G(P) = C_G(P)$. Now the Theorem of Burnside (7.2.1 on page 169) shows that G has a normal p-complement. This contradicts the simplicity of G. $\square$

---

[11]The hypothesis $(|G|, |A|) = 1$ in the non-cyclic case is essential since every group G with $Z(G) = 1$ acts fixed-point-freely on itself.

Chapter 10

The Embedding of p-Local Subgroups

Let p be a prime and G a group. A subgroup $M \leq G$ is a **p-local subgroup** of G if there exists a nontrivial p-subgroup $P \leq G$ such that $N_G(P) = M$. Clearly then

$$1 \neq P \leq O_p(M).$$

We have seen frequently—for example, in Grün's Theorem and in the Normal p-Complement Theorems of Frobenius and Thompson—that the structure of p-local subgroups is strongly related to the structure of G. This connection will be the main theme for last three chapters of this book.

In the first section of this chapter we investigate p-local subgroups by means of quadratic action which was introduced in the last chapter. In the second section we use the proof of the $p^a q^b$-Theorem of Burnside to demonstrate how these results can be applied. In the last section we introduce a method, the *amalgam method*, that allows us to investigate groups by means of suitable *coset graphs*.

A group M has **characteristic p** if

$$C_M(O_p(M)) \leq O_p(M).$$

According to 6.5.8 on page 144 this property is equivalent to

$$F^*(M) = O_p(M),$$

and for p-separable M to

$$O_{p'}(M) = 1;$$

see 6.4.3 on page 134. Let M be a proper subgroup of G and $p \in \pi(M)$.
Then M is called **strongly p-embedded** in G if

$$|M \cap M^g|_p = 1 \quad \text{for all } g \in G \setminus M.^1$$

We will derive statements about p-local subgroups of characteristic p of G
(in particular for $p = 2$) provided they are not strongly p-embedded in G.
Groups with a strongly 2-embedded subgroup have been classified by Bender
[29]. His result belongs to the fundamental theorems in group theory; see
the Appendix.

10.1   Primitive Pairs

In Section 6.6 we have called a proper subgroup M of G primitive if $M = N_G(A)$ for every nontrivial normal subgroup A of M. Suppose now that
M_1, M_2 are two primitive subgroups of G. Then for $\{i, j\} = \{1, 2\}$

$$\mathcal{P} \quad 1 \ne A \trianglelefteq M_i, \quad A \le M_1 \cap M_2 \quad \Rightarrow \quad N_{M_j}(A) = M_1 \cap M_2,$$

and this elementary property gives rise to the following generalization:

Let M_1, M_2 be two different—not necessarily primitive—subgroups of G.
Then (M_1, M_2) is a **primitive pair** of G if $\mathcal{P}$ holds for $\{i, j\} = \{1, 2\}$.

Let (M_1, M_2) be a primitive pair. Then (M_1, M_2) is **solvable** if M_1 and
M_2 are solvable; and (M_1, M_2) has **characteristic p** if M_1 and M_2 have
characteristic p and, in addition,

$$O_p(M_1) O_p(M_2) \le M_1 \cap M_2.$$

Note first:

10.1.1   *Let M be a group of characteristic p. Suppose that U is a subgroup
of M such that $U \trianglelefteq\trianglelefteq M$ or $O_p(M) \le U$. Then U has characteristic p.*

Proof. The case $O_p(M) \le U$ is obvious. In the other case the assertion
follows from 6.5.7 (b) on page 144. □

The next statements show how to get primitive pairs of characteristic p.

---

[1]For $n \in \mathbb{N}$ we denote by n_p the largest p-power dividing n.

10.1.2   *Let M_1 and M_2 be two different maximal p-local subgroups of G that both have characteristic p. Suppose that M_1 and M_2 have a common Sylow p-subgroup. Then (M_1, M_2) is a primitive pair of characteristic p.*

Proof. Let $1 \neq A \trianglelefteq M_i$ and $A \leq M_i \cap M_j$, $i \neq j$. By 10.1.1 $1 \neq O_p(A)$ ($\trianglelefteq M_i$), and the maximality of M_i gives

$$M_i = N_G(O_p(A)).$$

Hence $N_{M_j}(A) = M_i \cap M_j$, and this is $\mathcal{P}$.

Let S be a common Sylow p-subgroup of M_1 and M_2. Then

$$O_p(M_1)O_p(M_2) \leq S \leq M_1 \cap M_2. \qquad \square$$

10.1.3   *Let $p \in \pi(G)$. Suppose that every p-local subgroup of G has characteristic p and $O_p(G) = 1$. Then one of the following holds:*

(a)   *There exists a primitive pair of characteristic p in G.*

(b)   *Every maximal p-local subgroup of G is strongly p-embedded in G.*

Proof. Let M be a maximal p-local subgroup of G. Then $O_p(M) \neq 1$ and thus

$$N_G(M) \leq N_G(O_p(M)) = M < G,$$

in particular $M^g \neq M$ for all $g \in G \setminus M$. Moreover, also M^g is a maximal p-local subgroup.

Among all maximal p-local subgroups $L \leq G$ that are different from M we choose L such that $|M \cap L|_p$ is maximal.

We first treat the case $|M \cap L|_p \neq 1$. Let

$$T \in \mathrm{Syl}_p(M \cap L) \text{ and } U := N_G(T).$$

Since U is p-local there exists a maximal p-local subgroup H of G such that $U \leq H$. Obviously, either $H \neq L$ or $H \neq M$. We may assume that $H \neq M$ (the case $H = M$ follows with a symmetric argument, replacing L by M). Let $T \leq S \in \mathrm{Syl}_p M$. If $T < S$ we get from 3.1.10 on page 61

$$T < N_S(T) \leq H \cap M,$$

which contradicts the maximality of $|M \cap L|_p$. Thus, we have

$$(') \hspace{4cm} T \in \mathrm{Syl}_p\, M.$$

If also $T \in \mathrm{Syl}_p\, L$, then (a) follows from 10.1.2. Hence, we may assume that

$$T < S_1 \in \mathrm{Syl}_p\, L.$$

Then there exists $g \in S_1 \setminus T \subseteq G \setminus M$ such that $T^g = T$ and $M \neq M^g$ (3.1.10 on page 61). Now, again by 10.1.2, (M, M^g) is a primitive pair of characteristic p.

We have shown that (a) holds if $|M \cap L|_p \neq 1$. Hence, we may assume now that $|M \cap L|_p = 1$, whenever M and L are two different maximal p-local subgroups of G. But then $|M \cap M^g|_p = 1$ for every $g \in G \setminus M$, and M is strongly p-embedded in G. $\qquad\square$

A variation of the following theorem is called the theorem of Thompson-Wielandt.

10.1.4   Theorem (Bender [28]). *Let (M_1, M_2) be a primitive pair of G. Suppose that $F^*(M_1) \leq M_2$ and $F^*(M_2) \leq M_1$. Then there exists a prime p such that (M_1, M_2) has characteristic p.*

Proof. By our hypothesis

$$F^*(M_1)\, F^*(M_2) \trianglelefteq M_1 \cap M_2$$

and by 6.5.7 (b) on page 144

$$(') \hspace{3cm} F^*(M_1)\, F^*(M_2) \leq F^*(M_1 \cap M_2).$$

Hence, a component K of M_1 is also a component of $M_1 \cap M_2$ and normalizes $F^*(M_2)$. If $[F^*(M_2), K] = 1$, then $K \leq Z(F(M_2))$ (6.5.8), which contradicts $K' = K$. Thus, 6.5.2 on page 142 implies that $K \leq F^*(M_2) \leq M_1$, so

$$K \trianglelefteq\trianglelefteq F^*(M_2) \trianglelefteq M_2.$$

In particular, K is also a component of M_2. It follows that $E(M_1) \leq E(M_2)$ and with a symmetric argument $E(M_2) \leq E(M_1)$, i.e., $E(M_1) = E(M_2)$. The primitivity of (M_1, M_2) shows that $E(M_1) = E(M_2) = 1$ and thus

$$F^*(M_i) = F(M_i), \quad i = 1, 2.$$

In particular, by (') $F^*(M_1)F^*(M_2)$ is nilpotent. Let $p \in \pi(F(M_1))$. Since $O_p(M_1)$ centralizes every normal p'-subgroup of $M_1 \cap M_2$ we get from 6.1.4 on page 123

$$\pi(F(M_1)) = \pi(F(M_2)).$$

Assume now that (M_1, M_2) is a counterexample. Then there exists

$$q \in \pi(F(M_1)), \quad q \neq p.$$

Set

$$Y_1 := [M_1, O_p(M_2), O_p(M_2)] \quad \text{and} \quad Y_2 := [M_2, O_p(M_1), O_p(M_1)].$$

First we show that the case

(") $$Y_1 Y_2 \leq M_1 \cap M_2,$$

leads to a contradiction. In this case $O_p(M_1)$ is normalized by Y_2. Thus, we obtain a subnormal series

$$O_p(M_1) \trianglelefteq O_p(M_1)Y_2 \overset{1.5.5}{\trianglelefteq} O_p(M_1)[M_2, O_p(M_1)] \overset{1.5.5}{\trianglelefteq} M_2,$$

which shows that $O_p(M_1) \leq O_p(M_2)$. A symmetric argument, with $O_p(M_2)$ and Y_1 in place of $O_p(M_1)$ and Y_2, also gives $O_p(M_2) \leq O_p(M_1)$; so $O_p(M_1) = O_p(M_2)$. This contradicts the primitivity of (M_1, M_2).

Hence, it suffices to establish (") to show that G is not a counterexample. Note that

$$O_p(M_1) \leq C_{M_2}(O_{p'}(F(M_2))) \quad \text{and} \quad O_q(M_1) \leq C_{M_2}(O_{q'}(F(M_2))).$$

This implies that $X := [M_2, O_p(M_1)] \leq C_{M_2}(O_{p'}(F(M_2)))$ and

$$[X, O_q(M_1)] \leq C_{M_2}(O_{p'}(F(M_2))) \cap C_{M_2}(O_{q'}(F(M_2))) = Z(F(M_2))$$
$$\leq M_1 \cap M_2 \leq M_1.$$

It follows that

$$[X, O_q(M_1), O_p(M_1)] \leq O_p(M_1) \cap F(M_2) \leq O_p(M_2).$$

Thus $[O_p(M_1), O_q(M_1), X] = 1$, and the Three-Subgroups Lemma gives

$$[Y_2, O_q(M_1)] \leq [X, O_p(M_1), O_q(M_1)] \leq O_p(M_2).$$

This shows

$$Y_2 \leq N_{M_2}(O_q(M_1)O_p(M_2)) = N_{M_2}(O_q(M_1) \times O_p(M_2)).$$

Hence $Y_2 \leq N_{M_2}(O_q(M_1)) = M_1 \cap M_2$, and with a symmetric argument $Y_1 \leq M_1 \cap M_2$. This is ($''$). $\qquad\square$

In the following we investigate a primitive pair (M_1, M_2) of characteristic p of G. We set

$$B := O_p(M_1)\, O_p(M_2) \quad (\trianglelefteq M_1 \cap M_2).$$

For $i = 1, 2$ let S_i be a Sylow p-subgroup of M_i, which contains B. We further set

$$Z_i := \Omega(Z(S_i)), \quad V_i := \langle Z_i^{M_i}\rangle, \quad W_i := \langle V_j^{M_i}\rangle.$$

Note that $V_i \leq \Omega(Z(O_p(M_i)))$ and

$$(+) \qquad\qquad\qquad O_p(M_i/C_{M_i}(V_i)) = 1$$

(9.2.7 on page 235), and recall that M_i is not p-stable on V_i if any nontrivial subgroup of $M_i/C_{M_i}(V_i)$ acts quadratically on V_i. The investigation of (M_1, M_2) can be subdivided in three cases:

(I) $V_i \not\leq O_p(M_j)$ for some $i \in \{1, 2\}$, and $j \neq i$.

(II) $V_1 V_2 \leq O_p(M_1) \cap O_p(M_2)$, and W_i is non-Abelian for some $i \in \{1, 2\}$.

(III) W_1 and W_2 are Abelian.

10.1.5   Let (M_1, M_2) be a primitive pair of characteristic p of G. Then there exists $i \in \{1, 2\}$ such that one of the following holds:

(a)   The action of M_i on V_i or on $O_p(M_i)/\Phi(O_p(M_i))$ is not p-stable.

(b) W_i is elementary Abelian, and the action of M_i on W_i is not p-stable.

Proof. We treat the three cases (I), (II), and (III) separately.

Case (I): We choose the notation such that $V_1 \not\leq O_p(M_2)$. Since V_1 is normal in B we get

$$[O_p(M_2), V_1, V_1] \leq [V_1, V_1] = 1.$$

Hence, V_1 acts quadratically on the elementary Abelian p-group

$$W := O_p(M_2)/\Phi(O_p(M_2)).$$

Moreover $C_{M_2}(W) = O_p(M_2)$ by 8.2.9 on page 188 and thus

$$O_p(M_2/C_{M_2}(W)) = 1.$$

Since $V_1 \nleq O_p(M_2)$ this shows that the action of M_2 on W is not p-stable.

Case (II): We choose the notation such that W_2 is non-Abelian. Then there exists $x \in M_2$ such that

$$[V_1, V_1^x] \neq 1 \quad \text{and} \quad V_1^x \leq M_1.$$

The second property holds since $V_1^x \leq O_p(M_2) \leq M_1$. As V_1 and thus also V_1^x is normal in $O_p(M_2)$, we get

$$[V_1, V_1^x, V_1^x] \leq [V_1^x, V_1^x] = 1.$$

Hence, V_1^x acts nontrivially and quadratic on V_1. Now (+) shows that the action of M_1 on V_1 is not p-stable.

Case (III): In this case the Thompson subgroup $J(B)$ enters the stage. If $J(B) \leq O_p(M_1) \cap O_p(M_2)$, then

$$J(B) = J(O_p(M_i)) \trianglelefteq M_i, \ i = 1, 2,$$

which contradicts the primitivity of (M_1, M_2). We choose the notation such that

$$J(B) \nleq O_p(M_2).$$

Let

$$C_{M_2}(W_2) \leq D \trianglelefteq M_2 \quad \text{and} \quad D/C_{M_2}(W_2) = O_p(M_2/C_{M_2}(W_2)).$$

The primitivity of (M_1, M_2) shows that $C_{M_2}(W_2) \leq C_{M_2}(V_1) \leq M_1$ and thus $J(B) \trianglelefteq BC_{M_2}(W_2)$. It follows that

$$J(B) \cap C_{M_2}(W_2) \leq O_p(C_{M_2}(W_2)) \leq O_p(M_2),$$

so $[W_2, J(B)] \neq 1$. By 9.2.10 on page 237 there exists $A \in \mathcal{A}(B)$ such that

$$(') \qquad\qquad [W_2, A] \neq 1 = [W_2, A, A].$$

We assume now that the action of M_2 on W_2 is p-stable. Then $A \leq B \cap D$ and

$$B \cap D \trianglelefteq (B \cap D)C_{M_2}(W_2) \trianglelefteq\trianglelefteq D \trianglelefteq M_2,$$

so $A \leq B \cap D \leq O_p(M_2)$. Since $[W_2, A] \neq 1$ there exists $x \in M_2$ such that $[V_1^x, A] \neq 1$. Now

$$A \leq O_p(M_2) \leq M_1^x,$$

and $(')$ implies that A acts nontrivially and quadratically on V_1^x. Hence $(+)$ shows that the action of M_1^x on V_1^x and thus also the action of M_1 on V_1 is not p-stable.   □

10.1.6   Theorem. *Let (M_1, M_2) be a primitive pair of characteristic p of G. Then M_1 or M_2 has non-Abelian Sylow 2-subgroups. In particular, no group of odd order possesses a primitive pair of characteristic p.*

Proof. For $p \neq 2$ this follows from 10.1.5 and 9.4.5 on page 254. Let $p = 2$, and assume that the Sylow 2-subgroups of M_1 and M_2 are Abelian. Then $O_2(M_1) = O_2(M_2)$ and (M_1, M_2) is not primitive.   □

It should be pointed out that 10.1.6 is essential in the proof of the $p^a q^b$-Theorem in the next section, but none of the results coming now is used there.

In the case $p = 2$ every involution acts quadratically (see Section 9.1), so 10.1.5 does not give any information about the structure of M_1 and M_2. In this case one has to consider the "quality" of the quadratic action to get further information.

For the investigation of primitive pairs of characteristic 2 we need four lemmata that we will prove first.

10.1.7   *Let M be a p-separable group and A a p-subgroup of M satisfying*

$$\Phi(A) \leq O_p(M) \quad \text{and} \quad A \nleq O_p(M).$$

Then there exists $x \in O_{pp'}(M)$ such that for $L := \langle A, A^x \rangle$:

(a) $x \in O^p(L) \leq O_{pp'}(M).$

(b) $[O^p(L), A] = O^p(L)$.

(c) $|A/A \cap O_p(L)| = p$ and $[A \cap O_p(L), L] \leq O_p(M)$.

Proof. According to 6.4.11 on page 140 there exists a subgroup L with property (a) that is not a p-group. We choose L minimal among all such subgroups. Then (b) follows. Set

$$\overline{L} := L/O_p(L) \quad \text{and} \quad \overline{Q} := O_{p'}(\overline{L}),$$

so $\overline{L} = \overline{A}\overline{Q}$. Moreover, $\overline{A}$ is an elementary Abelian p-group since $\Phi(A) \leq O_p(M) \cap L \leq O_p(L)$. Let $\mathcal{B}$ be the set of maximal subgroups of A. By 8.3.4 on page 193

$$\overline{Q} = \langle C_{\overline{Q}}(\overline{U}) \mid U \in \mathcal{B} \rangle,$$

so $[C_{\overline{Q}}(\overline{U}), A] \neq 1$ for some $U \in \mathcal{B}$ since A acts nontrivially on $\overline{Q}$. The minimal choice of L gives $C_{\overline{Q}}(\overline{U}) = \overline{Q}$. This implies $U = A \cap O_p(L)$ and

$$[U, O^p(L)] \leq O_p(L) \cap O_{pp'}(M) \leq O_p(M),$$

and (c) follows.   □

10.1.8   *Let M be a group of characteristic 2 that possesses a section isomorphic S_3. Then M also possesses a section isomorphic S_4.*

Proof. Let M be a minimal counterexample. Since $O_2(S_3) = 1$ also $M/O_2(M)$ has a section isomorphic S_3. Let

$$O_2(M) \leq N \trianglelefteq X \leq M \quad \text{such that} \quad X/N \cong S_3.$$

The minimal choice of M gives $X = M$ (10.1.1). Let

$$\overline{M} := M/N \quad (\cong S_3)$$

and $D \in \mathrm{Syl}_3 M$. Then $\overline{D}$ ($\cong C_3$) is a normal subgroup of $\overline{M}$ that is inverted by every involution in $\overline{M}$. The Frattini argument yields $M = N_M(D)N$. Hence, there exist 2-elements that act nontrivially on the 3-group D. Let $t \in N_M(D)$ be such a 2-element. In addition, we choose the order of t minimal with that property. Then t acts as an involution on D. Thus, 8.1.8 on page 180 gives an element $d \in D$ such that

$$o(d) = 3 \text{ and } d^t = d^{-1},$$

so $\langle d, t \rangle / \langle t^2 \rangle \cong S_3$. The minimality of M shows that

$$M = O_2(M)\langle d, t \rangle, \quad t^2 \in O_2(M),$$

and together with 8.2.9 on page 188 and 8.4.2 on page 198

$$\Phi(O_2(M)) = 1 \text{ and } C_{O_2(M)}(d) = 1.$$

This implies that $t^2 = 1$. By 8.1.4 on page 177 there exists $1 \neq z \in C_{O_2(M)}(t)$. Set $V := \langle z, z^d, z^{d^2} \rangle$. Then $|V| \leq 8$, and V is normal in M. The case $|V| = 8$ contradicts $C_V(d) = 1$. Hence $V \cong C_2 \times C_2$ and $V\langle d, t \rangle \cong S_4$, so M is not a counterexample. $\square$

10.1.9   *Let M be a group that acts faithfully on the elementary Abelian 2-group V, and let A be an elementary Abelian 2-subgroup of M. Suppose that $C_M(O_{2'}(M)) \leq O_{2'}(M)$ and*

$$(*) \qquad\qquad |V/C_V(A)| < |A|^2.$$

Then M possesses a section isomorphic to S_3.

Proof. Among all elementary Abelian 2-subgroups that satisfy $(*)$, we choose A of minimal order.

Assume first that $|A| = 2$. Then $(*)$ implies that $A \in \mathcal{A}_V(M)$, and the conclusion follows from 9.3.7 on page 246.

Assume now that $|A| > 2$. The hypothesis $C_M(O_{2'}(M)) \leq O_{2'}(M)$ shows that A acts nontrivially on $O_{2'}(M)$. Let $Q \leq O_{2'}(M)$ be minimal such that $Q^A = Q$ and $[Q, A] \neq 1$. It follows from 8.5.2 on page 205 that

$$A_0 := C_A(Q)$$

is a maximal subgroup of A, and QA/A_0 acts faithfully on $C_V(A_0)$. The already treated case $|A| = 2$ —applied to the pair $(C_V(A_0), QA/A_0)$—gives the conclusion if

$$|C_V(A_0)/C_V(A)| < |A/A_0|^2 = 4.$$

This condition follows from the minimality of A since

$$|V/C_V(A)| < |A|^2 = 4|A_0|^2 \leq 4|V/C_V(A_0)|.$$ □

For the next lemma we need some additional notation. Let X be a group that acts on the elementary Abelian p-group Z.

$\mathcal{Q}(Z, X) := \{A \leq X \mid [Z, A, A] = 1 \neq [Z, A]\}$,

$q(Z, X) := 0$ if $\mathcal{Q}(Z, X) = \varnothing$, and otherwise

$q(Z, X) := \min\{e \in \mathbb{R} \mid |A/C_A(Z)|^e = |Z/C_Z(A)|, \ A \in \mathcal{Q}(Z, X)\}$.

10.1.10   *Let M be a group and V an elementary Abelian normal p-subgroup of M, and let $Z \leq V$ such that*

$$V = \langle Z^M \rangle \quad \text{and} \quad Z \trianglelefteq O_p(M).$$

Suppose that there exists $A \leq O_p(M)$ such that $[V, A, A] = 1$. Then

$$|A/C_A(V)|^q \leq |V/C_V(A)|,$$

where $q := q(Z, O_p(M))$.

Proof. Let $Z^M = \{Z_1, \ldots, Z_k\}$. Then Z_i, $i = 1, \ldots, k$, is normal in $O_p(M)$. We define the series

$$A := A_0 \geq \cdots \geq A_{i-1} \geq A_i \geq \cdots \geq A_k$$

putting

$$A_i := C_{A_{i-1}}(Z_i) \text{ for } i = 1, \ldots, k.$$

Then

$$A_k = C_A(V),$$

and the quadratic action of A on V gives

$$[Z_i, A_{i-1}, A_{i-1}] = 1 \text{ for } i = 1, \ldots, k.$$

If $[Z_i, A_{i-1}] = 1$, then $A_{i-1} = A_i$; and if $[Z_i, A_{i-1}] \neq 1$, then the definition of q implies

$$|A_{i-1}/A_i|^q \leq |Z_i/C_{Z_i}(A_{i-1})| = |Z_i C_V(A_{i-1})/C_V(A_{i-1})|$$
$$\leq |C_V(A_i)/C_V(A_{i-1})|.$$

It follows that

$$|A/C_A(V)|^q = \prod_{i=1}^{k} |A_{i-1}/A_i|^q \le \prod_{i=1}^{k} |C_V(A_i)/C_V(A_{i-1})| = |V/C_V(A)|. \quad \square$$

After these preparations we are now able to investigate solvable primitive pairs of characteristic 2. We show:

10.1.11   Theorem. *Let (M_1, M_2) be a solvable primitive pair of characteristic 2 of G. Then M_1 or M_2 possesses a section isomorphic to S_4.*

Proof. As in the proof of 10.1.5 we treat the cases (I), (II), and (III) given on page 266 separately. The notation is chosen as there. Because of 10.1.8, 10.1.9 and (+) on page 266 we may assume that for $i \in \{1, 2\}$

$(\times)$ $|A/C_A(V_i)|^2 \le |V_i/C_{V_i}(A)|$ for all $A \le B$ with $\Phi(A) \le C_B(V_i)$.

Case (I): Without loss we may assume that $V_1 \not\le O_2(M_2)$. We apply 10.1.7 with V_1 and M_2 in place of A and M. Then there exists a subgroup $L \le M$ and $x \in O_{22'}(M_2) \cap L$ such that

$$L = \langle V_1, V_1^x \rangle, \quad [V_1 \cap O_p(L), L] \le O_p(M_2),$$

and

(1) $|V_1 : V_1 \cap O_2(L)| = 2.$

Set

$$W := (V_1 \cap O_2(L))(V_1^x \cap O_2(L))$$

and

$$W_0 := V_1 \cap V_1^x \quad (\le W).$$

Clearly $W_0 \le Z(L)$, and thus

(2) $W_0 = Z(L) \cap W = C_{V_1^x \cap O_2(L)}(V_1) = C_{V_1 \cap O_2(L)}(V_1^x)$

since $x \in L$. Moreover

$$[W, V_1] \le V_1 \cap O_2(L) \le W,$$

as V_1 is normal in $O_2(M_2)V_1$. Similarly $[W, V_1^x] \leq W$, and thus

$$W \trianglelefteq L.$$

Now $[O_2(M_2), V_1] \leq V_1 \cap O_2(L) \leq W$ and $[O_2(M_2), V_1^x] \leq W$ imply $[O_2(M_2), L] \leq W$. The nontrivial action of $O^2(L)$ on $O_2(M_2)$ gives a nontrivial action of L on W and thus also on W/W_0 (8.2.2 on page 184). We now investigate the action of

$$A := V_1^x \cap O_2(L)$$

on V_1. Since

$$|A/C_A(V_1)| \overset{(2)}{=} |A/W_0| \overset{(1)}{=} \tfrac{1}{2}|V_1/W_0| \overset{(2)}{\geq} \tfrac{1}{2}|V_1/C_{V_1}(A)|$$

we get

$$|A/C_A(V_1)|^2 > |V_1/C_{V_1}(A)| \quad \text{or} \quad |A/W_0| = 2.$$

The first case contradicts ($\times$). In the second case $|W/W_0| = 4$. Since $O^2(L)$ acts nontrivially on W/W_0 we conclude that

$$L/C_L(W/W_0) \cong \mathrm{SL}_2(2) \cong S_3.$$

Now 10.1.8 shows that M_2 has a section isomorphic to S_4.

Case (II): We use a similar argument as in the proof of 10.1.5. As there we may assume that W_2 is non-Abelian. Then there exists $x \in M_2$ such that

$$[V_1, V_1^x] \neq 1 \quad \text{and} \quad V_1^x \leq O_2(M_2) \leq M_1 \cap M_1^x.$$

The symmetry between (V_1, M_1) and (V_1^x, M_1^x) allows to assume—possibly after interchanging the notation—that

$$|V_1/C_{V_1}(V_1^x)| \leq |V_1^x/C_{V_1^x}(V_1)|.$$

But this contradicts ($\times$) for $A = V_1^x$ and $V_i = V_1$.

Case (III): As in the proof of 10.1.5 we use the Thompson subgroup $J(B)$, and as there we may assume that $J(B) \not\leq O_2(M_2)$. Then there exists $A \in \mathcal{A}(B)$ such that

$$[W_2, A] \neq 1 = [W_2, A, A].$$

By 9.2.9 on page 236 $A \in \mathcal{A}_{V_1}(M_1)$ and

$$|A/C_A(V_1)| \geq |V_1/C_{V_1}(A)|$$

(this is $\mathcal{Q}_1'$ in 9.2 on page 237). Now $(\times)$ implies

(3) $$[V_1, A] = 1.$$

Assume that $A \le O_2(M_2)$. Since $[W_2, A] \ne 1$ there exists $x \in M_2$ such that $[V_1^x, A] \ne 1$. Note that

$$A \le O_2(M_2) = O_2(M_2)^x \le M_1^x,$$

and thus $[V_1, A^{x^{-1}}] \ne 1$ and $A^{x^{-1}} \le B$. Now $A^{x^{-1}} \in \mathcal{A}_{V_1}(B)$ and the definition of $\mathcal{A}_{V_1}(B)$ show that

$$|A^{x^{-1}}/C_{A^{x-1}}(V_1)| \ge |V_1/C_{V_1}(A^{x^{-1}})|,$$

which contradicts $(\times)$.

Assume now that $A \not\le O_2(M_2)$, and let L be as in 10.1.7 (with respect to A and $M = M_2$). Then

$$A_0 := A \cap O_2(L)$$

is a maximal subgroup of A. Set

$$Q := O^2(L), \quad U := \langle V_1^L \rangle, \quad \text{and} \quad \overline{U} := U/C_U(Q).$$

If $\overline{U} = 1$, then $Q \le C_G(V_1) \cap M_2 \le M_1 \cap M_2$; and thus

$$Q = [Q, A] \le O_2(M_1 \cap M_2),$$

since $A \le B \le O_2(M_1 \cap M_2)$. But Q is not a 2-group, a contradiction. Hence, we have $\overline{U} \ne 1$.

The subgroup V_1 is normalized by $O_2(L)$ since $O_2(L) \le A_0 O_2(M_2)$. In particular $O_2(L)U \le B$. We now apply 10.1.10 to (LU, V_1, U) in place of (M, Z, V). As $(\times)$ implies $q(V_1, O_2(LU)) \ge 2$, we get

$$|A_0/C_{A_0}(U)|^2 \le |U/C_U(A_0)|.$$

On the other hand, since $A \in \mathcal{A}(L)$ $(\subseteq \mathcal{A}_U(L))$

(+) $$|U/C_U(A)| \le |A/C_A(U)|.$$

Now $|A/A_0| = 2$ gives

$$|A/C_A(U)|^2 \le 2^2 |A_0/C_{A_0}(U)|^2 \le 4|U/C_U(A_0)| \le 4|U/C_U(A)|$$
$$\le 4|A/C_A(U)|;$$

so $|A/C_A(U)| \le 4$, and (+) yields either

(4) $A_0 = C_A(U)$ and $|U/C_U(A)| = 2$, or

(5) $|A_0/C_{A_0}(U)| = 2$ and $|U/C_U(A)| = 4$.

In case (4) we get that $|\overline{U}| = 4$ since $L = \langle A, A^x \rangle$ for some $x \in L$. Hence $L/C_L(\overline{U}) \cong S_3$, and 10.1.8 shows that M_2 possesses a section isomorphic to the symmetric group S_4.

Thus, we may assume now that we are in case (5). Let $C_U(Q) < W \le U$ such that L acts irreducibly on $\overline{W}$. Assume first that $|U/WC_U(A)| \ne 1$. Then $|W/C_W(A)| \le 2$, and the same argument as in case (4) (with W in place of U) shows that $|\overline{W}| = 4$ and $L/C_L(\overline{W}) \cong S_3$. Hence, M_2 possesses a section isomorphic to S_4.

Assume now that $U = WC_U(A)$, so $[U, Q] \le W$. Since $L = AQ$, $V_1^A = V_1$, and $U = \langle V_1^L \rangle$ this gives

(6) $\overline{U} = \overline{W}\,\overline{V}_1$.

Note that $O_2(L)$ and thus also A_0 act trivially on the L-chief factor $\overline{W}$. Hence (3) and (6) imply that $[\overline{U}, A_0] = 1$. Let

$$P := [\langle A_0^L \rangle, Q] \quad (\le O_2(L)).$$

As also $\langle A_0^L \rangle$ acts trivially on $\overline{U}$ we get $[U, P] \le C_U(Q)$, so

$$[U, P, P] = 1.$$

If $[U, P] = 1$, then PA_0 centralizes V_1. On the other hand PA_0 is normal in L $(= QA)$; so PA_0 also centralizes $\langle V_1^L \rangle = U$, which contradicts the fact that in case (5) $|A_0/C_{A_0}(U)| = 2$. Thus, we have

(7) $[U, P] \ne 1$.

As P acts quadratically on V_1 $(\le U)$ and is normal in L, we get $[U, \Phi(P)] = 1$. Hence $(\times)$ also holds for P in place of A. Now as before 10.1.10 yields

$$|P/C_P(U)|^2 \le |U/C_U(P)| \overset{(5)}{\le} 4^2.$$

It follows that $|P/C_P(U)| \le 4$. If Q acts trivially on $P/C_P(U)$, then Q also acts trivially on $\langle A_0^L \rangle/C_P(U)$ (8.2.2). But then $P = C_P(U)$, which contradicts (7). Thus, we have $P/C_P(U) \cong C_2 \times C_2$ and

$$L/C_L(P/C_P(U)) \cong S_3;$$

and again 10.1.8 gives a section of M_2 that is isomorphic to S_4. $\square$

We conclude this section combining 10.1.3 and 10.1.11:

10.1.12   Theorem. *Let G be a group of even order and $O_2(G) = 1$. Suppose that for every 2-local subgroup M of G:*

(1)   *M has characteristic 2 and is solvable.*

(2)   *M does not posses a section isomorphic to S_4.*

Then every maximal 2-local subgroup of G is strongly 2-embedded in G.   □

10.2   The $p^a q^b$-Theorem

In this section we prove:

10.2.1   Burnside's Theorem. *Every group of order $p^a q^b$ $(p, q \in \mathbb{P})$ is solvable.*

For the proof of this theorem Burnside used a short and elegant argument from the character theory of finite groups.[2] His result and that of Frobenius about the kernel of a Frobenius group (4.1.6 on page 80) established character theory as a tool in the investigation of finite groups. Sixty years passed before Bender [30], Goldschmidt [54], and Matsuyama [80] were able to give a proof of Burnside's result that is independent of character theory but much longer.

In the attempt to prove the theorem of Burnside without character theory one can hardly avoid concepts and notions we have already met in previous chapters:

- primitive maximal subgroups,

- the Fitting subgroup of a p-local subgroup,

- coprime and p-stable action.

Moreover, a further concept that will be central in the next chapter might have been guessed:

- the set of nontrivial q-subgroups of a group that are normalized by a given q'-subgroup.

---

[2]See [4], p. 321, or in a later presentation, for example [9].

In the 1960s all these concepts (and their generalizations) were put into the center of attention by the works of Thompson, Gorenstein, Glauberman and Bender; and their impact on the investigation of finite groups is responsible for much of the progress made in the last 40 years.

One may wonder how group theory would have developed, if Burnside had not found this beautiful character-theoretic proof and he and his contemporaries had studied the group-theoretic structure of the situation more intensively, instead.

We now begin with the proof of the theorem of Burnside. Let G be a counterexample of minimal order.

For $U \leq G$ a Sylow p- resp. q-subgroup of U is denoted by U_p resp. U_q.

By 1.1.6 on page 7 we have the factorization

$$G = G_p G_q.$$

The minimality of G implies that every proper subgroup of G and every factor group G/N, $1 \neq N \trianglelefteq G$, is solvable. Since G (as a counterexample) is not solvable we get from 6.1.2 on page 122 that G is a non-Abelian simple group. In particular

$$1 \neq U < G \quad \Rightarrow \quad N_G(U) \text{ is solvable.}$$

In the following we analyze the local structure of G. The essential tool will be 8.2.12 on page 189:

(1) *Let M be a maximal subgroup of G and P a p-subgroup of M. Then* $O_q(C_M(P)) \leq O_q(M)$.

Let $\mathcal{M}$ be the set of maximal subgroups of G and

$$\begin{aligned}
\mathcal{M}_p &:= \{M \in \mathcal{M} \mid M \text{ has characteristic } p\}, \\
\mathcal{M}_q &:= \{M \in \mathcal{M} \mid M \text{ has characteristic } q\}, \\
\mathcal{M}_0 &:= \mathcal{M} \setminus (\mathcal{M}_p \cup \mathcal{M}_q).
\end{aligned}$$

Note that

$$F(M) = O_p(M) \times O_q(M) \quad (M \in \mathcal{M}),$$

so

$$\begin{aligned}
M \in \mathcal{M}_p &\iff F(M) = O_p(M), \text{ and} \\
M \in \mathcal{M}_q &\iff F(M) = O_q(M).
\end{aligned}$$

(2)   Let $M \in \mathcal{M}$ and $G_p \leq M$. Then $M \in \mathcal{M}_p$.

Proof. Let $Q := O_q(M) \leq G_q$. Then

$$\langle Q^G \rangle = \langle Q^{G_p G_q} \rangle = \langle Q^{G_q} \rangle \leq G_q.$$

Thus, $\langle Q^G \rangle$ is a proper normal subgroup of G, and the simplicity of G implies that $Q = 1$.   □

(3)   Let $M \in \mathcal{M}_0$. Then M is the unique maximal subgroup of G containing $Z(F(M))$. In particular $C_G(a) \leq M$ for all $a \in Z(F(M))^{\#}$.

Proof. Let $A := Z(F(M))$. Since $M \in \mathcal{M}_0$ we get

$$A = A_p \times A_q, \quad A_p \neq 1 \neq A_q.$$

Moreover, the maximality of M gives $M = N_G(A_p) = N_G(A_q)$ and

$$A_q \leq O_q(C_M(A_p)) = O_q(C_G(A_p)).$$

Let $A \leq H \in \mathcal{M}$. Then also $A_q \leq O_q(C_H(A_p))$, and (1) implies $1 \neq A_q \leq O_q(H)$ and similarly (with p and q interchanged) $1 \neq A_p \leq O_p(H)$. Hence $H \in \mathcal{M}_0$ and

$$O_q(H) \leq C_G(A_p) \leq M \quad \text{and} \quad O_p(H) \leq C_G(A_q) \leq M,$$

so $F(H) \leq M$. With the roles of H and M interchanged we also get $F(M) \leq H$.

Either $H = M$ or (M, H) is a primitive pair. In the second case 10.1.4 shows that $M \in \mathcal{M}_p$ or $M \in \mathcal{M}_q$, which contradicts $M \in \mathcal{M}_0$.   □

(4)   Let $M \in \mathcal{M}_0$. Then there exist $x \in G \setminus M$ and $M_p \in \mathrm{Syl}_p M$ such that

$$M_p = M_p{}^x \quad (\leq M \cap M^x).$$

Proof. Choose G_p such that $M_p := G_p \cap M \in \mathrm{Syl}_p M$. By (2) $M_p < G_p$, so by 3.1.10 on page 61 there exists

$$x \in G_p \setminus M_p \subseteq G \setminus M$$

such that $M_p{}^x = M_p$.   □

(5) $\mathcal{M}_0 = \varnothing$.

Proof. Assume that $\mathcal{M}_0 \neq \varnothing$ and $M \in \mathcal{M}_0$. We choose the notation such that $p > q$. As in the proof of (3) let

$$A_p := Z(O_p(M)) \quad \text{and} \quad A_q := Z(O_q(M));$$

in addition let $x \in G \setminus M$ be as in (4). Then

$$A_p A_p{}^x \leq M \cap M^x.$$

Assume first that A_q and thus also $A_q{}^x$ is cyclic. The action of A_p on $A_q{}^x$ ($\trianglelefteq M^x$) is trivial since $p > q$ (2.2.5 on page 51). Thus, (3) implies that $A_q{}^x \leq M$, so $Z(F(M^x)) \leq M$. Now again (3) yields $M = M^x$, which contradicts $N_G(M) = M$ and $x \notin M$.

We have shown that A_q is not cyclic. By (4) there exists a Sylow subgroup M_q and an element $y \in G \setminus M$ such that

$$A_q A_q{}^y \leq M_q = M_q{}^y \leq M \cap M^y.$$

Hence $A_q \leq M^y$, and A_q acts on $P := A_p{}^y$ ($\trianglelefteq M^y$). Now 8.3.4 implies

$$P = \langle C_P(a) \mid a \in A_q{}^\# \rangle \overset{(3)}{\leq} M,$$

so $Z(F(M^y)) = A_p{}^y A_q{}^y \leq M$. As above (3) gives the contradiction $M = M^y$. $\qquad\square$

(6) *Let $M \in \mathcal{M}$ such that $Z(G_q) \cap M \neq 1$. Then $M \in \mathcal{M}_q$.*

Proof. Assume that M is a counterexample. By (5) $M \in \mathcal{M}_p$, and thus

$$C_M(O_p(M)) \leq O_p(M) =: P.$$

Let $P \leq G_p$. The maximality of M yields

$$Z := Z(G_p) \leq N_G(P) \leq M,$$

so $Z \leq P$. Let $Y := Z(G_q) \cap M$. Then $\langle Z^Y \rangle$ ($\leq P$) is a p-group. For $g \in G$ there exist $x \in G_p$, $y \in G_q$ such that $g = xy$. Hence

$$Z^{gY} = Z^{yY} = Z^{Yy};$$

so $\langle Z^{gY} \rangle = \langle Z^Y \rangle^g$, and $\langle Z^{gY} \rangle$ is a p-group. By the lemma of Matsuyama (6.7.8 on page 160) there exists $T \in \mathrm{Syl}_p\, G$ such that

$$R := \mathrm{wcl}_G(Z, T)$$

is normalized by Y. It follows that $\langle Y, T \rangle \leq N_G(R) < G$, and $\langle Y^T \rangle$ is a proper subgroup of G. On the other hand $G = G_q T$ and $Y^G = Y^T$, so $\langle Y^T \rangle$ is a normal subgroup of G. This contradicts the simplicity of G. $\square$

(7)   *Let L be a p-local subgroup of G.*

   (a) $L \cap Z(G_q) = 1$ *for all $G_q \in \mathrm{Syl}_q G$.*
   (b)   *L has characteristic p.*

Proof. Let $1 \neq R \leq G_p$ and $M \in \mathcal{M}$ such that $L = N_G(R)$ and $L \leq M$. In particular $Z(G_p) \leq L \leq M$.

(a) Assume that $L \cap Z(G_q) \neq 1$. Then $M \cap Z(G_q) \neq 1 \neq M \cap Z(G_p)$, and this contradicts (6).

(b) Assume that $Q := O_q(L) \neq 1$. Then $N_G(Q)$ is a q-local subgroup of G containing L and thus also $Z(G_p)$. This contradicts (a) (with the roles of p and q reversed). $\square$

(8) $|G|$ *is odd.*

Proof. In a counterexample let $q = 2$ and t be an involution in $Z(G_2)$ (3.1.11 on page 61). The theorem of Baer (6.7.5 on page 160) shows that there exists a p-element $y \neq 1$ in G such that $y^t = y^{-1}$. Hence, $L = N_G(\langle y \rangle)$ is a p-local subgroup of G containing t. But this contradicts (7a). $\square$

Now 10.1.6 on page 268 and (8) imply that G does not possess primitive pairs of characteristic p. But because of (7b) G also satisfies the hypothesis of 10.1.3 on page 263. Hence, for every maximal p-local subgroup M of G

$$|M \cap M^g|_p = 1 \quad \text{for all } g \in G \setminus M.$$

As we can choose M to contain G_p, we get $G_p \cap G_p{}^g = 1$ for $g \in G \setminus M$, so (1.1.6 on page 7)

$$|G_p|^2 = |G_p G_p{}^g| \leq |G|.$$

With a symmetric argument also $|G_q|^2 \leq |G|$. But this contradicts $|G| = |G_p||G_q|$ since $|G_p| \neq |G_q|$. $\square$

10.3   The Amalgam Method

In this section we present a method that is particularly suited investigation of primitive pairs (M_1, M_2) of characteristic p. This method was introduced by Goldschmidt [58] at the end of the 1970s and since then has become an integral part of the local structure theory of finite groups.[3] The name *amalgam method* refers to the fact that this method does not require a finite group but can be carried out already in the *amalgamated product* of the *finite* groups M_1 and M_2. In our presentation we do not use amalgamated products.

Let G be a group, and let P_1 and P_2 be two different subgroups of G. In this section we do not assume that G is a *finite* group, but only that the subgroups P_1 an P_2 are finite subgroups of G.

Let Γ be the set of right cosets of P_1 and P_2 in G. The elements of Γ are called **vertices**. For $\{1, 2\} = \{i, j\}$ two vertices $P_i x, P_j y \in \Gamma$ are **adjacent** if

$$P_i x \cap P_j y \neq \varnothing \quad \text{and} \quad P_i x \neq P_j y,$$

and in this case $\{P_i x, P_j y\}$ is called an **edge** of Γ. Then Γ is a **graph**, the **coset graph** of G with respect to P_1 and P_2.

Note that $i \neq j$ if $\{P_i x, P_j y\}$ is an edge and that $\{P_1, P_2\}$ is an edge since $1 \in P_1 \cap P_2$.

For $\alpha \in \Gamma$ let $\Delta(\alpha)$ be the set of all vertices adjacent to α.

The group G acts on Γ by right multiplication

$$g: \Gamma \to \Gamma \quad \text{with} \quad P_i x \mapsto P_i x g \quad (g \in G).$$

As usual we write α^g for the image of α under g, and we call α^g a vertex **conjugate** to α. As

$$P_i x \cap P_j y \neq \varnothing \quad \Longleftrightarrow \quad P_i x g \cap P_j y g \neq \varnothing,$$

g acts as an automorphism of *the graph* Γ, and this action gives rise to a homomorphism of G into $\operatorname{Aut}\Gamma$, the automorphism group of Γ.

We first collect some elementary properties of this action.

10.3.1   (a)   *G has two orbits on the vertices of Γ, and P_1 and P_2 are representatives of these orbits. Every vertex stabilizer G_α, $\alpha \in \Gamma$, is a G-conjugate of P_1 or P_2.*

---

[3]See [7].

(b) *G acts transitively on the edges of* Γ*. Every edge stabilizer in G is a
    G-conjugate of* $P_1 \cap P_2$.

(c) G_α *acts transitively on* $\Delta(\alpha)$*,* $\alpha \in \Gamma$*, in particular*

$$|\Delta(\alpha)| = |G_\alpha : G_\alpha \cap G_\beta| \text{ for } \beta \in \Delta(\alpha).$$

(d) $(P_1 \cap P_2)_G$ *is the kernel of the action of G on* Γ.[4]

Proof. (a) Note that for $P_i x \in \Gamma$ and $g \in G$:

$$P_i x g = P_i x \iff P_i g^{x^{-1}} = P_i \iff g \in P_i^x.$$

(b) Let $\{P_1 x, P_2 y\}$ be an edge, so there exists $z \in P_1 x \cap P_2 y$. Hence

$$P_1 x = P_1 z \quad \text{and} \quad P_2 y = P_2 z,$$

and the element z^{-1} conjugates the edge $\{P_1 x, P_2 y\}$ to $\{P_1, P_2\}$. According to (a) the stabilizer of $\{P_1 z, P_2 z\}$ is $P_1^z \cap P_2^z = (P_1 \cap P_2)^z$.

(c) By (a) we may assume that $\alpha = P_1$. Then

$$\Delta(\alpha) = \{P_2 y \mid P_2 y \cap P_1 \neq \varnothing\} = \{P_2 y \mid y \in P_1\}.$$

Thus P_1 is transitive on $\Delta(\alpha)$.

(d) By (a) any normal subgroup of G contained in $P_1 \cap P_2$ fixes every vertex of Γ. $\square$

An $(n+1)$-tuple $(\alpha_0, \alpha_1, \ldots, \alpha_n)$ of vertices is a **path of length** n from α_0 to α_n if

$$\alpha_i \in \Delta(\alpha_{i+1}) \text{ for } i = 0, \ldots, n-1 \quad \text{and} \quad \alpha_i \neq \alpha_{i+2} \text{ for } i = 0, \ldots, n-2.$$

Paths can be used to define the **distance** $d(\alpha, \beta)$ between vertices $\alpha, \beta \in \Gamma$. Here $d(\alpha, \beta) = \infty$ if there is no path in Γ from α to β, otherwise $d(\alpha, \beta)$ is the length of a shortest path from α to β.

The subset

$$\{\beta \in \Gamma \mid d(\alpha, \beta) < \infty\}$$

---

[4]$(P_1 \cap P_2)_G$ is the largest normal subgroup of G in $P_1 \cap P_2$.

is the **connected component** of Γ that contains α. Any two vertices of a connected component are joint by a path, and different connected components are disjoint. Γ is **connected** if Γ has only one connected component.

At first sight it is not transparent why these new objects and the language of graph theory should help to simplify the investigation of the structure of G (or better that of P_1 and P_2). The basic reason for this phenomena is the fact that the graph-theoretic notation allows us to describe the group theoretic properties that we investigate in a very easy way. Of course, the coming proofs should reveal this, but two things can be pointed out here already:

- Statement 10.3.2 below shows that Γ is connected if and only if G is generated by the two subgroups P_1 and P_2. This turns a fairly unhandy group theoretic property into an elementary graph-theoretic one that can be used easily in proofs and definitions, for example in 10.3.3 and the definition of a critical pair.

- By means of the above-defined distance, a large variety of normal subgroups of vertex stabilizers can be defined. For example, for $i \in \mathbb{N}$,

$$G_\alpha^{[i]} := \bigcap_{\substack{\delta \in \Gamma \\ d(\alpha,\delta) \leq i}} G_\delta.$$

The reader should try to define these normal subgroups for $\alpha = P_1$ (so $G_\alpha = P_1$) without the help of the graph Γ.

Of course, not all of these normal subgroups can be different since P_1 and P_2 are finite. In fact, one of the essential ideas of the amalgam method is to find out which of these subgroups are identical.

10.3.2 Γ *is connected if and only if* $G = \langle P_1, P_2 \rangle$.

Proof. Assume first that $G = \langle P_1, P_2 \rangle$. Let Δ be the connected component of Γ that contains P_1. Since P_1 and P_2 are adjacent also P_2 is in Δ. As different connected components are disjoint we get that

$$\Delta = \Delta^{\langle P_1, P_2 \rangle} = \Delta^G$$

and thus $\Delta = \Gamma$ by 10.3.1 (a).

Assume now that Γ is connected and set $G_0 := \langle P_1, P_2 \rangle$. Let

$$\Gamma_0 := \{P_1 x \mid x \in G_0\} \cup \{P_2 x \mid x \in G_0\}$$

be the coset graph of G_0 with respect to P_1 and P_2. As we have seen above Γ_0 is connected. Moreover $\Gamma = \Gamma_0$ implies $G = G_0$. Assume that $\Gamma \neq \Gamma_0$. Since Γ is connected there exists an edge $\{\alpha, \beta\}$ of Γ such that $\alpha \in \Gamma_0$ and $\beta \in \Gamma \setminus \Gamma_0$. By 10.3.1 (a) (applied to G_0 and Γ_0) G_α is in G_0. Since G_α is transitive on $\Delta(\alpha)$ (10.3.1 (c)) not only β but also every other element of $\Delta(\alpha)$ is in $\Gamma \setminus \Gamma_0$. Hence, in Γ_0 no vertex is adjacent to α. But then Γ_0 is not connected, a contradiction. $\square$

An essential tool in the investigation of coset graphs is the following elementary fact:

10.3.3   Let $G = \langle P_1, P_2 \rangle$ and $U \leq G_\alpha \cap G_\beta$. Suppose that $\{\alpha, \beta\}$ is an edge of Γ such that one of the following holds:

(1) $N_{G_\delta}(U)$ acts transitively on $\Delta(\delta)$ for $\delta \in \{\alpha, \beta\}$.

(1′) $U \trianglelefteq G_\alpha$ and $U \trianglelefteq G_\beta$.

Then U acts trivially on Γ.

Proof. Hypothesis (1′) together with 10.3.1 (c) implies Hypothesis (1). Thus, we may assume that (1) holds. Let

$$\Gamma_0 := \alpha^{N_G(U)} \cup \beta^{N_G(U)}.$$

Then U fixes every vertex in Γ_0. Let $\gamma \in \Gamma_0$, so there exists $g \in N_G(U)$ and $\delta \in \{\alpha, \beta\}$ such that $\gamma = \delta^g$. Then

$$\Delta(\delta^g) = \Delta(\gamma) \quad \text{and} \quad N_{G_\gamma}(U) = N_{G_\delta}(U)^g.$$

By (1) $N_{G_\gamma}(U)$ is transitive on $\Delta(\delta^g) = \Delta(\gamma)$. Moreover, one of the vertices in $\{\alpha^g, \beta^g\}$ is adjacent to γ and

$$\{\alpha^g, \beta^g\} \subseteq \Gamma_0.$$

It follows that $\Delta(\gamma) \subseteq \Gamma_0$. Since by 10.3.2 Γ is connected we conclude that $\Gamma = \Gamma_0$. Thus, U stabilizes every vertex in Γ. $\square$

We now present the amalgam method in action treating a particular situation that we will meet again in Chapter 12. For the rest of this section we assume:

$\mathcal{A}$ Let G be a group generated by two finite subgroups P_1 and P_2, and set
$T := P_1 \cap P_2$. Suppose that for $i = 1, 2$:

$\mathcal{A}_1$ $C_{P_i}(O_2(P_i)) \leq O_2(P_i)$.

$\mathcal{A}_2$ $T \in \mathrm{Syl}_2 P_i$.

$\mathcal{A}_3$ $T_G = 1$.

$\mathcal{A}_4$ $P_i/O_2(P_i) \cong S_3$.

$\mathcal{A}_5$ $[\Omega(Z(T)), P_i] \neq 1$.

The aim of our investigation is to show that $\mathcal{A}$ implies:

$\mathcal{B}$ $P_1 \cong P_2 \cong S_4$ or $P_1 \cong P_2 \cong C_2 \times S_4$.

In the following we assume $\mathcal{A}$. Let Γ be the coset graph of G with respect
to P_1 and P_2. According to 10.3.2 Γ is connected, and 10.3.1 (d) together
with $\mathcal{A}_3$ shows that G acts faithfully on Γ.

Let $\{\alpha, \beta\}$ be an edge of Γ. Since $\{\alpha, \beta\}$ is conjugate to the edge $\{P_1, P_2\}$
(10.3.1 (b)), the statements $\mathcal{A}_1, \ldots, \mathcal{A}_5$ also hold for G_α and G_β in place
of P_1 and P_2. In this sense we will apply $\mathcal{A}_1, \ldots, \mathcal{A}_5$ to arbitrary vertex
stabilizers G_α and edges $\{\alpha, \beta\}$.

10.3.4   *Let $\{\alpha, \beta\}$ be an edge of Γ.*

(a) $G_\alpha \cap G_\beta$ *has index 3 in G_α and is a Sylow 2-subgroup of G_α. In
particular $G_\alpha = \langle G_\alpha \cap G_\beta, t \rangle$ for all $t \in G_\alpha \setminus G_\beta$.*

(b) $|\Delta(\alpha)| = 3$ *and*

$$O_2(G_\alpha) = \bigcap_{\delta \in \Delta(\alpha)} (G_\alpha \cap G_\delta) \quad (= G_\alpha^{[1]}).$$

(c) G_α *acts 2-transitively on $\Delta(\alpha)$.*

Proof. (a) follows from $\mathcal{A}_4$ and (b), (c) from 10.3.1 (c), (a).   □

For $\alpha \in \Gamma$ let
$$
\begin{aligned}
Q_\alpha &:= O_2(G_\alpha), \\
Z_\alpha &:= \langle \Omega(Z(T)) \mid T \in \mathrm{Syl}_2 G_\alpha \rangle.
\end{aligned}
$$

10.3.5   *Let* $\alpha \in \Gamma$, $V \trianglelefteq G_\alpha$ *and* $T \in \mathrm{Syl}_2 \, G_\alpha$. *Suppose that*

$$\Omega(Z(T)) \leq V \leq \Omega(Z(Q_\alpha)) \quad and \quad |V : \Omega(Z(T))| = 2.$$

Then

$$V = C_V(G_\alpha) \times W \quad where \quad W := [V, G_\alpha].$$

Moreover $W \cong C_2 \times C_2$ *and* $C_{G_\alpha}(W) = Q_\alpha$, *i.e.,* $G_\alpha/C_{G_\alpha}(W) \cong S_3$.

Proof. Let $D \in \mathrm{Syl}_3 \, G_\alpha$. By 8.4.2 on page 198 we have the decomposition

$$V = C_V(D) \times W \quad with \quad W := [V, D].$$

$\mathcal{A}_5$ and $G_\alpha = DT$ imply $W \neq 1$ and thus $|W| \geq 4$. Let $d \in D^\#$. By our hypothesis

$$|V/\Omega(Z(T))| = 2 = |V/\Omega(Z(T^d))|.$$

Now $G_\alpha = \langle T, T^d \rangle$ shows that $|V/C_V(G_\alpha)| \leq 4$. It follows that $C_V(G_\alpha) = C_V(D)$ and $|W| = 4$. The other statements are consequences of $\mathcal{A}_4$. $\square$

10.3.6   *Let* $\{\alpha, \beta\}$ *be an edge of* Γ.

(a) $Z_\alpha \leq \Omega(Z(Q_\alpha))$.

(b) $Q_\alpha Q_\beta = G_\alpha \cap G_\beta \in Syl_2 \, G_\alpha$.

(c) $C_{G_\alpha}(Z_\alpha) = Q_\alpha$; *in particular, the Sylow 2-subgroups of* G_α *are non-Abelian.*

(d) $Z_\alpha Z_\beta$ *is normal in* G_α *if and only if there exists* $\gamma \in \Delta(\alpha) \setminus \{\beta\}$ *such that* $Z_\alpha Z_\beta = Z_\alpha Z_\gamma$.

Proof. (a) Let $T \in \mathrm{Syl}_2 \, G_\alpha$. Then $Q_\alpha \leq T$, and $\mathcal{A}_1$ implies that $\Omega(Z(Y)) \leq Z(Q_\alpha)$.

(b) By $\mathcal{A}_4$ and 10.3.4 Q_α and Q_β have index 2 in $G_\alpha \cap G_\beta$. Thus, it suffices to show that $Q_\alpha \neq Q_\beta$.

If $Q_\alpha = Q_\beta$, then $Q_\alpha = 1$ by 10.3.3 and 10.3.4 since G acts faithfully on Γ. This contradicts $\mathcal{A}_1$.

(c) By $\mathcal{A}_5$ the normal subgroup Z_α is not central in G_α. Thus, Z_α also is not central in $T \in \mathrm{Syl}_2\, G_\alpha$ since $G_\alpha = \langle T \mid T \in \mathrm{Syl}_2\, G_\alpha \rangle$.

By (a) $Q_\alpha \leq C_{G_\alpha}(Z_\alpha)$. If $Q_\alpha < C_{G_\alpha}(Z_\alpha)$, then $C_{G_\alpha}(Z_\alpha)$ contains a subgroup D of order 3, and $G_\alpha = DT$, $T \in \mathrm{Syl}_2\, G_\alpha$, by $\mathcal{A}_4$. But then $\Omega(Z(T))$ is central in G_α, which contradicts $\mathcal{A}_5$.

(d) If $Z_\alpha Z_\beta \trianglelefteq G_\alpha$, then $Z_\alpha Z_\beta = Z_\alpha Z_\gamma$ for all $\gamma \in \Delta(\alpha)$ since G_α is transitive on $\Delta(\alpha)$.

Assume now that $Z_\alpha Z_\beta = Z_\alpha Z_\gamma$ for some $\gamma \in \Delta(\alpha)$, $\gamma \neq \beta$. Then $Z_\alpha Z_\beta$ is normalized by $G_\alpha \cap G_\beta$ and $G_\alpha \cap G_\gamma$ and thus also by G_α $(\mathcal{A}_4)$. □

As a motivation for what will follow later we prove:

10.3.7   *Let $\{\alpha, \beta\}$ be an edge of Γ. Then the following statements are equivalent:*

(i)   *$\mathcal{B}$ holds.*

(ii)   *$Z_\alpha \nleq Q_\beta$.*

Proof. Assume that $\mathcal{B}$ holds. Then by 10.3.1 (a) for $\delta \in \{\alpha, \beta\}$

$$G_\delta \cong S_4 \qquad \text{and} \quad Q_\delta \cong C_2 \times C_2$$

or

$$G_\delta \cong S_4 \times C_2 \quad \text{and} \quad Q_\delta \cong C_2 \times C_2 \times C_2.$$

Hence $Z_\delta = Q_\delta$, and 10.3.6 (b) implies $Z_\alpha \nleq Q_\beta$.

Assume now that $Z_\alpha \nleq Q_\beta$. Let $\delta \in \{\alpha, \beta\}$ and set

$$T := Q_\alpha Q_\beta \quad \text{and} \quad E := Q_\alpha \cap Q_\beta.$$

10.3.6 (b) gives $T \in \mathrm{Syl}_2\, G_\delta$ and $|T/Q_\delta| = 2$. It follows that

$$(1) \qquad\qquad |Q_\alpha : E| = 2 = |Q_\beta : E|,$$

and

$$(2) \qquad\qquad T = Q_\beta Z_\alpha \quad \text{and} \quad Q_\alpha = E Z_\alpha.$$

By 10.3.6 (c) $[Z_\alpha, Z_\beta] \neq 1$ and thus also

$$Z_\beta \not\leq Q_\alpha.$$

Now a symmetric argument yields

(3) $T = Q_\alpha Z_\beta$ and $Q_\beta = E Z_\beta.$

Since Z_δ is an elementary Abelian subgroup of $Z(Q_\delta)$ we get from (2) and (3) together with 5.2.7 on page 106

$$\Phi(Q_\alpha) = \Phi(E) = \Phi(Q_\beta),$$

i.e., $\Phi(E)$ is characteristic in Q_δ. Hence, $\Phi(E)$ is normal in G_α and G_β. Now 10.3.3 shows that $\Phi(E)$ is trivial. We get

(4) Q_α and Q_β are elementary Abelian,

Moreover, $T = Q_\alpha Q_\beta$ implies

(5) $E = Z(T).$

Let $W_\delta := [Q_\delta, G_\delta]$. (1) allows to apply 10.3.5 to $V = Q_\delta$, so

(6) $Q_\delta = Z(G_\delta) \times W_\delta$ and $W_\delta \cong C_2 \times C_2.$

By (2) and (3) there exists an involution t_δ in $T \setminus Q_\delta$ that acts nontrivially on $O^2(G_\delta)/W_\delta$. Hence

$$X_\delta := O^2(G_\delta) \langle t_\delta \rangle \cong S_4.$$

Assume that $Z(G_\alpha) = 1$. Then $|T| = 8$, and $Z(G_\beta) = 1$ follows from (5) and (6). Thus, $G_\alpha = X_\alpha$ and $G_\beta = X_\beta$ are as in $\mathcal{B}$.

Assume that $Z(G_\alpha) \neq 1$. Then also $Z(G_\beta) \neq 1$, again by (5) and (6). On the other hand $Z(G_\beta) \cap Z(G_\alpha) = 1$ by 10.3.3. Since $Z(G_\alpha)$ and $Z(G_\beta)$ are in $Z(T) = E$ we get from (6) that $Z(G_\alpha) \cong C_2 \cong Z(G_\beta)$. This gives the second possibility in $\mathcal{B}$.   □

Let $\{\alpha, \beta\}$ be an edge of Γ. In order to prove that $\mathcal{A}$ implies $\mathcal{B}$ it suffices to show—according to 10.3.7—that the assumption $Z_\alpha \leq Q_\beta$ leads to a contradiction. In doing this the following parameter b plays a central role.

Let μ be a vertex. Since Z_μ acts faithfully on Γ there exists $\lambda \in \Gamma$ such that $Z_\mu \not\leq G_\lambda$, in particular $Z_\mu \not\leq Q_\lambda$. As Γ is connected $d(\mu, \lambda) < \infty$, so

$$b := \min\{d(\mu, \lambda) \mid \mu, \lambda \in \Gamma, \ Z_\mu \not\leq Q_\lambda\}$$

is an integer. Moreover $b \geq 1$ since $Z_\mu \leq Q_\mu$. A pair (α, α') of vertices is a **critical pair** if

$$Z_\alpha \not\leq Q_{\alpha'} \quad \text{and} \quad d(\alpha, \alpha') = b.$$

Hence, for vertices $\mu, \lambda \in \Gamma$ with $d(\mu, \lambda) < b$ the minimality of b yields

$$Z_\mu \leq Q_\lambda \quad \text{and} \quad Z_\lambda \leq Q_\mu.$$

According to 10.3.7 $b = 1$ is equivalent to $\mathcal{B}$.

In the following let (α, α') be a critical pair and γ a path of length b from α to α'. We enumerate the vertices of γ by

$$\gamma = (\alpha, \alpha + 1, \alpha + 2, \ldots, \alpha') \quad \text{or} \quad \gamma = (\alpha, \ldots, \alpha' - 2, \alpha' - 1, \alpha'),$$

i.e., $\alpha' - i = \alpha + (b - i)$ for $1 \leq i \leq b - 1$. In addition, we set

$$R := [Z_\alpha, Z_{\alpha'}].$$

10.3.8   (a) (α', α) *is also a critical pair.*

(b) $G_\alpha \cap G_{\alpha+1} = Z_{\alpha'} Q_\alpha$ *and* $G_{\alpha'-1} \cap G_{\alpha'} = Z_\alpha Q_{\alpha'}$.

(c) $R \leq Z(G_\alpha \cap G_{\alpha+1}) \cap Z(G_{\alpha'-1} \cap G_{\alpha'})$ *and*
        $R = [Z_\alpha, G_{\alpha+1} \cap G_\alpha] = [Z_{\alpha'}, G_{\alpha'-1} \cap G_{\alpha'}]$.

(d) $|R| = 2$.

(e) $Z_\alpha = [Z_\alpha, G_\alpha] \times \Omega(Z(G_\alpha))$ *and* $[Z_\alpha, G_\alpha] \cong C_2 \times C_2$.

(f) $|Z_\alpha : \Omega(Z(Y))| = 2$ *for* $Y \in \mathrm{Syl}_2 \, G_\alpha$.

Proof. The minimality of b implies

$$Z_\alpha \leq Q_{\alpha'-1} \leq G_{\alpha'-1} \cap G_{\alpha'} \quad \text{and} \quad Z_{\alpha'} \leq Q_{\alpha+1} \leq G_\alpha \cap G_{\alpha+1}.$$

Moreover, $Z_\alpha \not\leq Q_{\alpha'}$ shows that

$$G_{\alpha'-1} \cap G_{\alpha'} = Z_\alpha Q_{\alpha'}$$

since $Q_{\alpha'}$ has index 2 in $G_{\alpha'-1} \cap G_{\alpha'}$ (10.3.4 and $\mathcal{A}_4$). As Z_α and $Z_{\alpha'}$ are normal in G_α and $G_{\alpha'}$, respectively, we get that

$(')$ $\qquad\qquad\qquad\qquad R \leq Z_\alpha \cap Z_{\alpha'}.$

Now 10.3.6 (c) implies $R \neq 1$, thus also $Z_{\alpha'} \not\leq Q_\alpha$ and

$$G_\alpha \cap G_{\alpha+1} = Z_{\alpha'} Q_\alpha.$$

Hence (a) and (b) follow, and (c) is a consequence of $(')$ and 10.3.6 (a). In addition 10.3.6 (a), (c) show that

$$|Z_\alpha / C_{Z_\alpha}(Z_{\alpha'})| = |Z_{\alpha'} / Z_{\alpha'}(Z_\alpha)| = 2 \text{ and } C_{Z_\alpha}(Z_{\alpha'}) = \Omega(Z(G_\alpha \cap G_{\alpha+1})).$$

This implies (d) and (f), and 10.3.5 gives (e). $\square$

10.3.9   Let $\alpha - 1 \in \Delta(\alpha) \setminus \{\alpha + 1\}$. Suppose that $(\alpha - 1, \alpha' - 1)$ is not a critical pair. Then the following hold:

(a) $Z_\alpha Z_{\alpha+1} = Z_\alpha Z_{\alpha-1} \trianglelefteq G_\alpha.$

(b) $Q_\alpha \cap Q_\beta \trianglelefteq G_\alpha$ for all $\beta \in \Delta(\alpha).$

(c) α and α' are conjugate, and b is even.

Proof. Since $(\alpha - 1, \alpha' - 1)$ is not critical we get

$$Z_{\alpha-1} \leq Q_{\alpha'-1} \quad (\leq G_{\alpha'-1} \cap G_{\alpha'}),$$

In particular $b > 1$ and

$$Z_{\alpha-1} \overset{10.3.8}{\leq} Z_\alpha Q_{\alpha'} \overset{10.3.6}{=} Z_\alpha C_{G_{\alpha'}}(Z_{\alpha'}).$$

It follows that

$$[Z_{\alpha-1}, Z_{\alpha'}] \leq R \overset{10.3.6}{\leq} Z_\alpha.$$

Hence, $Z_{\alpha-1}Z_\alpha$ is normalized by $Z_{\alpha'}$ and $G_{\alpha-1} \cap G_\alpha$ and thus also by $\langle G_\alpha \cap G_{\alpha-1}, Z_{\alpha'} \rangle = G_\alpha$ (10.3.4). Now 10.3.1 (c) implies (a).

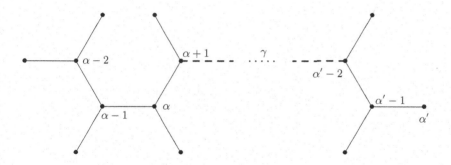

Fig. 1

Claim (b) follows from (a), 10.3.6 (c) and the transitivity of G_α on $\Delta(\alpha)$. Note that by 10.3.1 either $\alpha \in (\alpha')^G$ or $\alpha \in (\alpha'-1)^G$, so

$$\alpha \in (\alpha')^G \iff b \text{ is even}.$$

To prove (c) we may assume that α and $\alpha'-1$ and thus also G_α and $G_{\alpha'-1}$ are conjugate. Then (b) gives

$$Z_\alpha \le Q_{\alpha'-2} \cap Q_{\alpha'-1} = Q_{\alpha'-1} \cap Q_{\alpha'}.$$

This contradicts $Z_\alpha \nleq Q_{\alpha'}$. $\square$

10.3.10 *Suppose that there exists $\alpha - 1 \in \Delta(\alpha) \setminus \{\alpha + 1\}$ such that $(\alpha - 1, \alpha' - 1)$ is a critical pair. Then $b = 1$.*

Proof. Set $R_1 := [Z_{\alpha-1}, Z_{\alpha'-1}]$ and assume that $b > 1$. Then $Z_\alpha \le Q_{\alpha+1}$ and $Z_{\alpha'} \le Q_{\alpha'-1}$. As $(\alpha-1, \alpha'-1)$ is critical, 10.3.8 applies to $(\alpha-1, \alpha'-1)$ in place of (α, α'). Hence $|R_1| = 2$ and

$$R_1 = [Z_{\alpha-1}, G_{\alpha-1} \cap G_\alpha] \le Z(G_{\alpha-1} \cap G_\alpha) \cap Z(G_{\alpha'-2} \cap G_{\alpha'-1}),$$

in particular $R_1 \le Z(Q_{\alpha'-1})$ and thus $[R_1, Z_{\alpha'}] = 1$. By 10.3.8 (b) $Z_{\alpha'}$ and $G_{\alpha-1} \cap G_\alpha$ generate G_α, so

$$(1) \qquad\qquad\qquad R_1 \le Z(G_\alpha).$$

Let $\alpha - 2 \in \Delta(\alpha - 1) \setminus \{\alpha\}$ (see Fig. 1).

Next we show:

(2) $(\alpha - 2, \alpha' - 2)$ is a critical pair.

Assume that $(\alpha - 2, \alpha' - 2)$ is not critical. Then 10.3.9 (a), applied to $(\alpha - 1, \alpha' - 1)$ and $\alpha - 2$ in place of (α, α') and $\alpha - 1$, shows that $Z_{\alpha-1}Z_\alpha = Z_{\alpha-1}Z_\delta$ for all $\delta \in \Delta(\alpha - 1)$. With 10.3.1 (c) we get[5]

$$Z_{\alpha+1}Z_\alpha = Z_{\alpha+1}Z_{\alpha+2}.$$

The minimality of b yields $Z_{\alpha+1}Z_{\alpha+2} \leq Q_{\alpha'}$. But then also $Z_\alpha \leq Q_{\alpha'}$, and (α, α') is not a critical pair. This contradiction shows (2).

Set $R_2 := [Z_{\alpha-2}, Z_{\alpha'-2}]$. According to (2) $\alpha - 2$ and (α, α') satisfy the hypothesis. Hence, we also get for these vertices that $|R_2| = 2$ and

(3) $R_2 = [Z_{\alpha-2}, G_{\alpha-2} \cap G_{\alpha-1}] \leq Z(G_{\alpha-1})$

By 10.3.1 (c) there exist $y \in G_{\alpha-1}$ and $x \in G_\alpha$ such that

$$(\alpha - 2)^y = \alpha \quad \text{and} \quad (\alpha + 1)^x = \alpha - 1.$$

Hence

$$[Z_\alpha, G_\alpha \cap G_{\alpha-1}] = [Z_{\alpha-2}, G_{\alpha-2} \cap G_{\alpha-1}]^y = R_2{}^y \leq Z(G_{\alpha-1}),$$

and

$$R^x \stackrel{10.3.8\,(c)}{=} [Z_\alpha, G_\alpha \cap G_{\alpha+1}]^x = [Z_\alpha, G_\alpha \cap G_{\alpha-1}] = R_2{}^y \leq Z(G_{\alpha-1}).$$

It follows that

(4) $R \leq Z(G_{\alpha+1}).$

In addition, (1) and 10.3.3 give

(5) $R \cap R_1 = 1.$

Next we show:

(6) $b = 2.$

---
[5]Rotate around α such that $\alpha - 1$ goes to $\alpha + 1$.

Assume that $b > 2$. Then $Z_{\alpha'} \leq Q_{\alpha'-2}$, and by (3) and 10.3.6 (a) R_2 centralizes $Z_{\alpha'}$ and $G_{\alpha-1}$. Since $G_\alpha = \langle Z_{\alpha'}, G_\alpha \cap G_{\alpha-1} \rangle$ we conclude that R_2 centralizes $G_{\alpha-1}$ and G_α. Hence 10.3.3 yields $R_2 = 1$, which contradicts $|R_2| = 2$.

To treat the remaining case $b = 2$ we set

$$V_\alpha := \langle Z_\beta \mid \beta \in \Delta(\alpha) \rangle \quad (\unlhd G_\alpha)$$

and

$$V_{\alpha+1} := \langle Z_\beta \mid \beta \in \Delta(\alpha+1) \rangle \quad (\unlhd G_{\alpha+1}).$$

Note that $V_\alpha \leq Q_\alpha$ and $V_{\alpha+1} \leq Q_{\alpha+1}$ since $b > 1$. Moreover

$$Z_\alpha = \langle \Omega(Z(G_\alpha \cap G_{\alpha+1}))^{G_\alpha} \rangle \leq V_\alpha$$

since V_α is normal in G_α, similarly $Z_{\alpha+1} \leq V_{\alpha+1}$. Hence

(7) $$Z_\alpha Z_{\alpha+1} \leq V_\alpha \cap V_{\alpha+1}.$$

As $R_1 \leq Z(G_\alpha)$, the 2-transitive action of G_α on $\Delta(\alpha)$ (10.3.4 (c)) implies

$$V_\alpha' = R_1 \leq Z(G_\alpha).$$

We now derive a contradiction showing that V_α is Abelian: Since V_α is generated by involutions we get that V_α/R_1 is elementary Abelian, so

$$R_1 = \Phi(V_\alpha).$$

With the same argument (4) implies

$$R = \Phi(V_{\alpha+1}).$$

Let

$$\overline{V}_\alpha := V_\alpha/Z_\alpha.$$

From 10.3.8 (f) we get $|Z_\beta/Z_\alpha \cap Z_\beta| = 2$ for all $\beta \in \Delta(\alpha)$, so $|\overline{Z}_\beta| = 2$. In addition, $\overline{V}_\alpha$ is generated by the three subgroups $\overline{Z}_\beta$, $\beta \in \Delta(\alpha)$, thus

$$|\overline{V}_\alpha| \leq 8.$$

Set

$$W := V_\alpha \cap V_{\alpha+1} \quad (\unlhd G_\alpha \cap G_{\alpha+1}).$$

By (7) $Z_\alpha Z_{\alpha+1} \leq W$, and the definition of V_α gives

(8) $V_\alpha = \langle W^{G_\alpha} \rangle.$

Moreover by 5.2.7 on page 106

$$\Phi(W) \leq \Phi(V_\alpha) \cap \Phi(V_{\alpha+1}) = R_1 \cap R \overset{(5)}{=} 1.$$

Hence W is elementary Abelian, and $V_\alpha' \neq 1$ shows that $|V_\alpha/W| \geq 2$.

We investigate the action of G_α on $\overline{V}_\alpha$. The kernel of this action contains Q_α since $[G_{\alpha-1} \cap G_\alpha, Z_{\alpha-1}] = R_1 \leq Z_\alpha$. Set

$$\overline{V}_0 := [\overline{V}_\alpha, O^2(G_\alpha)].$$

If $\overline{V}_0 = 1$, then W is normal in G_α and $V_\alpha' = 1$. But this contradicts $V_\alpha' = R_1$ and $|R_1| = 2$.

Now let $\overline{V}_0 \neq 1$. Since $|\overline{V}_\alpha| \leq 8$ we get

(9) $|\overline{V}_0| = 4.$

Assume first that $|V_\alpha/W| = 2$. Let $x \in G_\alpha$ such that $W^x \neq W$. Then $V_\alpha = WW^x$ and thus $W \cap W^x = Z(V_\alpha)$ and $|V_\alpha/W \cap W^x| = 4$. Let $D \in \mathrm{Syl}_3 G_\alpha$. The nontrivial action of D on $\overline{V}_\alpha$ implies a nontrivial action of D on $V_\alpha/W \cap W^x$. Hence, all maximal subgroups of V_α that contain $W \cap W^x$, are D-conjugates of W. But then every element of $V_\alpha^{\#}$ is an involution and V_α is elementary Abelian. Again this contradicts $V_\alpha' = R_1$.

We have shown that
$$|V_\alpha/W| \geq 4.$$

Because of (7) and $|\overline{V}_\alpha| \leq 8$ we get

(10) $|\overline{V}_\alpha| = 8, \quad W = Z_\alpha Z_{\alpha+1} \quad \text{and} \quad |\overline{W}| = 2.$

As $Z_{\alpha'} \leq G_\alpha$ and $Z_{\alpha'} \not\leq Q_\alpha$ we get that $[\overline{V}_0, Z_{\alpha'}] \neq 1$. On the other hand, $b = 2$ and thus
$$[V_\alpha, Z_{\alpha'}] \leq [V_\alpha, V_{\alpha+1}] \leq W,$$

so $\overline{W} = [\overline{V}_0, Z_{\alpha'}]$. But then $\langle \overline{W}^{G_\alpha} \rangle = \overline{V}_0$, which contradicts (8), (9) and (10).   □

10.3.11   Theorem. *Suppose that $\mathcal{A}$ holds. Then either*

$$P_1 \cong P_2 \cong S_4 \quad or \quad P_1 \cong P_2 \cong C_2 \times S_4.$$

Proof. Assume that G is a counterexample. Among all (G, P_1, P_2, T) that satisfy $\mathcal{A}$ but not $\mathcal{B}$, we choose (G, P_1, P_2, T) such that $|T|$ is minimal. Then $b > 1$ (10.3.7) and $(\alpha - 1, \alpha' - 1)$ is not critical for all $\alpha - 1 \in \Delta(\alpha) \setminus \{\alpha + 1\}$ (10.3.10). Hence 10.3.9 implies:

(1) $\qquad\qquad b \equiv 0 \pmod 2 \quad \text{and} \quad X := Q_\alpha \cap Q_{\alpha+1} \trianglelefteq G_\alpha.$

Moreover by 10.3.3 (b) and 10.3.4 (a) $|Q_\alpha : X| = |Q_{\alpha+1} : X| = 2$. Let

$$D \in \mathrm{Syl}_3\, G_\alpha \quad \text{and} \quad \overline{G}_\alpha := G_\alpha / X.$$

Then $\overline{G}_\alpha$ is a group of order 12 and $\overline{Q}_\alpha$ a normal subgroup of order 2. It follows that $\overline{D}$ is also normal in $\overline{G}_\alpha$. Let $X \leq L \leq G_\alpha$ such that $\overline{L} = \overline{DQ}_{\alpha+1}$. We obtain:

(2a) L is a normal subgroup of index 2 in G_α,

(2b) $\overline{L} \cong S_3$,

(2c) $\mathrm{Syl}_2\, L = \{Q_\beta \mid \beta \in \Delta(\alpha)\}$,

(2d) $O_2(L) = X = Q_\alpha \cap Q_\beta$ for all $\beta \in \Delta(\alpha)$,

(2e) $Q_{\alpha+1} = Z_{\alpha'} O_2(L)$ (10.3.8 (b)),

(2f) $C_L(O_2(L)) \leq O_2(L)$.

For the proof of (2f) note that Z_α $(\trianglelefteq G_\alpha)$ is contained in $Q_{\alpha+1}$ and thus also in $O_2(L)$. Hence

$$C_L(O_2(L)) \leq C_L(Z_\alpha) \overset{10.3.6}{\leq} Q_\alpha \cap L \leq O_2(L).$$

According to $\mathcal{A}_4$ there exists an element $t \in G_{\alpha+1} \setminus Q_{\alpha+1}$ such that

$$\alpha^t = \alpha + 2 \quad \text{und} \quad t^2 \in Q_{\alpha+1}.$$

Thus $Q_{\alpha+1} = (Q_{\alpha+1})^t$ is a Sylow 2-subgroup of L $(\leq G_\alpha)$ and L^t $(\leq G_{\alpha+2})$. First we show:

(3) $\qquad\qquad\qquad O_2(L)$ is not elementary Abelian.

Assume that $O_2(L)$ is a counterexample. Then by (2b), (2d), $A_1 := O_2(L)$ and $A_2 := O_2(L^t)$ are two elementary Abelian subgroups of index 2 in $Q_{\alpha+1}$. If $A_1 = A_2$, then A_1 is normal in $\langle G_\alpha, G_{\alpha+2} \rangle$ and thus also in

$$\langle G_\alpha, G_\alpha \cap G_{\alpha+1}, G_{\alpha+1} \cap G_{\alpha+2} \rangle = \langle G_\alpha, G_{\alpha+1} \rangle = G.$$

But this contradicts $\mathcal{A}_3$ and (2f).

We have shown that $A_1 \neq A_2$. By (2b), (2c), and (2f) $Q_{\alpha+1}$ is non-Abelian; so

$$A := A_1 \cap A_2 = Z(Q_{\alpha+1}) \quad \text{and} \quad |Q_{\alpha+1}/A| = 4.$$

If $O^2(G_{\alpha+1})$ acts trivially on $Q_{\alpha+1}/A$, then

$$\langle G_\alpha, O^2(G_{\alpha+1}) \rangle \leq N_G(A_1),$$

which contradicts 10.3.3. Thus $O^2(G_{\alpha+1})$ acts transitively on $(Q_{\alpha+1}/A)^\#$. But then every element of $Q_{\alpha+1}^\#$ is an involution and $Q_{\alpha+1}$ is elementary Abelian, again a contradiction. This shows (3).

We now set

$$G_0 := \langle L, L^t \rangle,$$

and denote the largest normal subgroup of G_0 in $Q_{\alpha+1}$ by Q. Note that $G_0^t = G_0$ and thus also $Q^t = Q$. Next we show:

(4) $[Q, D] \neq 1$.

For the proof of (4) assume that $[Q, D] = 1$ and set

$$\widetilde{G}_0 := G_0/Q.^6$$

Because of $Q_{\alpha+1} \in \mathrm{Syl}_2\, L \cap \mathrm{Syl}_2\, L^t$ and (2) the quadruple

$$(\widetilde{G}_0, \widetilde{L}, \widetilde{L}^t, \widetilde{Q}_{\alpha+1})$$

satisfies the hypotheses $\mathcal{A}_2, \mathcal{A}_3, \mathcal{A}_4$. Moreover, since $[Q, D] = 1$ 10.3.6 (c) and 8.2.2 on page 184 imply

$$\widetilde{W} := [\widetilde{Z}_\alpha, \widetilde{D}] \neq 1 \quad (Q \leq W \leq O_2(L)),$$

and $C_{\widetilde{L}}(O_2(\widetilde{L})) \leq O_2(\widetilde{L})$; so also $\mathcal{A}_1$ and $\mathcal{A}_5$ hold.

---

[6]Tilde instead of bar convention.

Now the minimality of $|T|$ is used. Since $|Q_{\alpha+1}| < |T|$ we get

$$\widetilde{L} \cong S_4 \quad \text{or} \quad \widetilde{L} \cong C_2 \times S_4.$$

In particular, by 10.3.5

$$\widetilde{W} = [O_2(\widetilde{L}), O^2(\widetilde{L})] \not\leq O_2(\widetilde{L}^t),$$

and $\widetilde{W} \leq \widetilde{Z}_\alpha$ implies $Z_\alpha \not\leq O_2(L^t)$. Now (2b) gives

$$O_2(L) = (O_2(L) \cap O_2(L^t)) Z_\alpha.$$

Since $Z_\alpha \leq \Omega(Z(O_2(L)))$ we get

$$\Phi(O_2(L)) = \Phi(O_2(L) \cap O_2(L^t)),$$

and conjugation with t shows that $\Phi(O_2(L)) = \Phi(O_2(L^t))$. But now—as in the proof of step (3)—$\Phi(O_2(L))$ is normal in $\langle G_\alpha, G_{\alpha+2}\rangle = G$, and $\mathcal{A}_3$ yields $\Phi(O_2(L)) = 1$. This contradicts (3), and (4) is established.

(5) Let $\beta \in \Delta(\alpha)$ and $\gamma \in \Delta(\beta) \setminus \{\alpha\}$. Then $\langle Z_\alpha, Z_\gamma\rangle$ is not a normal in L.

We fix the notation $\Delta(\beta) = \{\alpha, \gamma, \delta\}$ and set

$$V_\beta := \langle Z_\alpha, Z_\gamma, Z_\delta\rangle \quad (\trianglelefteq G_\beta).$$

Every $x \in Q_\alpha \setminus Q_\beta$ interchanges γ and δ and normalizes L (see (2a)). If $\langle Z_\alpha, Z_\gamma\rangle$ is normal in L, then also $\langle Z_\alpha, Z_\delta\rangle = \langle Z_\alpha, Z_\gamma{}^x\rangle$ is normal in L. This implies that V_β is normal in L $(\not\leq G_\alpha \cap G_\beta)$, which contradicts 10.3.3.

(6) Let $b \geq 4$, $\alpha - 1 \in \Delta(\alpha) \setminus \{\alpha+1\}$ and $\alpha - 2 \in \Delta(\alpha-1) \setminus \{\alpha\}$. Then $(\alpha - 2, \alpha' - 2)$ is a critical pair.

Assume that $(\alpha - 2, \alpha' - 2)$ is not critical. Then $Z_{\alpha-2} \leq Q_{\alpha'-3} \cap Q_{\alpha'-2}$. Since by (1) $\alpha' - 2$ is conjugate to α we get

$$Z_{\alpha-2} \leq Q_{\alpha'-3} \cap Q_{\alpha'-2} \overset{(1)}{=} Q_{\alpha'-2} \cap Q_{\alpha'-1} \leq G_{\alpha'-1} \cap G_{\alpha'} \overset{10.3.8\,(b)}{=} Z_\alpha Q_{\alpha'},$$

so $[Z_{\alpha-2}, Z_{\alpha'}] \leq [Z_\alpha, Z_{\alpha'}] \leq Z_\alpha$. Hence $Z_{\alpha-2}Z_\alpha$ $(\leq Q_\alpha \cap Q_{\alpha-1})$ is normalized by $Z_{\alpha'}$ and also by $Q_{\alpha-1}$ $(\leq G_{\alpha-2} \cap G_\alpha)$. Now (2) implies that $Z_{\alpha-2}Z_\alpha$ is normal in L, which contradicts (5).

In the following let $\alpha - 1 \in \Delta(\alpha) \setminus \{\alpha + 1\}$ and $x \in L$ $(\leq G_\alpha)$ such that $(\alpha + 1)^x = \alpha - 1$. Then

$$\alpha - 2 := (\alpha + 2)^x$$

is adjacent to $\alpha - 1$ and different from α and $\alpha + 2$.

Let $b \geq 4$. By (6) $(\alpha - 2, \alpha' - 2)$ is critical. Hence 10.3.8 implies

$$R_2 := [Z_{\alpha-2}, Z_{\alpha'-2}] \leq Z(G_{\alpha-2} \cap G_{\alpha-1}) \cap Z_{\alpha'-2}.$$

In addition, $b \geq 4$ implies $Z_{\alpha'} \leq Q_{\alpha'-2}$, thus also $[R_2, Z_{\alpha'}] = 1$. Now (2) gives $[R_2, L] = 1$, and

$$R_2 \leq Z(G_{\alpha+2} \cap G_{\alpha+1})$$

since $x \in L$.

As also (α', α) is a critical pair (10.3.8 (a)) there exists $\alpha' + 2$ such that $d(\alpha', \alpha' + 2) = 2$ and $(\alpha' + 2, \alpha + 2)$ is critical. Assume that $b > 4$. Then $Z_{\alpha'+2} \leq Q_{\alpha'-2}$ and thus

$$[R_2, Z_{\alpha'+2}] = 1$$

since $R_2 \leq Z_{\alpha'-2}$. Hence

$$G_{\alpha+2} \cap G_{\alpha+3} \overset{10.3.8\,(b)}{=} Q_{\alpha+2} Z_{\alpha'+2}$$

is centralized by R_2. It follows that $R_2 \leq Z(G_{\alpha+2})$ and also $R_2 \leq Z(G_{\alpha-2})$ after conjugation with $x \in L$. This contradicts the action of $Z_{\alpha'-2}$ on $Z_{\alpha-2}$, see 10.3.8 (b), (e), and (f).

We have shown:

(7) $b \leq 4,$

We will now derive a final contradiction showing that (7) and (4) contradict each other.

As $Q \leq O_2(L^t) \leq Q_{\alpha+2}$ we have

$(')$ $[Q, Z_{\alpha+2}] = 1.$

We now distinguish the two cases $Z_{\alpha+2} \not\leq O_2(L)$ and $Z_{\alpha+2} \leq O_2(L)$.

In the first case $Q_{\alpha+1} = O_2(L) Z_{\alpha+2}$ and

$$L = \langle Z_{\alpha+2}^L \rangle O_2(L) = C_L(Q) O_2(L).$$

This implies $O^2(L) \leq C_L(Q)$ since Q is normal in L. In particular $[Q, D] = 1$, which contradicts (4).

Thus we are in the case $Z_{\alpha+2} \leq O_2(L)$. Then $Z_{\alpha+2} \leq Q_\alpha$, and (7) and (1) show that $b = 4$. Hence (6) implies

$$Z_{\alpha+2} \not\leq Q_{\alpha-2} = Q_{(\alpha+2)^x},$$

and L^{tx} is a normal subgroup of index 2 in $G_{\alpha-2}$. The subgroup

$$\langle (Z_{\alpha+2})^{L^{tx}} \rangle \quad (\leq G_0)$$

contains a Sylow 3-subgroup D_2 of $G_{\alpha-2}$. Now as above $(')$ and $Q \trianglelefteq G_0$ show that $[Q, D_2] = 1$. This contradicts (4) since D_2 is a G_0-conjugate of the Sylow 3-subgroup D of G_α. $\qquad\square$

We conclude this section with two examples of groups satisfying $\mathcal{A}$ and thus also $\mathcal{B}$. They are also examples for the two alternatives of $\mathcal{B}$.

(1) Let G be the symmetric group S_6 and

$$a := (1\,2), \quad b := (1\,2)\,(3\,4)\,(5\,6),$$

and let

$$P_1 := C_G(a), \quad P_2 := C_G(b).$$

Then for $x \in G$

$$x \in P_1 \iff \{1,2\}^x = \{1,2\} \iff \{3,4,5,6\}^x = \{3,4,5,6\}.$$

Thus

$$P_1 = \langle a \rangle \times G_{1,2} \cong C_2 \times S_4, {}^{7}$$
$$O_2(P_1) = \langle a \rangle \times \langle (3\,4)\,(5\,6) \rangle \times \langle (3\,5)\,(4\,6) \rangle,$$

and

$$T := O_2(P_1)\,\langle (3\,4) \rangle \in \mathrm{Syl}_2\, P_1.$$

Similarly for $x \in G$ and $\Omega := \{(12),(34),(56)\}$

$$x \in P_2 \iff \Omega^x = \Omega.$$

Hence P_2 acts on Ω. The kernel of this action is

$$N := \langle (1\,2) \rangle \times \langle (3\,4) \rangle \times \langle (5\,6) \rangle,$$

---

[7] $G_{1,2} := \{x \in G \mid 1^x = 1 \text{ and } 2^x = 2\}.$

and

$$P_2/N \cong S_\Omega \cong S_3.$$

It follows that $N = O_2(P_2)$ and

$$N \langle (3\,5)\,(4\,6) \rangle = T$$

is a Sylow 2-subgroup of P_2. Note further that $P_1 \neq P_2$ and $|P_1 : T| = 3 = |P_2 : T|$, so $P_1 \cap P_2 = T$. In addition

$$|G : \langle P_1, P_2 \rangle| \leq \frac{|G|}{|P_1 P_2|} = \frac{6!}{3 \cdot 48} = 5$$

and thus $G = \langle P_1, P_2 \rangle$ since P_1 is not contained in the simple group A_6 (see also 3.1.2 on page 57). This shows that the triple $(\langle P_1, P_2 \rangle, P_1, P_2)$ satisfies $\mathcal{A}$.

It should be noted that the triple $(A_6, P_1 \cap A_6, P_2 \cap A_6)$ is an example for the other alternative in $\mathcal{B}$.

(2) Let $G := \mathrm{GL}_3(2)$ be the group of invertible 3×3-matrices over $\mathbb{F}_2$. Let P_1 be the set of all $x \in G$ of the form

$$x = \begin{pmatrix} a & b & c \\ 0 & d & e \\ 0 & f & g \end{pmatrix},$$

and P_2 be the set of all $x \in G$ of the form

$$x = \begin{pmatrix} a & b & c \\ d & e & f \\ 0 & 0 & g \end{pmatrix},$$

$a, b, c, d, e, f, g \in \mathbb{F}_2$. Then P_1 and P_2 are subgroups of G. The mappings

$$\varphi_1 : P_1 \to \mathrm{SL}_2(2) \text{ such that } x \mapsto \begin{pmatrix} d & e \\ f & g \end{pmatrix},$$

$$\varphi_2 : P_2 \to \mathrm{SL}_2(2) \text{ such that } x \mapsto \begin{pmatrix} a & b \\ d & e \end{pmatrix}$$

are epimorphisms with kernels

$$\mathrm{Ker}\ \varphi_1 = \left\{ \begin{pmatrix} 1 & b & c \\ 0 & 1 & 0 \\ 0 & 0 & 1 \end{pmatrix} \mid b, c \in \mathbb{F}_2 \right\} \cong C_2 \times C_2,$$

$$\mathrm{Ker}\ \varphi_2 = \left\{ \begin{pmatrix} 1 & 0 & c \\ 0 & 1 & f \\ 0 & 0 & 1 \end{pmatrix} \mid c, f \in \mathbb{F}_2 \right\} \cong C_2 \times C_2.$$

As Ker φ_i $(i = 1, 2)$ has a complement in P_i that acts faithfully on Ker φ_i, we get

$$P_1 \cong S_4 \cong P_2.$$

Moreover

$$P_1 \cap P_2 = \left\{ \begin{pmatrix} 1 & a & b \\ 0 & 1 & c \\ 0 & 0 & 1 \end{pmatrix} \mid a, b, c \in \mathbb{F}_2 \right\}$$

is a subgroup of order 8 and thus a Sylow 2-subgroup of P_1 and P_2. Since

$$|G : \langle P_1, P_2 \rangle| \leq \frac{|G|}{|P_1 P_2|} = \frac{168}{72} < 3$$

and the order of G is not divisible by 16 we get $G = \langle P_1, P_2 \rangle$. Hence, the triple $(\langle P_1, P_2 \rangle, P_1, P_2)$ satisfies $\mathcal{A}$.

Chapter 11

Signalizer Functors

In previous chapters it became quite clear that normalizers of nontrivial p-subgroups (i.e., p-local subgroups) are of particular importance for the structure of finite groups. In this chapter we introduce another important concept that in a certain sense is dual to the concept of *normalizing*.

Let G be a group and A a subgroup of G. While the concept of the *normalizer* of A deals with subgroups U satisfying $A^U = A$, we are now interested in subgroups U satisfying $U^A = U$.

This dualization of the notion of a normalizer is one of the fundamental ideas in the proof of the theorem of Feit-Thompson. It was then Gorenstein who developed the general concept of a signalizer functor [60], one of the great achievements in group theory.

The goal of this chapter is to give a proof of the Completeness Theorem of Glauberman [52]. Two important special cases of this theorem ($r(A) \geq 4$ and $p = 2$) had been proved earlier by Goldschmidt ([56], resp. [55]). Another proof was given by Bender [31]. Later this proof was generalized by Aschbacher [1] to obtain a new proof of the Completeness Theorem.

In this chapter, A is always a p-group, while the A-invariant subgroups under consideration usually are p'-subgroups. Therefore we will frequently use elementary facts about coprime action, as given in 8.2.2, 8.2.3, 8.2.7, and 8.3.4. We will refer to such properties using the abbreviation (cp).

We find it appropriate to regard A not as a subgroup of G but as a group that acts on G.

11.1   Definitions and Elementary Properties

In the following p is a prime and A is a noncyclic elementary Abelian p-group that acts on the group G. Note that for every $a \in A$ the fixed-point groups $C_G(a)$ are A-invariant since A is Abelian.

Let U be an A-invariant p'-subgroup of G. Then 8.3.4 on page 193 implies that

$$(+) \qquad\qquad U = \langle\, C_G(a) \cap U \mid a \in A^\# \,\rangle,$$

and this gives rise to the following generalization.

Let θ be a mapping that associates with every $a \in A^\#$ an A-*invariant* and *solvable* p'-subgroup of $C_G(a)$.[1] This subgroup is denoted by $\theta(C_G(a))$, so

$$\theta(C_G(a)) := a^\theta.$$

By $\mathcal{W}_\theta(A)$ we denote the set of solvable A-invariant p'-subgroups $U \leq G$ satisfying

$$C_G(a) \cap U \leq \theta(C_G(a)) \text{ for all } a \in A^\#.$$

In other words, for $U \in \mathcal{W}_\theta(A)$

$$(+') \qquad\qquad U = \langle\, \theta(C_G(a)) \cap U \mid a \in A^\# \rangle.$$

Moreover, it is clear from the definition that every A-invariant subgroup of U is again in $\mathcal{W}_\theta(A)$.

The mapping θ is a **solvable A-signalizer functor** on G if

$$\mathcal{S} \qquad\qquad \theta(C_G(a)) \cap C_G(b) \leq \theta(C_G(b)) \text{ for all } a, b \in A^\#,$$

or equivalently

$$\mathcal{S}' \qquad\qquad \theta(C_G(a)) \in \mathcal{W}_\theta(A) \text{ for all } a \in A^\#.$$

The solvable A-signalizer functor θ is **complete** if $\mathcal{W}_\theta(A)$ contains a unique maximal element.[2] If θ is complete, then the unique maximal element of $\mathcal{W}_\theta(A)$ is denoted by $\theta(G)$.

Let θ be a solvable A-signalizer functor on G and set

$$E := \langle \theta(C_G(a)) \mid a \in A^\# \rangle.$$

---

[1] Most results of this section also hold without the solvability requirement.

[2] With respect to inclusion.

Property $(+)$ implies that θ is complete if and only if $E \in \mathbb{N}_\theta(A)$. Thus if θ is complete, then $E = \theta(G)$.

The goal of this chapter is to show that θ is complete if $r(A) \geq 3$—this is the Completeness Theorem of Glauberman.

For an A-invariant subgroup H of G

$$\theta_H : a \mapsto \theta_H(C_H(a)) := \theta(C_G(a)) \cap H \quad (a \in A^{\#})$$

is the **restriction** of θ to H. Here

$$\mathbb{N}_{\theta_H}(A) = \{U \in \mathbb{N}_\theta(A) \mid U \leq H\}.$$

If θ is complete, then also θ_H is complete and

$$\theta_H(H) = \theta(G) \cap H.$$

If θ_H is complete, we set $\theta(H) := \theta_H(H)$.

Condition $\mathcal{S}'$ says that for every $a \in A^{\#}$ the restriction $\theta_{C_G(a)}$ is complete with $\theta(C_G(a))$ being the unique maximal element. Evidently also $\theta_{C_G(B)}$ is complete for every nontrivial subgroup $B \leq A$, and

$$\theta(C_G(B)) = \theta(C_G(a)) \cap C_G(B) = \bigcap_{b \in B^{\#}} \theta(C_G(b)) \quad (a \in B^{\#}).$$

Before we continue with properties of signalizer functors we present a typical example:

11.1.1 *Let p be a prime and*

$$\theta : a \mapsto O_{p'}(C_G(a)) \quad (a \in A^{\#}).$$

(a) *Suppose that $C_G(a)$ is solvable for all $a \in A^{\#}$. Then θ is a solvable A-signalizer functor on G.*

(b) *Suppose that G is solvable. Then θ is complete and $\theta(G) = O_{p'}(G)$.*

Proof. (a) For $a, b \in A^{\#}$ 8.2.12 on page 189 gives

$$O_{p'}(C_G(a)) \cap C_G(b) \leq O_{p'}(C_{C_G(b)}(a)) \leq O_{p'}(C_G(b))$$

Thus $\mathcal{S}$ holds.

(b) Again by 8.2.12

$$\theta(C_G(a)) = O_{p'}(G) \cap C_G(a),$$

so $O_{p'}(G) \in \mathcal{W}_\theta(A)$ and

$$O_{p'}(G) = \langle \theta(C_G(a)) \mid a \in A^{\#} \rangle \in \mathcal{W}_\theta(A).$$

Hence θ is complete and $\theta(G) = O_{p'}(G)$. $\square$

In the following θ is a solvable A-signalizer functor on G. We set:

$\quad C_a := \theta(C_G(a))$ for $a \in A^{\#}$,

$\quad C_B := \theta(C_G(B))$ for $1 \neq B \leq A$,

$\quad \mathcal{W}_\theta^*(A)$: the set of maximal elements of $\mathcal{W}_\theta(A)$,

and for a set π of primes

$\quad \mathcal{W}_\theta(A, \pi) := \{U \in \mathcal{W}_\theta(A) \mid U \text{ is a } \pi\text{-group}\}$,

$\quad \mathcal{W}_\theta^*(A, \pi)$: set of maximal elements of $\mathcal{W}_\theta(A, \pi)$.

11.1.2   *Let $X, Y \in \mathcal{W}_\theta(A)$ such that $XY = YX$. Suppose that*

(1) $Y \leq N_G(X)$,   *or*

(1') XY *is solvable.*

Then $XY \in \mathcal{W}_\theta(A)$.

Proof. Also in case (1) XY is solvable (see 6.1.2 on page 122). Thus, in both case XY is an A-invariant solvable p'-group. Hence, 8.2.11 on page 188 implies

$$C_{XY}(a) = C_X(a)\,C_Y(a) \leq C_a \text{ for all } a \in A^{\sharp},$$

so $XY \in \mathcal{W}_\theta(A)$. $\square$

11.1.3   *Let N be an A-invariant normal p'-subgroup of G and $\overline{G} := G/N$. Then the mapping*

$$\overline{\theta} \colon a \mapsto \overline{\theta}(C_{\overline{G}}(a)) := \overline{C}_a \quad (a \in A^{\#})$$

is a solvable A-signalizer functor on $\overline{G}$, and $\overline{\mathcal{W}_\theta(A)} \subseteq \mathcal{W}_{\overline{\theta}}(A)$.

Proof. Let $a, b \in A^{\#}$ and $M := NC_a$, so

$$\overline{M} = \bar{\theta}(C_{\overline{G}}(a)) \quad \text{and} \quad C_{\overline{M}}(b) \overset{\text{(cp)}}{=} \overline{C_M(b)}.$$

It follows that

$$C_M(b) \overset{8.2.11}{=} C_N(b) C_{C_a}(b) = C_N(b) (C_a \cap C_G(b)) \leq N \theta(C_G(b)),$$

and thus

$$\bar{\theta}(C_{\overline{G}}(a)) \cap C_{\overline{G}}(b) = C_{\overline{M}}(b) = \overline{C_M(b)} \leq \bar{\theta}(C_{\overline{G}}(b)).$$

Similarly for $U \in \mathbb{N}_\theta(A)$

$$C_{\overline{U}}(a) \overset{\text{(cp)}}{=} \overline{C_U(a)} = \overline{C_a \cap U} \leq \bar{\theta}(C_{\overline{G}}(a)),$$

and $\overline{U} \in \mathbb{N}_{\bar{\theta}}(A)$. $\qquad\square$

11.1.4 *Assume in 11.1.3, in addition, that $N \in \mathbb{N}_\theta(A)$. Then $\overline{\mathbb{N}_\theta(A)} = \mathbb{N}_{\bar{\theta}}(A)$. In particular θ is complete if and only if $\bar{\theta}$ is complete.*

Proof. Let $N \leq U \leq G$ such that $\overline{U} \in \mathbb{N}_{\bar{\theta}}(A)$. According to 11.1.3 it suffices to show that $U \in \mathbb{N}_\theta(A)$.

For $a \in A^{\#}$

$$\overline{C_U(a)} \overset{\text{(cp)}}{=} C_{\overline{U}}(a) \leq \bar{\theta}(C_{\overline{G}}(a)) = \overline{C}_a = C_a N/N$$

and thus

$$C_U(a) \leq NC_a \cap C_G(a) = C_N(a) C_a \leq C_a$$

since $N \in \mathbb{N}_\theta(A)$; so $U \in \mathbb{N}_\theta(A)$. $\qquad\square$

We now set

$$\pi(\theta) := \bigcup_{a \in A^{\#}} \pi(C_a).$$

11.1.5 *Let $U \in \mathbb{N}_\theta(A)$. Then $\pi(U) \subseteq \pi(\theta)$.*

Proof. By (cp) for every $q \in \pi(U)$ there exists an A-invariant Sylow q-subgroup Q of U. Moreover, since A is noncyclic there exists $a \in A$ such that $C_Q(a) \neq 1$ (see 8.3.4). As $C_Q(a) \leq C_U(a) \leq C_a$ we get $q \in \pi(\theta)$.   □

It is clear that the restriction of θ to A-invariant subgroups and the solvable A-signalizer functor $\overline{\theta}$ (as in 11.1.3) can be used in proofs by induction. But in this context another variation of θ that reduces the number of primes in $\pi(\theta)$ is more important.

Let π be a set of primes such that $p \notin \pi$.[3] We apply 8.2.6 (d) to the solvable and A-invariant p'-group $C_a = \theta(C_G(a))$ $(a \in A^\#)$. Then C_a contains a unique maximal AC_A-invariant π-subgroup,[4] and this subgroup we denote by $\theta_\pi(C_G(a))$.

11.1.6   *The mapping*

$$\theta_\pi \colon a \mapsto \theta_\pi(C_G(a)) \quad (a \in A^\sharp)$$

is a solvable A-signalizer functor on G satisfying

$$\pi(\theta_\pi) \subseteq \pi \quad \text{and} \quad \{U \in \mathsf{M}_\theta(A, \pi) \mid U^{C_A} = U\} \subseteq \mathsf{M}_{\theta_\pi}(A).$$

Proof. Let $a, b \in A^\sharp$. Then $\theta_\pi(C_G(a)) \cap C_G(b)$ is an AC_A-invariant π-subgroup of C_b. Thus, $\theta_\pi(C_G(a)) \cap C_G(b)$ is contained in the unique maximal AC_A-invariant π-subgroup $\theta_\pi(C_G(b))$. This shows that θ_π is a solvable A-signalizer functor on G.

Clearly $\pi(\theta_\pi) \subseteq \pi$, and the other property again follows from the uniqueness of $\theta_\pi(C_G(a))$.   □

In the proof of the next lemma we use an argument that is also useful in other situations.

11.1.7   *Let A be an elementary Abelian p-group with $r(A) \geq 3$ that acts on the group X, and let $p \neq q \in \pi(X)$. Suppose that Q_1 and Q_2 are two A-invariant q-subgroups of X such that, for $D := Q_1 \cap Q_2$,*

$$Q_1 \neq D \neq Q_2.$$

Then there exists $a \in A^\#$ such that

$$N_G(D) \cap C_{Q_i}(a) \nleq D \text{ for } i = 1, 2.$$

---

[3] In this context observe that 1 is the only π-subgroup for $\pi = \varnothing$.

[4] AC_A-invariant means A-invariant and C_A-invariant.

Proof. Since $Q_1 \neq D \neq Q_2$ we get

$$D < N_{Q_i}(D) =: N_i \quad \text{for} \quad i \in \{1, 2\};$$

and 8.3.4 on page 193 shows that N_i is generated by the subgroups $C_{N_i}(B)$, $B \leq A$, and $r(A/B) \leq 1$. In particular, for $i = 1, 2$, there exists a maximal subgroup B_i of A such that

$$C_{N_i}(B_i) \not\leq D.$$

As $r(A) \geq 3$ we get $B_1 \cap B_2 \neq 1$. Now choose $1 \neq a \in B_1 \cap B_2$. □

Our first major result is:

11.1.8   Transitivity Theorem. *Let θ be a solvable A-signalizer functor on G and $q \in \pi(\theta)$. Suppose that $r(A) \geq 3$. Then the elements in $\mathcal{W}_\theta^*(A, q)$ are conjugate under C_A.*

Proof. Assume that the assertion is false. Among all pairs of elements of $\mathcal{W}_\theta^*(A, q)$ that are not conjugate under C_A we choose Q_1 and Q_2 such that

$$D := Q_1 \cap Q_2$$

is maximal. Set

$$N := N_G(D) \quad \text{and} \quad N_a := N \cap C_a \quad (a \in A^\#).$$

By 11.1.7 there exists $a \in A^\#$ such that

$$N_a \cap Q_1 \not\leq D \quad \text{and} \quad N_a \cap Q_2 \not\leq D.$$

Moreover, N_a is an A-invariant p'-group. Thus by 8.2.3 (b) and (c) there exists an element $c \in C_{N_a}(A)$ $(\leq C_A)$ such that

$$E := \langle (N_a \cap Q_1)^c, N_a \cap Q_2 \rangle$$

is an A-invariant q-subgroup of N_a. As D and E are in $\mathcal{W}_\theta(A, q)$, by 11.1.2 also $DE \in \mathcal{W}_\theta(A, q)$. Hence, there exists $Q_3 \in \mathcal{W}_\theta^*(A, q)$ containing DE, so

$$D < D(N_a \cap Q_1)^c \leq Q_1{}^c \cap Q_3 \quad \text{and} \quad D < D(N_a \cap Q_2) \leq Q_2 \cap Q_3.$$

The maximal choice of D implies that $Q_1{}^c$ and Q_3 are conjugate under C_A as well as Q_2 and Q_3. But then also Q_1 and Q_2 are conjugate under C_A, a contradiction. □

We note two corollaries of the Transitivity Theorem:

11.1.9   Let $q \in \pi(\theta)$ and $Q \in \mathbb{N}^*_\theta(A, q)$. Suppose that $r(A) \geq 3$. Then the following hold:

(a)   For every $H \in \mathbb{N}_\theta(A)$ there exists $c \in C_A$ such that $Q^c \cap H$ is an A-invariant Sylow q-subgroup of H.

(b) $C_Q(B)$ is an A-invariant Sylow q-subgroup of C_B for every $1 \neq B \leq A$.

Proof. (a) Every A-invariant Sylow q-subgroup Q_1 of H is in $\mathbb{N}_\theta(A, q)$ and thus contained in an element $Q_2 \in \mathbb{N}^*_\theta(A, q)$. By 11.1.8 there exists $c \in C_A$ such that $Q_2 = Q^c$, and $Q^c \cap H = Q_1$ follows.

(b) follows from (a) (with $H := C_B$) since $C_A \leq C_B$. $\square$

11.1.10   If $|\pi(\theta)| \leq 1$ and $r(A) \geq 3$, then θ is complete.

Proof. The case $\pi(\theta) = \varnothing$ gives $\theta(G) = 1$. Assume that $\pi(\theta) = \{q\}$. Then $\mathbb{N}_\theta(A) = \mathbb{N}_\theta(A, q)$, and there exists $Q \in \mathbb{N}^*_\theta(A)$ such that $C_A \leq Q$. By 11.1.8 Q is the only element in $\mathbb{N}^*_\theta(A, q)$. $\square$

We conclude this section with an example that shows that one cannot drop the hypothesis $r(A) \geq 3$ in the Completeness Theorem of Glauberman (11.3.2).

Let q be an odd prime and V an elementary Abelian group of order q^2 with generators v, w, i.e.,

$$V = \langle v \rangle \times \langle w \rangle \cong C_q \times C_q.^5$$

Let $x, t, z \in \operatorname{Aut} V$ be defined by

$$
\begin{aligned}
(v^x, w^x) &:= (v, vw), \\
(v^t, w^t) &:= (v^{-1}, w), \\
(v^z, w^z) &:= (v^{-1}, w^{-1}),
\end{aligned}
$$

---

[5]V is—written additively—a vector space over $\mathbb{F}_q$ with basis v, w.

and let U be the subgroup of $\operatorname{Aut} V$ generated by x, t, z. Then

$$[t, z] = [z, x] = 1 \quad \text{and} \quad x^t = x^{-1}.$$

Let H be the semidirect product of U with V. We identify U and V with the corresponding subgroups of H. Then

$$G := V\langle x \rangle$$

is a non-Abelian normal subgroup of order q^3 and $A := \langle t, z \rangle$ an elementary Abelian subgroup of order 4 in H satisfying:

$(')$ $C_G(A) = 1,$

$('')$ $G = \langle x, w \rangle = \langle C_G(z), C_G(t) \rangle,$

$(''')$ $\langle v \rangle^A = \langle v \rangle \leq C_G(tz).$

We now define

$$\theta(C_G(t)) := C_G(t), \quad \theta(C_G(z)) := C_G(z), \quad \text{and} \quad \theta(C_G(tz)) := 1.$$

By $(')$ θ is a solvable A-signalizer functor on G.

Assume that θ is complete on G. Then by $('')$ G is the maximal element of $\mathbb{W}_\theta(A)$. But then $C_G(tz) = \theta(C_G(tz)) = 1$, which contradicts $(''')$. Hence, θ is not complete.

11.2   Factorizations

As in the previous section, A is a noncyclic elementary Abelian p-group that acts on the group G, and θ is a solvable A-signalizer functor on G.

In earlier chapters we have learned that global properties of groups can be deduced from local properties carried by their p-local subgroups.

In the case of signalizer functors we follow a similar strategy. Now the *carriers* of local information are not the p-local subgroups, but the θ-local subgroups: the normalizers of nontrivial subgroups of $\mathbb{W}_\theta(A)$.

We introduce the following notation:

For $q \in \pi(\theta)$ set $\theta_{q'} := \theta_{\pi(\theta) \setminus \{q\}}$, where the signalizer functor on the right-hand side is defined as in 11.1.6. Then θ is said to be **locally complete** on G if

- $\theta_{N_G(U)}$ is complete for all nontrivial $U \in \mathbb{N}_\theta(A)$, and

- $\theta_{q'}$ is complete for every $q \in \pi(\theta)$.

Actually, this notion is rather unnecessary since it will turn out in the next section that for $r(A) \geq 3$ locally complete solvable A-signalizer functors are complete. But we have introduced this notion for two reasons: Firstly it should again emphasize the importance of local properties in group theory. Secondly it can be used to partition a long proof into independent intermediate results.

In this section we deal with the case $\pi(\theta) \neq \{2,3\}$, in particular with the case $p = 2$. We will show that in this case every locally complete solvable A-signalizer functor θ with $r(A) \geq 3$ is complete.

11.2.1 *Let G be a p'-group and let X and Y be A-invariant subgroups of G. Suppose that*

(1) $C_G(a) = C_X(a)\, C_Y(a)$ *for all $a \in A^{\#}$, and*

(2) X *is $C_G(A)$-invariant.*

Then $G = XY$.

Proof. Let q be a prime divisor of $|G|$ (for $G = 1$ there is nothing to prove). First assume that G is a nontrivial q-group, so $Z(G) \neq 1$. Since A is noncyclic there exists $a \in A^{\#}$ such that

$$N := C_{Z(G)}(a) \neq 1.$$

It is evident that N is A-invariant and that $\overline{G} := G/N$ satisfies the hypothesis (with respect to $\overline{X}, \overline{Y}$). Using induction on $|G|$ we may assume that $\overline{G} = \overline{XY}$, so

$$G = XYN = XNY = XC_G(a)Y \stackrel{(1)}{=} XY.$$

We now deduce the general case from the case just treated.

Let $a \in A^{\#}$. According to (cp), applied to $C_Y(a)$, Y, and G, there exists an A-invariant Sylow q-subgroup of G such that

$$Q \cap Y \in \mathrm{Syl}_q Y \quad \text{and} \quad Q \cap C_Y(a) \in \mathrm{Syl}_q C_Y(a).$$

Moreover, since X, $C_X(a)$ and $C_G(A)$ are $C_G(A)$-invariant we get from 8.2.5 on page 186 that

$$X \cap Q \in \mathrm{Syl}_q X, \ C_X(a) \cap Q \in \mathrm{Syl}_q C_X(a) \text{ and } C_Q(a) \in \mathrm{Syl}_q C_G(a).$$

As $X \cap Y$ and $C_{X \cap Y}(a)$ both are $C_Y(A)$-invariant, the same argument also yields

$$X \cap Y \cap Q \in \mathrm{Syl}_q X \cap Y \text{ and } C_{X \cap Y}(a) \cap Q \in \mathrm{Syl}_q C_{X \cap Y}(a).$$

Now

$$
\begin{aligned}
|C_Q(a)| &= |C_G(a)|_q \overset{(1)}{=} |C_X(a) \, C_Y(a)|_q \\
&= |C_X(a)|_q \, |C_Y(a)|_q \, |C_{X \cap Y}(a)|_q^{-1} \\
&= |C_X(a) \cap Q| \, |C_Y(a) \cap Q| \, |C_{X \cap Y}(a) \cap Q|^{-1} \\
&= |C_{X \cap Q}(a) \, C_{Y \cap Q}(a)|,
\end{aligned}
$$

and thus $C_Q(a) = C_{X \cap Q}(a) C_{Y \cap Q}(a)$. Since also $Q \cap X$ is $C_Q(A)$-invariant the above proved case gives $Q = (Q \cap X)(Q \cap Y)$. It follows that

$$|Q| = |X \cap Q||Y \cap Q||X \cap Y \cap Q|^{-1} = |X|_q |Y|_q |X \cap Y|_q^{-1} = |XY|_q.$$

Hence $|Q|$ divides $|XY|$ for every $q \in \pi(G)$, and $G = XY$. $\qquad\square$

In the following the notation is chosen as in Section 11.1. In particular

$$C_a := \theta(C_G(a)) \text{ for } a \in A^\# \quad \text{and} \quad C_B := \theta(C_G(B)) \text{ for } 1 \neq B \leq A.$$

11.2.2 Let θ be locally complete on G and $M \in \mathbb{N}_\theta^*(A)$. Suppose that $|\pi(\theta)| = 2$ and there exists an A-invariant subgroup $F \leq F(M)$ such that

$$O_q(F) \neq 1 \text{ for every } q \in \pi(\theta).$$

Then M is the unique element in $\mathbb{N}_\theta^*(A)$ containing F.

Proof. Let $\pi(\theta) = \{q, r\}$. We proceed by way of contradiction. Let (F, M) be a counterexample such that F is maximal, and set

$$F_q := O_q(F), \ F_r := O_r(F) \text{ and } N_q := N_G(F_q).$$

Note that θ_{N_q} is complete since $F_q \in \mathbb{N}_\theta(A)$. We first show:

($'$) M is the unique element in $\mathbb{M}_\theta^*(A)$ containing $\theta(N_q)$, $q \in \pi(\theta)$.

Let

$$\theta(N_q) \leq L \in \mathbb{M}_\theta^*(A).$$

Assume first that $F_q = O_q(M)$. Then $M \leq \theta(N_q)$, and the maximality of M yields $\theta(N) = M = L$.

Assume now that $F_q < O_q(M)$ and set

$$\widetilde{F} := N_{O_q(M)}(F_q) \times O_r(M).$$

Then $\widetilde{F}$ is A-invariant and $F < \widetilde{F}$. Hence $(\widetilde{F}, M)$ satisfies the hypothesis but is not a counterexample. Thus also in this case $M = L$, and ($'$) is proved.

Now let $F \leq H \in \mathbb{M}_\theta^*(A)$, so

$$N_H(F_q) \leq \theta(N_q) \overset{(')}{\leq} M.$$

It follows that $F_r \leq O_{q'}(N_H(F_q))$, and 8.2.13 on page 190 implies that $F_r \leq O_{q'}(H) = O_r(H)$.

The same argument with the roles of r and q reversed also shows that $F_q \leq O_{r'}(H) = O_q(H)$. Thus $F \leq F(H)$, and the pair (F, H) satisfies the hypothesis. Now either ($'$) also applies to (F, H), or (F, H) is not a counterexample. In both cases $\theta(N_q) \leq H$ since $F \leq \theta(N_q)$. Hence ($'$) shows that $H = M$. But then (F, M) is not a counterexample.   □

The following remark describes a situation we will meet in the next proofs.

11.2.3   *Let G be a p'-group. Suppose that*

$$\theta(C_G(a)) = C_G(a) \quad \text{for all } a \in A^\#.$$

Then the following hold:

(a) $\mathbb{M}_\theta(A)$ *is the set of all A-invariant solvable subgroups of G. In particular, every A-invariant Sylow subgroup is in $\mathbb{M}_\theta(A)$.*

(b) θ *is complete if and only if $G = \theta(G)$.*

(c) *Let θ be locally complete on G. Then*

$$1 \neq U \in \mathbb{N}_\theta(A) \quad \Rightarrow \quad N_G(U) \in \mathbb{N}_\theta(A). \qquad \square$$

In the following we investigate a factorization of G:

$$G = KQ \quad \text{with } K, Q \in \mathbb{N}_\theta(A).$$

Then G is a p'-group, and for all $a \in A^\#$

$$C_G(a) \overset{8.2.11}{=} C_K(a)C_Q(a) = \theta(C_G(a)).$$

Thus, we are in the situation of 11.2.3.

11.2.4 *Let θ be locally complete but not complete on G, $q \in \pi(\theta)$, and*

$$G = KQ \quad \text{with } K \in \mathbb{N}_\theta(A, q') \text{ and } Q \in \mathbb{N}_\theta(A, q).$$

(a) *Q does not normalize any nontrivial q'-subgroup of G.*

(b) *$F(U) \leq Q$ for all $Q \leq U \in \mathbb{N}_\theta(A)$.*

(c) *Let Q be Abelian. Then $U \leq N_G(Q)$ for all $Q \leq U \in \mathbb{N}_\theta(A)$.*

Proof. Note that $Q \in \mathrm{Syl}_q G$ and $\mathrm{Syl}_r K \subseteq \mathrm{Syl}_r G$ for $r \in \pi(K)$. Moreover, $G = KQ$ shows that

(1) $\mathrm{Syl}_r G = \bigcup\limits_{g \in Q} \mathrm{Syl}_r K^g$ for $r \in \pi(G) \setminus \{q\}$.

Since θ is locally complete but not complete, no nontrivial normal subgroup of G is in $\mathbb{N}_\theta(A)$. In particular

(2) $\bigcap\limits_{g \in G} K^g = \bigcap\limits_{g \in Q} K^g = 1$.

Let X be a q'-subgroup that is normalized by Q. According to (cp) for every $r \in \pi(X)$ there exists a Q-invariant Sylow r-subgroup of X. Thus to prove (a) we may assume that X itself is an r-group. Then (1) implies that $X \leq K$, so by (2)

$$X \leq \bigcap\limits_{g \in Q} K^g = 1.$$

Claim (b) is a direct consequence of (a) since $Q \in \mathrm{Syl}_q G$.

For the proof of (c) recall from 6.1.4 on page 123 that

$$F(U) \overset{(b)}{\le} Q \le C_U(F(U)) \le F(U)$$

since Q is Abelian, so $Q = F(U)$.   □

11.2.5 *Let θ be locally complete on G, $q \in \pi(\theta)$, and $Q \in \mathbb{M}_\theta(A, q)$. Suppose that Q is Abelian, $r(A) \ge 3$, and*

$$G = KQ \quad for \quad K := \theta_{q'}(G).$$

Then θ is complete and $\theta(G) = G$.

Proof. Note that the remark after 11.2.3 shows that G is a p'-group such that

$$C_a = C_G(a) \quad and \quad C_B = C_G(B) \quad for \ a \in A^\#, \ 1 \ne B \le A.$$

In particular, 11.2.3 applies to G. Hence, $\mathbb{M}_\theta(A)$ is the set of A-invariant solvable subgroups of G, and θ is complete if and only if G is solvable.

Thus, we may assume that G is not solvable. Moreover, as θ is locally complete, also the normalizers of A-invariant solvable subgroups are solvable, i.e.,

(1) $U \in \mathbb{M}_\theta(A) \Rightarrow N_H(U) \in \mathbb{M}_\theta(A)$.

The $\{p^a q^b\}$-Theorem of Burnside (10.2.1) shows that $|\pi(G)| \ge 3$, so $|\pi(K)| \ge 2$. Hence, as K is solvable, there exist $r, r_0 \in \pi(K)$ such that

$$1 \ne O_{r_0}(K) \le O_{r'}(K).$$

We fix the following notation:

$$Q_a := Q \cap C_a, \ a \in A^\#, \ Q_B := Q \cap C_B, \ 1 \ne B \le A,$$

$$L := N_G(Q), \quad and \quad K_0 := O_{r_0}(K).$$

Suppose that $Q \le \theta_{r'}(G)$. Then 11.2.4 (c) yields $\theta_{r'}(G) \le L$. In particular $1 \ne O_{r'}(K) \le L$. The factorization of G gives

$$O_{r'}(K) \le \bigcap_{g \in K} L^g = \bigcap_{g \in G} L^g =: D.$$

Hence, D is a nontrivial normal subgroup of G in $\mathbb{M}_\theta(A)$. But then G is solvable by (1), a contradiction. We have shown:

(2) $Q \not\leq \theta_{r'}(G)$.

Let Q^* be a nontrivial A-invariant subgroup of Q. Then

$$N_G(Q^*) \in \mathcal{M}_\theta(A)$$

by (1), and $Q \leq N_G(Q^*)$ since Q is Abelian. Hence 11.2.4 (c) implies:

(3) $N_G(Q^*) \leq L$ and in particular $N_G(Q^*)$ is q-closed.

Now let U be an A-invariant subgroup of G such that $U \cap Q \neq 1$. Then (3) shows that $N_U(Q \cap U) \leq L$ and $U \cap Q \in \mathrm{Syl}_q U$. Thus:

(4)   Let U be an A-invariant subgroup such that $U \cap Q \neq 1$. Then $U \cap Q \in \mathrm{Syl}_q U$.

Let B be a nontrivial subgroup of A. By 11.1.9 $Q_B \in \mathrm{Syl}_q C_B$ since $C_A \leq C_B$. Hence

$$C_B = O_{q'}(C_B)N_{C_B}(Q_B)$$

since C_B is solvable and Q_B is Abelian. Now (3) implies:

(5) $C_B = O_{q'}(C_B)(C_B \cap L)$ for all $1 \neq B \leq A$.

Set

$$\mathcal{B} := \{B \leq A \mid |A/B| = p \text{ and } Q_B \neq 1\}.$$

In the following let $B \in \mathcal{B}$. Next we show:

(6)   Let $T \in \mathrm{Syl}_r O_{q'}(C_A)$. Then $T \not\leq L$.

Assume that $T \leq L$. According to 8.2.6 on page 186 there exists an A-invariant Hall r'-subgroup H of C_B containing Q_B. Moreover, the A-invariant Hall r'-subgroups of C_B are conjugate under C_A, so $H \cap O_{q'}(C_A)$ is a Hall r'-subgroup of $O_{q'}(C_A)$ and

$$O_{q'}(C_A) = T(H \cap O_{q'}(C_A)).$$

Together with (5) this gives

$$C_A = (H \cap O_{q'}(C_A))(C_A \cap L).$$

Since $C_A \cap L$ normalizes Q_B we get

$$X := \langle Q_B{}^{C_A} \rangle \leq \langle Q_B{}^{H} \rangle \leq H.$$

In particular X is an AC_A-invariant r'-subgroup of C_B, so

$$Q_B \leq \theta_{r'}(G) \quad \text{for all } B \in \mathcal{B}.$$

But now 8.3.4 on page 193 shows that $Q \leq \theta_{r'}(G)$ since A is noncyclic. This contradicts (2), and this contradiction proves (6).

(7)   Let $B \in \mathcal{B}$ and V be an A-invariant Sylow r-subgroup of $O_{q'}(C_B)$. Then $[V, Q_B] \neq 1$.

Assume that $[V, Q_B] = 1$. Then (3) implies $V \leq C_B \cap L$. On the other hand, again by 8.2.6 on page 186, $V \cap C_A$ is an A-invariant Sylow r-subgroup of $O_{q'}(C_A)$. This contradicts (6).

We now set

$$K_B := \bigcap_{x \in Q_B} K^x.$$

As $O_{q'}(C_B) O_{q'}(C_a) \leq K$ and $Q_B \leq C_a$ for $a \in B^\sharp$, we obtain:

(8) $O_{q'}(C_B) O_{q'}(C_a) \leq K_B$ for $a \in B^\sharp$.

Next we show:

(9) $O_{r'}(K) \leq K_B$.

Let V be an AQ_B-invariant Sylow r-subgroup of $O_{q'}(C_B)$ and set

$$W := [V, Q_B] \quad \text{and} \quad X := O_{r'}(K).$$

By (8) W normalizes X, so 8.2.7 and 8.3.4 give:

(+) $X = C_X(W)[X, W], \quad C_X(W) = \langle C_X(W) \cap C_a \mid a \in B^\# \rangle,$

and

(++) $[X, W] = \langle [C_X(a), W] \mid a \in B^\# \rangle.$

Let $a \in B^\#$. By (5) and (8)

$$W \leq O_{q'}(C_a) \leq K_B.$$

Thus also

$$[C_a, W] \leq O_{q'}(C_a) \leq K_B,$$

and $(++)$ implies that

$$[X, W] \leq K_B.$$

Hence, by $(+)$ and (8) it suffices to prove

$$C_X(W) \cap C_a \leq O_{q'}(C_a).$$

Let

$$S := C_G(W) \cap C_a.$$

Then S and SQ_B are AQ_B-invariant, so by (4) $S \cap Q_a \in \mathrm{Syl}_q S$. If $S \cap Q_a \neq 1$, then (3) yields $W \leq L$ and thus $[W, Q_B] = 1$. But this contradicts (7). Hence S is a q'-group, and

$$[S, Q_B] \overset{(cp)}{=} [S, Q_B, Q_B] \overset{(5)}{\leq} O_{q'}(C_a) \overset{(8)}{\leq} K_B.$$

But then $(S \cap X)[S, Q_B]$ is a Q_B-invariant subgroup of K and thus in K_B. It follows that $S \cap X = C_X(W) \cap C_a \leq K_B$, and (9) is proved.

Recall that $1 \neq K_0 = O_{r_0}(K) \leq O_{r'}(K)$. Hence by (9)

$$K_0 \leq K_B \quad \text{and thus} \quad K_0 \leq O_{r_0}(K_B).$$

It follows that $\langle K_0^{Q_B} \rangle$ is an AQ_B-invariant r_0-subgroup of G.

Among all AQ_B-invariant r_0-subgroups of G that contain K_0 and are generated by conjugates of K_0 we choose R maximal. Then R is an A-invariant solvable subgroup, so by (1)

$$M := N_G(R) \in \mathbb{M}_\theta(A).$$

Set $Q_0 := M \cap Q$. By (4) $Q_0 \in \mathrm{Syl}_q M$ since $1 \neq Q_B \leq M$. For every $g \in G$ there exist $x \in K$, $y \in Q$ such that $g = xy$. Hence

$$\langle K_0^{gQ_0} \rangle = \langle K_0^{yQ_0} \rangle = \langle K_0^{Q_0} \rangle^y,$$

as $K_0 \trianglelefteq K$ and Q is Abelian. In particular,

$$\langle K_0^{gQ_0} \rangle \text{ is an } r_0\text{-group for every } g \in G.$$

Let $g \in G$ such that $K_1 := K_0^g \leq M$. Then the lemma of Matsuyama $(6.7.8$ on page $160)$, with (M, K_1, Q_0) in place of (G, Z, Y), gives a Sylow r_0-subgroup T_1 of M such that $K_1 \leq T_1$ and

$$R_1 := \mathrm{wcl}_M(K_1, T_1)$$

is normalized by Q_0. Now 6.4.4 on page 134 shows that

$$R_1 \leq N := O_{q'}(M).$$

We have shown that $\mathrm{wcl}_G(K_0, M) \leq N$. The coprime action of AQ_B on N gives an A-invariant Sylow r_0-subgroup T of M such that

$$R \leq T \cap N \quad \text{and} \quad (T \cap N)^{Q_B} = T \cap N.$$

The maximal choice of R implies

$$R = \mathrm{wcl}_G(K_0, T \cap N) = \mathrm{wcl}_G(K_0, T).$$

It follows that $N_G(T) \leq N_G(R) = M$, and $T \in \mathrm{Syl}_{r_0} G$. We have shown:

(10)   For every $B \in \mathcal{B}$ there exists an A-invariant $T \in \mathrm{Syl}_{r_0} G$ such that $K_0 \leq T$ and $\mathrm{wcl}_G(K_0, T)$ is invariant under AQ_B.

We now derive a final contradiction: Let B and T be as in (10). Moreover, as above,

$$R := \mathrm{wcl}_G(K_0, T), \quad M := N_G(R), \quad \text{and} \quad Q_0 := Q \cap M \in \mathrm{Syl}_q M.$$

Then $1 \neq R \leq O_{r_0}(M)$ and thus $Q \not\leq M$ by 11.2.4. Since

$$Q = \prod_{B_1 \in \mathcal{B}} Q_{B_1}$$

there exists $B_1 \in \mathcal{B}$ with

$$Q_1 := Q_{B_1} \not\leq M.$$

We now apply (10) with B_1 in place of B. Then there exists an A-invariant Sylow r_0-subgroup T_1 of G such that

$$R_1 := \mathrm{wcl}_G(K_0, T_1)$$

is AQ_1-invariant. According to the coprime action of A on G and on M, there exists $c \in C_A$ such that $T_1^c = T$; so $R_1^c = R$ and $Q_1^c \leq M$, and there exists $c' \in C_A \cap M$ such that

$$Q_1^{cc'} \leq Q_0.$$

It follows that
$$Q_1^{cc'} \leq C_{Q_0}(B_1) \leq Q_1$$
and $Q_1 = Q_1^{cc'} \leq M$. This contradicts the choice of $B_1 \in \mathcal{B}$.   □

Next we prove a version of 3.2.9 on page 67 that suits the situation of this section.

11.2.6 Let θ be locally complete on G and $q \in \pi(\theta)$, and let $L, M \in \mathcal{M}_\theta(A)$, $W \in \mathcal{M}_\theta(A, q)$, and $U := \theta(N_G(W))$. Suppose that $W \leq M$,
$$M = O_{q'}(M)(L \cap M), \quad \text{and} \quad U = O_{q'}(U)(U \cap M).$$
Then there exists $c_1 \in C_A \cap M$, such that
$$U = O_{q'}(U)\,(U \cap L^{c_1}).$$

Proof. The factorization $M = O_{q'}(M)(L \cap M)$ and (cp) imply that $L \cap M$ contains an A-invariant Sylow q-subgroup of M. Hence, there exists $c_1 \in C_A \cap M$ such that
$$W \leq (L \cap M)^{c_1} = L^{c_1} \cap M \quad \text{and} \quad M = O_{q'}(M)(L^{c_1} \cap M).$$
With
$$(M, q, O_{q'}(M), L^{c_1} \cap M, W) \text{ in place of } (G, p, N, H, P),$$
then 3.2.9 on page 67 shows that
$$U \cap M = (U \cap O_{q'}(M))(U \cap (L^{c_1} \cap M))$$
and thus
$$U = O_{q'}(U)(U \cap O_{q'}(M))(U \cap (L^{c_1} \cap M)).$$
Since the third factor normalizes the second one, this second factor is contained in $O_{q'}(U)$; and the claim follows.   □

11.2.7 Let θ be locally complete on G, $r(A) \geq 3$ and $q \in \pi(\theta)$. Suppose that there exists $M \in \mathcal{M}_\theta(A)$ satisfying:

(∗) For every $W \in \mathcal{M}_\theta(A, q)$, $W \neq 1$, there exists $c \in C_A$ such that
$$\theta(N_G(W)) = O_{q'}(\theta(N_G(W)))\,(N_G(W) \cap M^c).$$

Then θ is complete on G.

Proof. Let Q_0 be an A-invariant Sylow q-subgroup of $O_{q'q}(M)$. Since $q \in \pi(\theta)$ there exists $1 \neq W \in \mathcal{U}_\theta(A, q)$. Hence, $(*)$ implies that $q \in \pi(M)$, and $Q_0 \neq 1$ since M is solvable.

We fix the following notation:

$$Q := Z(Q_0), \quad L := \theta(N_G(Q)), \quad \text{and} \quad K := \theta_{q'}(G).$$

The Frattini argument and 8.2.11 imply

(1) $M = O_{q'}(M)(M \cap L)$ and $M \cap C_a = (O_{q'}(M) \cap C_a)(M \cap L \cap C_a)$ for $a \in A^\sharp$.

We show:

(2) $C_a = O_{q'}(C_a)(C_a \cap L)$ for $a \in A^\sharp$.

Let W_a be an A-invariant Sylow q-subgroup of $O_{q'q}(C_a)$. If $W_a = 1$, then (2) holds. Thus, we may assume that $W_a \neq 1$. Set $U := \theta(N_G(W_a))$. Then

$$C_a = O_{q'}(C_a)(U \cap C_a)$$

and by $(*)$

$$U = O_{q'}(U)(U \cap M^c) \text{ for some } c \in C_A.$$

In particular $W_a \leq O_q(U) \leq M^c$. We now apply 11.2.6 to

$$M^c \overset{(1)}{=} O_{q'}(M^c)(M^c \cap L^c).$$

Then there exists $c_1 \in C_A \cap M^c$ such that

$$U = O_{q'}(U)(U \cap L^{cc_1}).$$

Hence

$$C_a = O_{q'}(C_a)(U \cap C_a) \overset{8.2.11}{=} O_{q'}(C_a)(O_{q'}(U) \cap C_a)(U \cap L^{cc_1} \cap C_a)$$
$$= O_{q'}(C_a)(C_a \cap L^{cc_1}).$$

This implies (2) since $cc_1 \in C_A$.

As in 11.2.5 let

$$\mathcal{B} = \{B \leq A \mid |A/B| = p \text{ and } Q_B \neq 1\}$$

and pick $B \in \mathcal{B}$. We set

$$Q_B := C_Q(B) \quad \text{and} \quad K_B := \langle O_{q'}(C_a) \mid a \in B^\# \rangle.$$

Note that $K_B \leq K$ and K_B is AC_A-invariant. Hence (2) implies

(3) $C_a = (C_a \cap K_B)(C_a \cap L)$ and $C_a \cap K = (C_a \cap K_B)(K \cap L \cap C_a)$
        for all $a \in B^\sharp$.

As $C_K(a) = K \cap C_a$ and K_B is $C_K(A)$-invariant, we get from 11.2.1

(4) $K = K_B(K \cap L)$ for $B \in \mathcal{B}$.

Since Q_B normalizes K_B this gives

$$Q_B K = Q_B K_B(L \cap K) = K_B Q_B(L \cap K)$$

$$\subseteq K_B Q(L \cap K) = K_B(L \cap K)Q \overset{(4)}{=} KQ,$$

and

$$QK = \prod_{B \in \mathcal{B}} Q_B K \subseteq KQ.$$

It follows that

$$G_0 := KQ = QK$$

is an A-invariant p'-subgroup of G. Thus 11.2.5 applied to G_0 yields:

(5) θ_{G_0} is complete and $\theta(G_0) = G_0$.

Let $G_1 := [Q, G_0]Q$, so $G_1 = [Q, K]Q$ since Q is Abelian. For $B \in \mathcal{B}$ we
get

$$[K, Q] = [Q, K] \overset{(4)}{=} [Q, (K \cap L)K_B] \overset{1.5.4}{\leq} [Q, K_B]Q$$

since $[Q, K \cap L] \leq Q$. Thus:

(6) $G_1 = [Q, K_B]Q$ for all $B \in \mathcal{B}$.

Note that $C_L(B)$ normalizes K_B and thus by (6) also G_1. Since $L = \langle C_L(B) \mid B \in \mathcal{B} \rangle$ we get that $L \leq N_G(G_1)$. On the other hand, G_1 is an A-invariant normal subgroup of $G_0 \in \mathcal{N}_\theta(A)$; so

$$G_1 \in \mathcal{N}_\theta(A) \text{ and } \langle K, L \rangle \leq N_G(G_1).$$

The local completeness of θ gives

$$\langle K, L \rangle \in \theta(N_G(G_1)),$$

and with (2) we conclude that

$$E := \langle C_a \mid a \in A^\sharp \rangle \leq \langle K, L \rangle \in \mathcal{N}_\theta(A).$$

Hence θ is complete. $\square$

11.2.8 *Suppose that θ is locally complete on G, $r(A) \geq 3$ and $\pi(\theta) \neq \{2,3\}$. Then θ is complete on G.*

Proof. For $|\pi(\theta)| \leq 1$ the claim follows from 11.1.10. Hence, we may assume that there exists $q \in \pi(\theta)$ such that

$$q \geq 5 \quad \text{and} \quad \mathcal{M}_\theta(A, q) \neq \{1\}.$$

Let $S \in \mathcal{M}_\theta^*(A, q)$ and $Q := W(S)$, where $W(S)$ is the nontrivial characteristic subgroup of S defined in Section 9.4. We set

$$M := \theta(N_G(W(S))).$$

Then the completeness of θ follows from 11.2.7 if we can show:

$(*)$ For every $W \in \mathcal{M}_\theta(A, q)$, $W \neq 1$, there exists $c \in C_A$ such that

$$\theta(N_G(W)) = O_{q'}(\theta(N_G(W)))(N_G(W) \cap M^c).$$

For the proof of $(*)$ let $W \in \mathcal{M}_\theta(A, q)$ be a counterexample,

$$U := \theta(N_G(W)), \quad \text{and} \quad T \in \operatorname{Syl}_q U \text{ with } T^A = T.$$

In addition, we choose the counterexample W such that $|T|$ is maximal. Then $T \neq 1$ and $T \in \mathcal{M}_\theta(A, q)$. Hence

$$L := \theta(N_G(W(T)))$$

exists since θ is locally complete. Pick $S^* \in \mathcal{M}_\theta^*(A, q)$ such that $T \leq S^*$. From 9.4.6 on page 255 we get

(1) $U = O_{q'}(U)(U \cap L).$

Assume first that $T = S^*$. According to the Transitivity Theorem there exists $c \in C_A$ with $S^c = T$. Hence $M^c = L$ (see 9.4.1), and (1) shows that W is not a counterexample.

Assume now that $T < S^*$. Then

$$T < N_{S^*}(W(T)) \leq M.$$

The maximal choice of T gives property $(*)$ for $W(T)$ in place of W, so

$$L = O_{q'}(L)(L \cap M^c) \text{ for some } c \in C_A.$$

Hence 11.2.6 (with (L, M^c) in place of (M, L)) implies that

$$U = O_{q'}(U)(L \cap M^{cc_1}), \quad c_1 \in C_A,$$

and W is not a counterexample. This final contradiction proves $(*)$. $\qquad \square$

An important special case of the Completeness Theorem, the theorem of Goldschmidt [55], is now a consequence of 11.2.8.

11.2.9 *Let $p = 2$ and θ be a solvable A-signalizer functor on G with $r(A) \geq 3$. Then θ is complete on G.*

Proof. Let (G, A, θ) be a counterexample such that $|G| + |\pi(\theta)|$ is minimal.

Assume that N is a nontrivial normal subgroup of G contained in $\mathcal{W}_\theta(A)$. Then $\overline{\theta}$, defined as in 11.1.3, is a solvable A-signalizer functor on $\overline{G} := G/N$, and the minimality of $|G| + |\pi(\theta)|$ shows that $\overline{\theta}$ is complete. But now by 11.1.4 also θ is complete, a contradiction.

Hence, no nontrivial normal subgroup of G is contained in $\mathcal{W}_\theta(A)$. In particular $N_G(U) < G$ for $1 \neq U \in \mathcal{W}_\theta(A)$. Now again by the minimality of $|G| + |\pi(\theta)|$ θ is locally complete on G. Thus 11.2.8 shows that θ also is complete, a contradiction. $\qquad \square$

11.3 The Completeness Theorem of Glauberman

As before A is an noncyclic elementary Abelian p-group that acts on the group G. The notation is chosen as in Sections 11.1 and 11.2.

We will need the following remark:

11.3.1 *Let G be a solvable p'-group and $q \in \pi(G)$. Suppose that U is a q'-subgroup of G such that $[U, C_G(B)]$ is a q'-group for every maximal subgroup B of A. Then $U \leq O_{q'}(G)$.*

Proof. We may assume that $O_{q'}(G) = 1$ and show that $U = 1$. Let $Q := O_q(G)$. Then

$$Q \overset{(cp)}{=} \langle C_Q(B) \mid B \leq A, \; |A/B| = p \rangle,$$

so $[Q, U] = 1$ since $[U, C_Q(B)] \leq Q$. It follows that

$$U \leq C_G(Q) \overset{6.4.3}{\leq} Q$$

and thus $U = 1$. $\qquad\qquad\qquad\qquad\qquad\qquad\qquad\qquad\qquad\qquad\qquad\qquad\qquad$ $\square$

11.3.2   Completeness Theorem of Glauberman. *Let θ be a solvable A-signalizer functor on G. Suppose that $r(A) \geq 3$. Then θ is complete.*

Proof. We proceed by induction on $|G| + |\pi(\theta)|$. Let (G, A, θ) be a counterexample such that $|G| + |\pi(\theta)|$ is minimal. As in the proof of 11.2.9:

(1) θ is locally complete.

Thus 11.2.8 implies:

(2) $\pi(\theta) = \{2, 3\}$.

In the following let q and r be the two primes in $\pi(\theta)$, so

$$\pi(\theta) = \{q, r\} = \{2, 3\}.$$

We set

$$\mathbb{M} := \mathbb{M}_\theta(A), \;\; \mathbb{M}^* := \mathbb{M}_\theta^*(A), \;\; \mathbb{M}_s := \mathbb{M}_\theta(A, s), \;\; \mathbb{M}_s^* := \mathbb{M}_\theta^*(A, s) \;\;\; (s \in \pi(\theta)).$$

The Transitivity Theorem 11.1.8 yields

$$\mathbb{M}_q^* = S^{C_A} \;\; \text{and} \;\; \mathbb{M}_r^* = R^{C_A} \;\; \text{for} \;\; S \in \mathbb{M}_q^* \;\; \text{and} \;\; R \in \mathbb{M}_r^*,$$

which we will use frequently.

First we prove:

(3)   *Let $S \in \mathbb{M}_q^*$. Then S is contained in a unique element of $\mathbb{M}^*$.*

Proof. Clearly $S \neq 1$ since $q \in \pi(\theta)$. Let

$$S \leq M_1 \cap M_2 \quad \text{for} \quad M_1, M_2 \in \mathsf{M}^*.$$

We show that $M_1 = M_2$. For $i = 1, 2$ there exists an A-invariant Sylow r-subgroup R_i of M_i such that $M_i = SR_i$. Let $R_i \leq \widehat{R}_i \in \mathsf{M}^*_r$ and $c \in C_A$ with $\widehat{R}_1^c = \widehat{R}_2$. From 11.1.9 we get that $C_A \cap \widehat{R}_i \in \mathrm{Syl}_r C_A$ and $C_A \cap S \in \mathrm{Syl}_q C_A$, so $C_A = (\widehat{R}_i \cap C_A)(S \cap C_A)$. Hence, we can choose c in $S \cap C_A$. It follows that

$$R_1{}^c S = (R_1 S)^c = (SR_1)^c = SR_1{}^c.$$

Thus

$$M^* := S\langle R_1{}^c, R_2 \rangle$$

is an A-invariant subgroup and $\langle R_1{}^c, R_2 \rangle \leq \widehat{R}_2$, in particular $\langle R_1{}^c, R_2 \rangle \in \mathsf{M}$. The $p^a q^b$-Theorem shows that M^* is solvable. Hence 11.1.2 yields $M^* \in \mathsf{M}$. Since

$$c \in S \leq M_2 \leq M^*$$

the maximality of M_1 and M_2 implies that

$$M_2 = M^* = M_1{}^c = M_1. \qquad \square$$

(4)   *Let $S \in \mathsf{M}^*_q$ and $S \leq M \in \mathsf{M}^*$. Then S contains an A-invariant subgroup $Q \neq 1$ such that the following hold:*

     (4a) $\theta(N_G(Q))$ *is not Thompson factorizable (with respect to q).*

     (4b) $\theta(N_G(Q)) \cap C_A \not\leq M$, *and in particular $C_A \not\leq M$.*

Proof. Note first that 11.2.7 and $S^{C_A} = \mathsf{M}^*_q$ show that S contains an A-invariant subgroup $Q \neq 1$ such that

$$U := \theta(N_G(Q)) \neq O_{q'}(U)(U \cap M).$$

In addition, we choose Q such that $|U \cap M|_q$ is maximal. Let T be an A-invariant Sylow q-subgroup of $U \cap M$. After conjugation in M we may assume that $T \leq S$. By (3) $T < S$ and thus

$$T < N_S(T) \leq N_G(J(T)).$$

The choice of Q implies

$$U_1 := \theta(N_G(J(T))) = O_{q'}(U_1)(U_1 \cap M).$$

On the other hand $Q \leq T$ and thus $\Omega(Z(S)) \leq \Omega(Z(T)) \leq J(T)$, so again by (3)

$$O_{q'}(U_1) \leq \theta(C_G(\Omega(Z(T)))) \leq \theta(C_G(\Omega(Z(S)))) \leq M.$$

Hence $U_1 \leq M$ follows. We have shown:

$$(*) \qquad E := \langle \theta(C_G(\Omega(Z(T)))), \theta(N_G(J(T))) \rangle \leq M.$$

In particular $\theta(N_G(T)) \leq M$ and thus

$$T \in \mathrm{Syl}_q U.$$

If U is Thompson factorizable, then

$$U = O_{q'}(U)\, N_U(J(T))\, C_U(\Omega(Z(T))) = O_{q'}(U)(U \cap M),$$

which contradicts the choice of U. Hence U has property (4a). Now 9.3.10 (a) on page 247 and (cp) give

$$U = O_{q'}(U)\langle E \cap U, C_U(A) \rangle \overset{(*)}{=} O_{q'}(U)\langle M \cap U, C_U(A) \rangle.$$

It follows that $C_U(A) = C_A \cap U \not\leq M$, and this is (4b). $\square$

(5)   *There exist elementary Abelian subgroups of order 9 and 8 in C_A.*

Proof. Let $U := \theta(N_G(Q))$ be as in (4). We apply 9.3.10 (b) to $U/O_{q'}(U)$. Because of (cp) there exist subgroups $W \leq D \leq C_A$ such that

$$W \cong C_q \times C_q \quad \text{and} \quad D/C_D(W) \cong \mathrm{SL}_2(q).$$

Let $q = 3$. Then W has the desired property. Moreover, there exists an element of order 6 in $C_A/O_{2'}(C_A)$ since $D/C_D(W) \cong \mathrm{SL}_2(3)$ contains such an element (see 8.6.10 on page 219 and note that $O_{2'}(\mathrm{SL}_2(3)) = 1$).

Now let $q = 2$ and $\overline{C}_A := C_A/O_{2'}(C_A)$. Then every 3-element $d \in D \setminus C_D(W)$ acts fixed-point-freely on W. Hence, $\overline{C}_A$ satisfies the hypothesis of

8.5.6 on page 209, and C_A also contains an elementary Abelian subgroup of order 8. □

In the following $\mathcal{B}$ denotes the set of all maximal subgroups of A,[6] and B always denotes an element of $\mathcal{B}$. For $q \in \pi(\theta)$ we set

$$K := \theta_{q'}(G),$$

$$K_\vee(B) := \langle O_{q'}(C_a) \mid a \in B^\# \rangle, \ B \in \mathcal{B},$$

$$K_\wedge(B) := \bigcap_{a \in B^\#} O_{q'}(C_a), \ B \in \mathcal{B},$$

$$K_\mathcal{B} := \langle K_\wedge(B) \mid B \in \mathcal{B} \rangle.$$

All four subgroups are C_A-invariant, so they are in K. In addition

$$K_\wedge(B_1) \leq O_{q'}(C_a) \leq K_\vee(B) \quad \text{for all } B, B_1 \in \mathcal{B} \text{ and } a \in (B \cap B_1)^\#.$$

In particular

$$K_\wedge(B) \leq K_\mathcal{B} \leq K_\vee(B) \quad \text{for all } B \in \mathcal{B}.$$

(6)   (6a) $K_\mathcal{B} \cap H \leq O_{q'}(H)$ *for all* $H \in \mathsf{M}$.

   (6b)   *If* $H \in \mathsf{M}$ *such that* $H \cap C_a \leq O_{q'}(C_a)$ *for all* $a \in A^\#$, *then* $H \leq K_\mathcal{B}$.

   (6c)   *If* $F \leq C_A$ *is a noncyclic Abelian* q-*subgroup, then*

$$K_\mathcal{B} = \langle O_{q'}(\theta(C_G(f))) \mid f \in F^\# \rangle.$$

   In particular $\theta(C_G(F)) \leq N_G(K_\mathcal{B})$.

Proof. (a) Note that $K_\vee(B) \cap H$ is an $H \cap C_B$-invariant q'-subgroup of H. Since $K_\mathcal{B} \cap H \leq K_\vee(B) \cap H$ we get that $[K_\mathcal{B} \cap H, H \cap C_B]$ is a q'-subgroup for every $B \in \mathcal{B}$. Hence (a) follows from 11.3.1.

(b) We have

$$H \cap C_B = C_H(B) = \bigcap_{a \in B^\#} (H \cap C_a) \leq K_\wedge(B),$$

so $H = \langle H \cap C_B \mid B \in \mathcal{B} \rangle \leq K_\mathcal{B}$.

---

[6]This is different from the notation of the the previous section.

(c) Let $f \in F^{\sharp}$. Then $\langle f \rangle$ is A-invariant and thus $\langle f \rangle \in \mathsf{M}$. Set $C_f := \theta(C_G(f))$. Since $K_{\mathcal{B}}$ is a C_A-invariant q'-subgroup

$$K_{\mathcal{B}} = \langle K_{\mathcal{B}} \cap C_f \mid f \in F^{\sharp} \rangle \overset{(a)}{\leq} \langle O_{q'}(C_f) \mid f \in F^{\sharp} \rangle.$$

Conversely, for $a \in A^{\sharp}$ 8.2.12 gives

$$C_a \cap O_{q'}(C_f) \leq O_{q'}(C_{C_a}(f)) \leq O_{q'}(C_a),$$

and this implies (c). $\square$

(7) $K_{\mathcal{B}} = 1$ *for all* $q \in \pi(\theta)$.

Proof. Assume that $K_{\mathcal{B}} \neq 1$ and set

$$N := \theta(N_G(K_{\mathcal{B}})).$$

Then $C_A \leq N$, and (3) and (4b) imply

$(')$ $S \not\leq N$ for all $S \in \mathsf{M}_q^*$.

By (5) there exists $E \leq C_A$ $(\leq N)$ such that

$$E \cong \begin{cases} C_q \times C_q & \text{for } q = 3 \\ C_q \times C_q \times C_q & \text{for } q = 2. \end{cases}$$

Let Q be an A-invariant Sylow q-subgroup of N with $E \leq Q \leq N$.

First we show that the statement

$('')$ Every Q-invariant q'-subgroup of $\mathsf{M}_{q'}$ is contained in in $K_{\mathcal{B}}$.

does not hold.

For doing this let Q_1 be an A-invariant Sylow q-subgroup of $\theta(N_G(Q))$. Then $K_0 := \langle K_{\mathcal{B}}^{Q_1 \cap C_B} \rangle$ is Q-invariant. On the other hand,

$$K_{\mathcal{B}} \leq K_{\vee}(B) \leq K \quad \text{and} \quad Q_1 \cap C_B \leq C_B \leq \theta(N_G(K_{\vee}(B)));$$

so K_0 is a Q-invariant q'-subgroup in M.

Assume now that $('')$ holds. Then $K_0 = K_{\mathcal{B}}$ and thus $Q_1 \cap C_B \leq N$. It follows that
$$Q_1 = \langle Q_1 \cap C_B \mid B \in \mathcal{B} \rangle \leq N,$$
and $Q = Q_1$ since $Q \leq Q_1$. This shows that $Q \in \mathcal{W}_q^*$, which contradicts $(')$.
To derive a contradiction it suffices now to verify $('')$. Let $U \neq 1$ be a Q-invariant subgroup in $\mathcal{W}_{q'}$. First assume that $q = 2$, so $|E| = 8$. Then
$$U \stackrel{(cp)}{=} \langle C_U(F) \mid F \leq E, \ |F| = 4 \rangle \stackrel{(6c)}{\leq} N,$$
and thus even $U \leq O_{q'}(N)$ since $Q \in \mathrm{Syl}_q N$ (see 6.4.4 on page 134).
Let $F \leq E$ with $|F| = 4$. Then
$$C_U(F) \leq O_{q'}(N) \cap \theta(N_G(F)) \leq O_{q'}(\theta(C_G(F))),$$
and 8.2.12 implies for all $f \in F^\sharp$
$$C_U(F) \leq O_{q'}(\theta(C_G(F))) \leq O_{q'}(\theta(C_G(f))) \stackrel{(6c)}{\leq} K_{\mathcal{B}}.$$
It follows that
$$U = \langle C_U(F) \mid F \leq E, \ |F| = 4 \rangle \leq K_{\mathcal{B}},$$
and $('')$ holds if $q = 2$.

Assume now that $q = 3$, so $|E| = 9$ and $E \leq C_A \leq C_a$ for every $a \in A^\sharp$. Let S_0 be an A-invariant Sylow q-subgroup of $C_{C_a}(E)$. Then $E \leq Z(S_0)$ and thus

(z) $\hspace{3cm} Z(S_0) \in \mathrm{Syl}_q C_{C_a}(S_0).$

Moreover, (6c) implies that $S_0 \leq N$. Hence, there exists $d \in C_A$ such that $S_0 \leq Q^d$ and thus
$$S_0 \leq C_a \cap Q^d = (C_a \cap Q)^d.$$
The q'-group $T := (U \cap C_a)^d \leq C_a$ is normalized by the q-group S_0. Let
$$\overline{C}_a := C_a / O_{q'}(C_a) \quad \text{and} \quad \overline{X} := O_q(\overline{C}_a).$$
The semidirect product $S_0 T$ acts on the q-group $\overline{X}$, and
$$C_{\overline{X}}(S_0) \stackrel{(cp)}{=} \overline{C_X(S_0)} \stackrel{(z)}{\leq} \overline{Z(S_0)} \cap \overline{X} \text{ and } [\overline{Z(S_0)} \cap \overline{X}, T] \leq \overline{X} \cap \overline{T} = 1.$$
Now 8.5.3 on page 205 shows that $[\overline{X}, T] = 1$. It follows that $T \leq O_{q'}(C_a)$ and thus $U \cap C_a \leq O_{q'}(C_a)$ since $d \in C_a$. Now (6b) (with $H = U$) implies $('')$. $\hspace{1cm}\square$

(8)   Let $M \in \mathcal{M}^*$ with $C_A \leq M$. Then $F(M) = O_q(M)$ for some $q \in \pi(\theta)$.

Proof. Let M be a counterexample, so $O_2(M) \neq 1 \neq O_3(M)$ by (2). Moreover, let $E \leq C_A$ be an elementary Abelian q-subgroup as in (5), so $r(E) \geq 2$. Then $C_{O_q(M)}(E) \neq 1$, and there exists $e \in E^{\#}$ such that $O_{q'}(M) \cap \theta(C_G(e)) \neq 1$. It follows that

$$C_{O_2(M)}(e) \neq 1 \neq C_{O_3(M)}(e).$$

Hence, by 11.2.2 M is the unique subgroup in $\mathcal{M}^*$ that contains $C_{F(M)}(e)$. In particular $\theta(C_G(e)) \leq M$. This shows that

$$O_{q'}(M) \cap \theta(C_G(e)) \leq O_{q'}(\theta(C_G(e))) \overset{(6c)}{\leq} K_{\mathcal{B}} \overset{(7)}{=} 1,$$

a contradiction. $\square$

We now derive a final contradiction that shows that (G, A, θ) is not a counterexample. Let $S \in \mathcal{M}_2^*$. There exists $B \in \mathcal{B}$ such that

$$Z_B := Z(S) \cap C_B \neq 1.$$

On the other hand, by (7) $K_{\mathcal{B}} = 1$ for $q = 3$. Hence, there exists $b \in B^{\#}$ with

$$Z_B \nleq O_{3'}(C_b) = O_2(C_b).$$

Let $M \in \mathcal{M}^*$ such that $C_b \leq M$. Then also $C_A \leq M$, and 11.1.9 implies that $S \cap M \in \mathrm{Syl}_2 M$. In particular

$$O_2(M) \leq S \quad \text{and} \quad [O_2(M), Z_B] = 1.$$

But $Z_B \nleq O_2(M)$ since $O_2(M) \cap C_b \leq O_2(C_b)$, so (8) and 6.4.4 on page 134 give

$$F(M) = O_3(M) \quad \text{and} \quad [O_3(M), Z_B] \neq 1.$$

Since $O_3(M) = \langle C_{\widetilde{B}} \cap O_3(M) \mid \widetilde{B} \in \mathcal{B} \rangle$ there exists $\widetilde{B} \in \mathcal{B}$ such that

$$O_3(M) \cap C_{\widetilde{B}} \nleq C_G(Z_B).$$

Hence

(+) $[C_{\widetilde{B}}, Z_B]$ is not a 2-group.

With the same argument for $T \in \mathcal{W}_3^*$ there exist subgroups $D, \widetilde{D} \in \mathcal{B}$ such that $Z(T) \cap C_D \neq 1$ and

$$(++) \qquad\qquad [C_{\widetilde{D}}, Z(T) \cap C_D] \text{ is not a 3-group.}$$

We now use the fact that $r(A) \geq 3$. Then $\widetilde{B} \cap \widetilde{D} \neq 1$, so there exists $1 \neq w \in \widetilde{B} \cap \widetilde{D}$. Let

$$C_w \leq H \in \mathcal{W}^*.$$

According to (8) either $F(H) = O_2(H)$ or $F(H) = O_3(H)$, and 11.1.9 (a) shows that

$$S \cap H \in \mathrm{Syl}_2 H \text{ and } T \cap H \in \mathrm{Syl}_3 H.$$

Assume that $F(H) = O_2(H)$. Then $O_2(H) \leq S$, and thus

$$Z_B \leq C_H(O_2(H)) \overset{6.4.4}{\leq} O_2(H).$$

Hence $[C_{\widetilde{B}}, Z_B]$ is a 2-group since $C_{\widetilde{B}} \leq H$. This contradicts $(+)$.

If $F(H) = O_3(H)$, then an analogous argument with T in place of S shows that $[C_{\widetilde{D}}, Z(T) \cap C_D]$ is not a 3-group. This contradicts $(++)$. $\qquad\square$

We will use the Completeness Theorem in the next chapter. Here we give an elementary consequence:

11.3.3 *Let A be an elementary Abelian p-group with $r(A) \geq 3$ that acts on the p'-group G. Suppose that*

$$C_G(a) \text{ is solvable for all } a \in A^{\#}.$$

Then G is solvable.

Proof. Define

$$\theta(C_G(a)) := C_G(a), \quad a \in A^{\#}.$$

By 11.1.1 θ is a solvable A-signalizer functor on G, and this signalizer functor is complete by the above Completeness Theorem of Glauberman. Now 8.3.4 on page 193 implies that G is the maximal element of $\mathcal{W}_\theta(A)$. In particular, G is solvable. $\qquad\square$

Chapter 12

N-Groups

In this chapter we will demonstrate how to use the methods and results established in the previous chapters. Our target is to investigate the structure of N-groups. Here an **N-group** is a group G that satisfies:

$\mathcal{N}$ G has even order, and every 2-local subgroup of G is solvable.

This definition differs slightly from that used in the literature, where an N-group G satisfies $\mathcal{N}$ not only for 2 but for every prime in $p \in \pi(G)$; i.e., every local subgroup of G is solvable. All nonsolvable groups satisfying this stronger condition have been classified by Thompson [94]. Later his result was generalized by Gorenstein and Lyons [61], Janko [73], and F. Smith [83] to groups that satisfy $\mathcal{N}$. In the 1970s Thompson's proof became a pattern for the classification of the finite simple groups.

A complete treatment of N-groups would be far beyond the reach of this book. Therefore, in this chapter we will assume the following additional hypothesis:

$\mathcal{Z}$ $\qquad C_G(\Omega(Z(S))) \leq N_G(S)$ for $S \in \mathrm{Syl}_2 G$.

A group satisfying $\mathcal{Z}$ and $\mathcal{N}$ is said to be a ZN-group.

Property $\mathcal{Z}$ implies that $C_G(\Omega(Z(S)))$ and thus also $N_G(\Omega(Z(S)))$ is 2-closed. Hence, $\mathcal{Z}$ is equivalent to

$$N_G(\Omega(Z(S))) \leq N_G(S) \text{ for } S \in \mathrm{Syl}_2 G.$$

For example, simple N-groups in which the normalizer of a Sylow 2-subgroup is a maximal subgroup have this property.

The investigation of ZN-groups shows the pattern of proof that is typical for many classification problems:

- Reduction to groups of local characteristic 2.

- Determination of the 2-local structure.

- Identification of the groups by means of their 2-local structure.

We will carry out the first two steps for ZN-groups. For the identification we will refer to the corresponding literature.

In this chapter a strongly 2-embedded subgroup of G (for the definition see page 262) will be called **strongly embedded** in G.

A group G has **local characteristic 2** if G has even order and

$$\mathcal{L} \quad C_L(O_2(L)) \leq O_2(L) \text{ for all 2-local subgroups } L \text{ of } G.$$

In other words, a group of even order has local characteristic 2 if every 2-local subgroup has characteristic 2. Note further that the condition $C_L(O_2(L)) \leq O_2(L)$ implies the apparently stronger condition

$$C_G(O_2(L)) \leq O_2(L)$$

since $L = N_G(Q)$ for some nontrivial 2-subgroup $Q \leq O_2(L)$ and

$$C_G(O_2(L)) \leq C_G(Q)) \leq L.$$

If G is an N-group of even order, then by 6.4.4 (a) on page 134 $\mathcal{L}$ is equivalent to:

$$O_{2'}(L) = 1 \text{ for all 2-local subgroups } L \text{ of } G.$$

In this chapter we prove:

Theorem 1. *Let G be an ZN-group with $O_{2'}(G) = 1 = O_2(G)$, and let $S \in \mathrm{Syl}_2\, G$ and $Z := \Omega(Z(S))$. Then one of the following holds for*

$$H := O^2(G) \quad and \quad R := S \cap H.$$

(a)   *H contains a strongly embedded subgroup.*

(*b*)   *R* is a dihedral or semidihedral group.

(*c*) $Z \cap R \cong C_2$ and $Z \cap R$ is weakly closed in R with respect to H.

(*d*) $\Omega(R) = Z \cong C_2 \times C_2$.

In all four cases of Theorem 1 other theorems are available that allow to determine the groups in question:

- The theorem of Bender about groups having a strongly embedded subgroup.

- The theorem of Gorenstein-Walter about groups having dihedral Sylow 2-subgroups.

- The theorem of Alperin-Brauer-Gorenstein about groups having semi-dihedral Sylow 2-subgroups,

- The Z^*-Theorem of Glauberman (in case (c)).

- The theorem of Goldschmidt about groups having a strongly closed Abelian 2-subgroup (in case (d)).

We will state these results in the Appendix since they are not only of use in our special situation but fundamental for the classification of the finite simple groups in general.

The first step in the proof of Theorem 1 (Section 12.1) describes those groups that are not of local characteristic 2. This can be done using a slightly weaker hypothesis:

$\mathcal{C}$ *G* has even order, and $C_G(t)$ is solvable for every involution $t \in G$.

A group satisfying $\mathcal{Z}$ and $\mathcal{C}$ is said to be a *ZC*-group. Using the Completeness Theorem of Glauberman we prove in Section 12.1:

Theorem 2. *Let G be a ZC-group with $O_{2'}(G) = 1 = O_2(G)$. Then one of the cases (a), (c), and (d) in Theorem 1 holds, or $O^2(G)\Omega(Z(S))$ has local characteristic 2.*

The second part of the proof of Theorem 1 (Sections 12.2 and 12.3) investigates the 2-local structure of groups of local characteristic 2. After a lengthy analysis of the 2-local structure that mainly uses theorems from Chapters 9 and 10, one ends up with an astonishingly elementary structure:

Theorem 3. *Let G be an ZN-group of local characteristic 2 with $O_2(G) = 1$. Then G possesses a strongly embedded subgroup, or there exist two maximal 2-local subgroups M_1 and M_2 of $O^2(G)$ such that*

$$(*) \qquad M_1 \cong S_4 \cong M_2 \quad and \quad M_1 \cap M_2 \in \operatorname{Syl}_2 M_i, \, i = 1, 2.$$

It should be pointed out that strongly embedded subgroups show up in Theorem 2 as well as in Theorem 3. This is typical for the impact of strongly embedded subgroups in classification problems, and it demonstrates the fundamental importance of the theorem of Bender. In Theorem 2 strongly embedded subgroups occur as normalizers of nontrivial subgroups of odd order, in Theorem 3 as normalizers of nontrivial 2-subgroups.

Finally an elementary argument shows that $(*)$ of Theorem 3 implies case (b) of Theorem 1, which concludes the proof of Theorem 1.

12.1   An Application of the Completeness Theorem

In this section we investigate the relation between the existence of nontrivial signalizer functors and the existence of strongly embedded subgroups.

Before we turn to the proof of Theorem 2 we formulate two independent Lemmata.

12.1.1   Thompson's Transfer Lemma. *Let G be a group and $S \in \operatorname{Syl}_2 G$. Suppose that there exists a maximal subgroup $U \leq S$ and an involution $t \in S$ such that $t^G \cap U = \varnothing$. Then t is not contained in $O^2(G)$.*

Proof. The group G acts by right multiplication on the set Ω of cosets Ug, $g \in G$. Let $n := |\Omega| = |G : U|$ and φ be the homomorphism from G to S_n that describes this action of G on Ω. Then

$$n = 2|G : S| \quad and \quad |G : S| \text{ is odd.}$$

For $Ug \in \Omega$

$$(Ug)t = Ug \iff gtg^{-1} \in U.$$

The hypothesis $U \cap t^G = \varnothing$ shows that the involution $t^\varphi \in S_n$ has no fixed point on $\{1, \ldots, n\}$, so t^φ is the product of $\frac{n}{2}$ transpositions. As $\frac{n}{2} = |G : S|$ is odd, t^φ is not in A_n and thus t is not in $N := A_n^{\varphi^{-1}}$. In addition, $|S_n/A_n| = 2$ implies $|G/N| = 2$, so $O^2(G) \leq N$ and $t \notin O^2(G)$. $\square$

12.1.2 Let G be a group satisfying $\mathcal{C}$. Suppose that $O_{2'}(C_G(t)) = 1$ for every involution $t \in G$. Then G is of local characteristic 2.

Proof. By way of contradiction we may assume that there exists a 2-local subgroup $L \leq G$ such that

$$C_L(O_2(L)) \nleq O_2(L).$$

Then 6.5.8 on page 144 implies that

$$F^*(L) \neq O_2(L).$$

Let t be an involution in $Z(O_2(L))$, so $F^*(L) \leq C_G(t)$. By our hypothesis $C_G(t)$ is solvable. Thus $F^*(L) = F(L)$ and

$$O_{2'}(L) \neq 1.$$

An application of 8.2.13 on page 190, with

$$(2, L, C_G(t), O_{2'}(L)) \quad \text{in place of} \quad (p, N_G(P), L, U),$$

gives $1 \neq O_{2'}(L) \leq O_{2'}(C_G(t))$, a contradiction. $\square$

We now begin with the proof of Theorem 2 and consider the following situation:

$\mathcal{S}$

G is a ZC-group with $O_2(G) = 1 = O_{2'}(G)$,

$H := O^2(G)$,

$S \in \mathrm{Syl}_2\, G$, $Z := \Omega(Z(S))$ and $T := S \cap H \in \mathrm{Syl}_2\, H$.

Let $\mathcal{B}(G)$ be the set of maximal[1] Abelian 2-subgroups of G that contain an elementary Abelian subgroup of order 8. Recall from 11.1.1 on page 305 that for $B \in \mathcal{B}$

$$\theta_B \colon a \mapsto O_{2'}(C_G(a)), \quad a \in \Omega(B)^\sharp,$$

is a solvable $\Omega(B)$-signalizer functor on G. The Completeness Theorem of Glauberman (11.3.2 on page 326) shows that θ_B is complete. We denote the maximal element of $\mathcal{V}_{\theta_B}(\Omega(B))$ by $\theta_B(G)$. It is evident from the definition of a signalizer functor that for $R := \theta_B(G)$:

$C_R(a) = O_{2'}(C_G(a))$ for $a \in B^\sharp$;

$C_R(B_0) = O_{2'}(C_G(B_0))$ for $1 \neq B_0 \leq B$;

$R = \langle O_{2'}(C_G(a)) \mid a \in B^\# \rangle$;

$R^g = \theta_{B^g}(G)$ for every $g \in G$, and in particular $N_G(B) \leq N_G(R)$.

12.1.3   *Suppose that $\mathcal{S}$ holds. Let $B \in \mathcal{B}(G)$ and R be a B-invariant $2'$-subgroup of G. Then*

$$R \leq \theta_B(G).$$

Proof. From 11.3.3 on page 333 we get that R is solvable. Hence, according to the definition of $\theta_B(G)$, it suffices to show that

$$C_R(b) \leq O_{2'}(C_G(b)) \text{ for every } b \in B^\sharp.$$

Let $b \in B^\sharp$ and $X := C_R(b)$. After suitable conjugation we may assume that

(1) $B \leq C_G(b) \cap S \in \mathrm{Syl}_2\, C_G(b)$.

Since $B \in \mathcal{B}(G)$ we have

(2) $Z \leq C_S(B) = B$.

This shows that $R^Z = R$; so

(3) $X \overset{8.2.7}{=} C_X(Z)\,[X, Z],$

---

[1] With respect to inclusion.

and

$$[X, Z] \overset{6.4.4}{\leq} R \cap O_{2'2}(C_G(b)) \leq O_{2'}(C_G(b)).$$

Thus, it remains to prove that $Q := C_X(Z) \leq O_{2'2}(C_G(b))$. Property $\mathcal{Z}$ implies that $S^Q = S$. This gives $QB = Q \times B$ since Q is B-invariant. Now the $P \times Q$-Lemma and (2) show that

$$[S, Q] = 1.$$

In particular $[Q, S \cap C_G(b)] = 1$. Hence 6.4.4 (b) on page 134 yields $Q \leq O_{2'}(C_G(b))$, and we are done. $\qquad\square$

12.1.4 *Suppose that $\mathcal{S}$ holds. Let $A, B \in \mathcal{B}(G)$ such that $A, B \leq S$. Then $\theta_A(G) = \theta_B(G)$. In particular $N_G(S) \leq N_G(\theta_B(G))$ for all $B \in \mathcal{B}(G)$ with $B \leq S$.*

Proof. Let $R := \theta_B(G)$ and $M := N_G(R)$. Then

$$B \leq S \cap M =: S_0,$$

and thus also $B \leq S_0 \leq N_G(R^x)$ for $x \in N_S(S_0)$. Now 12.1.3 gives $R = R^x$ and $x \in M \cap S = S_0$. Hence $N_S(S_0) = S_0$ and thus $S = S_0 \leq M$. In particular $A \leq M$, and again by 12.1.3

$$R = \theta_B(G) \leq \theta_A(G).$$

A symmetric argument, with the roles of A and B reversed, gives $\theta_A(G) \leq \theta_B(G)$ and thus $\theta_A(G) = \theta_B(G)$.

As $N_G(S)$ acts on $\mathcal{B}(S)$, we also get

$$\theta_{B^g}(G) = \theta_B(G)^g = \theta_B(G) \text{ for } g \in N_G(S). \qquad\square$$

12.1.5 *Suppose that $\mathcal{S}$ holds. Then one of the following holds:*

(a) *Every involution of TZ is contained in an element of $\mathcal{B}(G)$.*

(b) $Z \cong C_2$, *and all involutions of H are H-conjugates of the involution in Z.*

(c) $\Omega(T) = Z \cong C_2 \times C_2$.

(d) $Z \cap T \cong C_2$, and $Z \cap T$ is weakly closed in T with respect to H.

Proof. Let $\mathcal{N}(S)$ be the set of nontrivial elementary Abelian normal subgroups of S. Suppose that there exists $X \in \mathcal{N}(S)$ such that $r(X) \geq 3$. Then

$$r(C_X(t)) \geq 2$$

for every involution t of S, since t acts quadratically on X (Example (b) on page 225 and 9.1.1 (b)). Hence $|C_X(t)\langle t \rangle| \geq 8$, and $C_X(t)\langle t \rangle$ is contained in an element of $\mathcal{B}(G)$. This is (a). Thus we may assume:

(1) $r(X) \leq 2$ for all $X \in \mathcal{N}(S)$.

Suppose that $r(X) = 1$ for all $X \in \mathcal{N}(S)$. Then 5.3.9 on page 116 shows that S contains a cyclic maximal subgroup. Hence, also T contains a cyclic maximal subgroup U. Moreover $Z \leq U$ if $U \neq 1$, since $r(Z) = 1$. Now (b) follows with Thompson's Transfer Lemma (12.1.1) applied to H and T. Thus we may assume:

(2)   There exists $V \in \mathcal{N}(S)$ such that $V \cong C_2 \times C_2$.

Let $S_0 := C_S(V)$. Then S_0 is a subgroup of index at most 2 in S. Suppose that $V < \Omega(S_0)$. Then every involution of S_0 is contained in an element of $B \in \mathcal{B}(G)$. If $V = Z$, then $S = S_0$ and (a) follows. In the other case

$$V \neq Z, \quad |Z| = |S/S_0| = 2,$$

and $Z \leq T$ since T is normal in S. Now 12.1.1 shows that every involution in $T \setminus S_0$ is conjugate to an involution in S_0. Thus, again (a) holds. Hence we may assume:

(3) $V = \Omega(S_0)$, and in particular $Z \leq V$.

Suppose that there exists $B \in \mathcal{B}(G)$ such that $B \leq S$. Then by (2) and (3) $B \nleq S_0$ and $V \leq B \cap S_0$. But now $B \leq C_S(V) = S_0$, a contradiction. We have shown:

(4) $\mathcal{B}(G) = \varnothing$.

Suppose that $Z \not\cong C_2$, so $Z = V$ and $S_0 = S$. Set $Z_0 := Z \cap T$. If $Z \leq T$, then (3) implies (c). In the other case $Z_0 = \Omega(T) \cong C_2$, and (d) holds. Since $Z_0 \neq 1$ we may assume now:

(5) $Z \cong C_2$ and $Z \leq T$.

In particular, we have $Z < V$ and thus

$$|S : S_0| = 2.$$

From now on we assume that (d) does not hold. Then there exists $g \in H$ such that $Z \neq Z^g \leq S$. Let

$$W := ZZ^g \quad (\cong C_2 \times C_2), \quad M := N_G(W), \quad \text{and} \quad D := S \cap S^g.$$

By Hypothesis $\mathcal{Z}$

$$C_G(W) = N_G(S) \cap N_G(S^g) \quad \text{and} \quad C_S(W) = D.$$

Moreover (4) yields

(6) $W = \Omega(D)$.

Since $D \neq S$ also $D < N_S(W)$ and $D < N_{S^g}(W)$. Hence $M/C_G(W)$ is not 2-closed. It follows that

$$M/C_M(W) \cong S_3 \quad \text{and} \quad (S \cap M)/D \cong C_2.$$

In particular, all involutions in W are conjugate under $O^2(M)$.

Suppose first that $S \leq M$. Then D is a maximal subgroup of S, and by 12.1.1 every involution of T is conjugate in H to an involution of $D \cap T$. But by (6) this latter involution is in $W \cap T$ and thus an $O^2(M)$-conjugate of the involution in Z. This implies (b) since $O^2(M) \leq H$.

Suppose now that $S \not\leq M$. Then there exists $x \in N_S(S \cap M)$ such that $W^x \neq W$. Thus $|(S \cap M) : D| = 2$ and (6) imply

$$S \cap M = W^x D = W D^x \quad \text{and} \quad D = W(D \cap D^x).$$

It follows that

$$\Phi(D) = \Phi(D \cap D^x).$$

Assume that $\Phi(D) \neq 1$. Then $W \cap \Phi(D) \neq 1$, and $W \cap \Phi(D)$ is M-invariant since $D = O_2(M)$. The transitive action of M on $W^{\#}$ gives

$$W \leq \Phi(D) = \Phi(D \cap D^x) \leq D^x,$$

which contradicts $\Omega(D^x) = W^x \neq W$.

We have shown that $\Phi(D) = 1$ and thus $D = W$. On the other hand,

$$[V, W] \leq Z \leq W,$$

since $|V/Z| = 2$ and $V \trianglelefteq S$. Hence

$$V \leq S \cap M = WW^x.$$

As WW^x is non-Abelian of order 8, it is a dihedral group of order 8 with W and W^x being the only two elementary Abelian subgroups of order 4. It follows that either $V = W$ or $V = W^x$. Since x is in S both cases yield $V = W = W^x$, a contradiction.   □

12.1.6 *Suppose that S and one of the cases* (a) *and* (b) *in* 12.1.5 *hold. Then H possesses a strongly embedded subgroup, or HZ has local characteristic* 2.

Proof. We may assume that HZ does not have local characteristic 2. Then by 12.1.2 there exists an involution $t \in HZ$ such that $O_{2'}(C_{HZ}(t)) \neq 1$, so also

$$O_{2'}(C_G(t)) \neq 1.$$

Suppose first that case (b) of 12.1.5 holds. Then t is conjugate to the involution $z \in Z$, and $C_G(z) = N_G(S)$ since $\mathcal{Z}$ holds and $|Z| = 2$. It follows that

$$R := O_{2'}(N_G(S)) = O_{2'}(C_G(z)) \neq 1.$$

Let $M := N_H(R)$. Then $M \neq H$ since $O_{2'}(G) = 1$. We want to show that M is strongly embedded in H.

Assume that there exists $g \in H \setminus M$ such that $M \cap M^g$ has even order. Then $M \cap M^g$ contains an involution v, and after suitable conjugation in M we may assume that

$$v \in S \cap S^g.$$

Since $[R, S] = 1$ we get

$$[R, v] = 1 = [R^g, v],$$

so $\langle R, R^g \rangle \leq C_G(v)$. As $R = O_{2'}(C_G(z))$ and $\langle Z^g, Z \rangle \leq C_G(v)$ we can apply 8.2.13 on page 190 to $C_G(v)$ and get

$$\langle R, R^g \rangle \leq O_{2'}(C_G(v)) \overset{(b)}{=} O_{2'}(N_G(S^y)) = R^y \text{ for some } y \in G.$$

Hence $R = R^g$ and $g \in M$. This contradiction shows that M is strongly embedded in H.

Suppose now that case (a) of 12.1.5 holds. Then there exists $B \in \mathcal{B}(G)$ such that $t \in B$. As $C_G(t)$ is solvable, we get

$$(1) \qquad\qquad 1 \neq O_{2'}(C_G(t)) \leq \theta_B(G).$$

We now set

$$R := \theta_B(G) \quad \text{and} \quad M := N_G(R)$$

and show that $M \cap H$ is strongly embedded in H.

As above $O_{2'}(G) = 1$ shows that $M \cap H \neq H$. According to 12.1.4 we may assume, after suitable conjugation, that

$$(2) \qquad\qquad B \leq S \leq N_G(S) \leq M.$$

Let $g \in G \setminus M$. If there exists $A \in \mathcal{B}(G)$ such that $A \leq M \cap M^g$, then again by 12.1.3 and 12.1.4 $R = R^g$ and $g \in M$. Thus, we have:

(3)　$A \not\leq M \cap M^g$ for all $A \in \mathcal{B}(G)$ and $g \in G \setminus M$. In particular, $M \cap M^g$ does not contain a Sylow 2-subgroup of G.

Next we show:

(4)　$Z \not\leq M \cap M^g$ for all $g \in G \setminus M$.

Assume that (4) is false. Then there exists $g \in G \setminus M$ such that

$$Z \leq M \cap M^g =: D.$$

By (2) $S \in \mathrm{Syl}_2 M$, so

$$Z \leq S^{gh} \text{ for some } h \in M^g.$$

In particular $[Z, Z^{gh}] = 1$, and by $\mathcal{Z}$ and (2) $ZZ^{gh} \le D$ and $Z \ne Z^{gh}$. Since also $gh \in G \setminus M$ we may assume that

$$ZZ^g \le S \cap S^g \le D.$$

Let
$$W := ZZ^g,$$

so $r(W) \ge 2$ since $Z \ne Z^g$. If $r(W) \ge 3$, then there exists $A \in \mathcal{B}(G)$ with $W \le A$. But then $\mathcal{Z}$ implies $A \le S \cap S^g \le D$, which contradicts (3). Thus, we have

$$W \cong C_2 \times C_2 \quad \text{and} \quad |Z| = 2.$$

Now, as in the proof of 12.1.5, (2) yields

$$N_G(W)/C_G(W) \cong S_3.$$

In particular, all elements of $W^{\#}$ are conjugate. Hence $\mathcal{Z}$ implies for $a \in W^{\sharp}$

$$C_G(a) = N_G(S^y) \text{ for some } y \in G.$$

Since $R^W = R$ and $W \le S^y$ we get

$$[C_R(a), W] \le R \cap S^y = 1 \text{ for all } a \in W^{\sharp}.$$

Now 8.3.4 on page 193 shows that

$$R \le C_R(W) \overset{\mathcal{Z}}{\le} O_{2'}(C_G(Z)) \cap C_G(Z^g).$$

Another application of 8.2.13 yields

$$R \le O_{2'}(C_G(Z^g)) \overset{12.1.4}{\le} R^g$$

and thus $R = R^g$ and $g \in M$, a contradiction. Hence (4) is proved.

To derive a final contradiction we now assume that $M \cap H$ is not strongly embedded in H. Then there exists $g \in H \setminus M$ such that $H \cap M \cap M^g$ has even order. Note first that by the Frattini argument $M = N_G(M)$ since $N_G(S) \le M$, so $M \ne M^g$. Let

$$Q \in \mathrm{Syl}_2 \, H \cap M \cap M^g \text{ and } D := S \cap S^g.$$

After conjugation in M we may assume that $Q \le D$. By (2) and (3) there exists a 2-element y of $M^g \setminus D$ such that $Q^y = Q$, and by 8.1.4 on page 177 there exists an involution $w \in Z(Q)$ such that $w^y = w$. Hence

(5) $C := C_G(w) \nleq M$.

There exists $x \in G$ such that

$$S \cap C \leq S^x \cap C \in \mathrm{Syl}_2\, C.$$

Since $Z \leq S \cap C$ and $S^x \cap C \leq M^x$, (4) implies that $x \in M$. Thus, after conjugation in M, we may assume that

$$S \cap C \in \mathrm{Syl}_2\, C.$$

As we are in case (a) of 12.1.5, there exists $A \in \mathcal{B}(G)$ with $A \leq S \cap C$. Now 12.1.3 and 12.1.4 imply $O_{2'}(C) \leq R$, so $O_{2'}(C)(S \cap C) \leq M$. But then 6.4.4 on page 134 shows that

$$Z \leq O_{2'2}(C) \leq C \cap M.$$

Hence, Z is in M^x for all $x \in C$, and (4) yields $C \leq M$. This contradicts (5).   □

The cases (c), (d) in 12.1.5 correspond to the cases (c), (d) of Theorem 1; and the cases (a), (b) have been treated in 12.1.6. Hence, Theorem 2 is proved.

12.2 $J(T)$-Components

In this section G is a ZN-group of local characteristic 2, and T is a non-trivial 2-subgroup of G.

By $J(X)$ we denote the Thompson subgroup of the group X with respect to the prime 2. Let $\mathcal{L}(T)$ be the set of subgroups $L \leq G$ satisfying:

$$T \in \mathrm{Syl}_2\, L, \quad J(T) \nleq O_2(L), \quad \text{and} \quad C_G(O_2(L)) \leq O_2(L).$$

The last condition implies that $O_{2'}(L) = 1$ and

$$Z(S) \leq Z(T) \leq Z(O_2(L)) \text{ for } T \leq S \in \mathrm{Syl}_2\, G.$$

In particular L is a ZN-group.

For $L \in \mathcal{L}(T)$ we use the following notation:

$$V := \langle \Omega(Z(T))^L \rangle \quad (\leq \Omega(Z(O_2(L)))),$$

$$C := C_L(V),$$

$$\overline{L} := L/C \text{ and } \widetilde{L} := L/O_2(L).^2$$

12.2.1 *Let L be a 2-local subgroup of G and $T \in \mathrm{Syl}_2 L$.*

$$L \in \mathcal{L}(T) \iff J(T) \nleq O_2(L).$$

Proof. As mentioned on page 336 L satisfies

$$C_G(O_2(L)) \le O_2(L)$$

since G is of local characteristic 2. □

12.2.2 *Let $L \in \mathcal{L}(T)$.*

(a) $\widetilde{C} = O_{2'}(\widetilde{C})$ *and* $\widetilde{C}\widetilde{T} = \widetilde{C} \times \widetilde{T}$.

(b) $J(T) \nleq C$.

Proof. (b) follows from (a) since $J(T) \nleq O_2(L)$. Let $T \le S \in \mathrm{Syl}_2 G$. As mentioned earlier

$$Z := \Omega(Z(S)) \le \Omega(Z(T)) \le V.$$

Now $\mathcal{Z}$ implies

$$C \le C_G(Z) \cap L \le N_L(S) \le N_L(T).$$

Hence C is 2-closed, and (a) follows. □

According to 12.2.2 (b) and 9.2.12 on page 238 $L \in \mathcal{L}(T)$ is not Thompson factorizable, so we are allowed to apply the results of Section 9.3. From 9.3.8 we get:

12.2.3 *Let $L \in \mathcal{L}(T)$. Then there exist subgroups $E_1, \ldots, E_r$ of L such that the following hold:*

---

[2]We will use the bar and tilde convention.

(a) $C \leq E_i \trianglelefteq\trianglelefteq L$,

(b) $\{E_1, \ldots, E_r\}^L = \{E_1, \ldots, E_r\}$,

(c) $\overline{J(L)} = \overline{E}_1 \times \cdots \times \overline{E}_r$,

(d) $V = [V, E_1] \times \cdots \times [V, E_r] \times C_V(\overline{J(L)})$,

(e) $[V, E_i] \cong C_2 \times C_2$ *and* $\overline{E}_i \cong \mathrm{SL}_2(2)$ *for* $i = 1, \ldots, r$. $\square$

Next we introduce the notion of a $J(T)$-component. It can be seen as an attempt to describe the structure of the groups E_i of 12.2.3 independent of the particular choice of $L \in \mathcal{L}(T)$. This then allows us to investigate the embeddings of such $J(T)$-components into different elements of $\mathcal{L}(T)$. The ultimate goal (in Section 12.3) will be to show that a suitably chosen $J(T)$-component is contained in a unique maximal 2-local subgroup of G.

A subgroup $K \leq G$ is a **$J(T)$-component** if the following hold:

$\mathcal{K}_1$ $K = O^2(K) = [K, J(T)]$ and $K/O_2(K) \cong C_3$,

$\mathcal{K}_2$ $J(T) = J(\widehat{T})$ for $J(T) \leq \widehat{T} \in \mathrm{Syl}_2(KJ(T))$,

$\mathcal{K}_3$ $W_K \cong C_2 \times C_2$ for $W_K := [\Omega(Z(O_2(K))), K]$.

The set of $J(T)$-components of G we denote by $\mathcal{K}(T)$. For $L \leq G$

$$\mathcal{K}_L(T) := \{K \in \mathcal{K}(T) \mid K \leq L\},$$

$$\mathcal{K}_0(T) := \{K \in \mathcal{K}(T) \mid J(T) = J(T_0) \text{ for } T \leq T_0 \in \mathrm{Syl}_2 N_G(W_K)\}.$$

The first observation is elementary but useful:

12.2.4   *Let* $K \in \mathcal{K}(T)$ *and* Q *be a subgroup of* G *satisfying*

$$KJ(T) \leq N_G(Q) \quad \text{and} \quad Q \leq N_G(J(T)).$$

Then $Q \leq N_G(K)$.

Proof. By our hypothesis

$$\langle J(T)^{QK} \rangle = \langle J(T)^K \rangle \stackrel{\mathcal{K}_1}{=} KJ(T),$$

so Q normalizes $KJ(T)$ and thus, again by $\mathcal{K}_1$, also $O^2(KJ(T)) = K$. $\square$

12.2.5   *Let* $L \in \mathcal{L}(T)$ *and* $\Omega := \{E_1, ..., E_r\}$ *be the set of subnormal subgroups of* L *given in* 12.2.3. *Then there exists a bijection*

$$\rho : \mathcal{K}_L(T) \to \Omega \; \text{ such that } \; K = O^2([O^2(K^\rho), J(T)])$$

for $K \in \mathcal{K}_L(T)$. *Moreover*

$$K \trianglelefteq\trianglelefteq L \quad \text{and} \quad W_K = [V, K^\rho].$$

Proof. Let $K \in \mathcal{K}_L(T)$. By 12.2.2 (a) C normalizes T and thus also $J(T)$. Hence 12.2.4 (with $Q := C$) gives

(1) $K \trianglelefteq KC.$

In particular $K \trianglelefteq KO_2(L)$ and

$$K \overset{\mathcal{K}_1}{=} [K, J(T)] \le [KC, J(T)] \le K[C, J(T)] \overset{12.2.2}{\le} KO_2(L).$$

Now $\mathcal{K}_1$ implies

(2) $K = O^2([KC, J(T)]).$

Again by $\mathcal{K}_1$

(3) $\overline{K} \le O^2(\overline{E}_1) \times \cdots \times O^2(\overline{E}_r),$

so $O_2(\overline{K}) = 1$ and

$$O_2(K) \le C \cap O_2(K) \overset{(1)}{\le} O_2(C) \le O_2(L).$$

Now $\mathcal{K}_1$ and 12.2.2 (a) give

(4) $\overline{K} \cong C_3$ and $[V, K] = [V, K, K] = W_K \cong C_2 \times C_2.$

Hence, by 12.2.3 there exists exactly one $i \in \{1, \dots, r\}$ such that the projection of $\overline{K}$ on $\overline{E}_i$ is nontrivial, so

(5) $KC = O^2(E_i)C.$

This shows that there exists a mapping

$$\rho : \mathcal{K}_L(T) \to \Omega \text{ such that } \overline{K} = O^2(\overline{K^\rho}).$$

Moreover, for $E_i = K^\rho$ we have

$$K \stackrel{(2)}{=} O^2([KC, J(T)]) \stackrel{(5)}{=} O^2([O^2(E_i)C, J(T)]) \stackrel{12.2.2\,(a)}{\leq} [O^2(E_i), J(T)]O_2(L).$$

As $K = O^2(K)$, this shows that $K = O^2([O^2(E_i), J(T)]) = O^2(K^\rho, J(T)])$.

By (1) and (5) K is subnormal in L, and by (4) $[V, K] = W_K$. Thus, it remains to show the bijectivity of ρ. The injectivity follows from (2).

For proving the surjectivity of ρ we fix $E \in \Omega$ and set

$$K_0 := O^2([O^2(E_j)C, J(T)]) \text{ and } W_0 := [\Omega(Z(O_2(K_0))), K_0].$$

It suffices to show that K_0 is a $J(T)$-component.

Clearly K_0 satisfies $\mathcal{K}_2$ since $T \in \mathrm{Syl}_2 L$. Set $X := O^2(E)C$. By 12.2.2 $\widetilde{X}$ is a $2'$-group and

$$|\widetilde{X}/C_{\widetilde{X}}(J(T))| = 3.$$

Hence 8.4.4 on page 198 implies that

$$\widetilde{K}_0 = [\widetilde{X}, J(T)] \cong C_3.$$

In addition, $K_0/[K_0, J(T)]$ is a 2-group since $K_0 \leq [K_0, J(T)]O_2(L)$. Thus, the definition of K_0 gives

$$K_0 = O^2(K_0) = [K_0, J(T)],$$

and $\mathcal{K}_1$ holds for K_0.

It remains to show that K_0 satisfies $\mathcal{K}_3$. Again by the definition of K_0 we have $[V, K_0] \leq W_0$. Hence, it suffices to show that $|W_0| \leq 4$.

Among all $A \in \mathcal{A}(T)$ satisfying $[K_0, A] \not\leq O_2(K_0)$ we choose A such that $C_A(W_0)$ is maximal. There exists $d \in K_0$ such that $\langle A, A^d \rangle$ contains a Sylow 3-subgroup D of K_0. By 8.4.2 on page 198, D acts fixed-point-freely on W_0 since $K_0 = DO_2(K_0)$. It follows that

(6) $C_{W_0}(A) \cap C_{W_0}(A^d) = 1$ and $W_0 = [W_0, A][W_0, A^d]$.

Set $A_0 := C_A([W_0, A])[W_0, A]$ and $A_1 := C_A(K_0/O_2(K_0))$. Then $|A/A_1| = 2$, and 9.2.3 on page 233 implies

(7) $A_0 \in \mathcal{A}(T)$ and $[W_0, A_0] \neq 1$.

As
$$C_A(W_0)[W_0, A] \leq C_{A_0}(W_0)$$
the maximality of $|C_{W_0}(A)|$ yields either $A_0 \leq A_1$ or $A_0 = A$. In the first case
$$A_0^d O_2(K_0) = A_0 O_2(K_0) \leq C_T([W_0, A]) \cap C_T([W_0, A^d]),$$
and (6) implies $[W_0, A_0] = 1$, which contradicts (7). In the second case the same argument shows $[W_0, A_1] = 1$, and thus $|A/C_A(W_0)| = 2$. Now the maximality of $|A|$ gives $C_{W_0}(A) = W_0 \cap A$ and

$$|A| \geq |W_0 C_A(W_0)| = |C_A(W_0)||W_0/C_{W_0}(A)|.$$

Thus
$$|W_0/C_{W_0}(A)| = |W_0/[W_0, A]| = 2,$$
and $|W_0| \leq 4$ follows from (6).   □

The next two results are consequences of 12.2.5:

12.2.6   Let $L \in \mathcal{L}(T)$, $K \in \mathcal{K}_L(T)$, and $Z(T) \cap W_K \neq 1$. Then $T \leq N_L(K)$.

Proof. By 12.2.5 there exists a subgroup E_i as in 12.2.3 such that

$$W_K = [V, E_i] \quad \text{and} \quad K = O^2([O^2(E_i), J(T)]).$$

Since by 12.2.3 (b), (d)

$$[V, E_i] \cap [V, E_i^x] = 1 \text{ for } x \in L \setminus N_L(E_i).$$

The hypothesis $Z(T) \cap W_K \neq 1$ implies that $T \leq N_L(E_i)$, and the claim follows.   □

12.2.7   Let $K \in \mathcal{K}_0(T)$. Then the following hold:

(a) $N_G(K) = N_G(W_K)$.

(b) K is a $J(T^g)$-component for all $g \in G$ with $J(T^g) \leq N_G(W_K)$.

Proof. (a) Let $L := N_G(W_K)$ and $T \leq T_0 \in \mathrm{Syl}_2 L$. Then $J(T) = J(T_0)$ by the definition of $\mathcal{K}_0(T)$, so K is also a $J(T_0)$-component. It follows that $J(T_0) \nleq O_2(L)$, and by 12.2.1 $L \in \mathcal{L}(T_0)$.

We apply 12.2.3 and 12.2.5 with the notation given there. Then there exists $E \in \Omega$ with

$$K = O^2([O^2(E), J(T)]) \text{ and } W_K = [V, E].$$

Thus, by 12.2.3 (b) and (e)

$$E \trianglelefteq L \text{ and } \overline{J(L)} = \overline{E}.$$

On the other hand, the Frattini argument yields

$$L = N_L(J(T))J(L) = N_L(J(T))EC,$$

and by 12.2.2 $C \leq N_L(J(T))$. Now (a) follows.

(b) Let $J(T^g) \leq N_G(W_K)$. Then $J(T^g)$ is a $N_G(W_K)$-conjugate of $J(T)$, and (b) follows from (a). $\square$

We are now able to prove the main result of this section.

12.2.8 *Let* $K \leq L \in \mathcal{L}(T)$ *such that* $K \trianglelefteq\trianglelefteq \langle K, J(T) \rangle$. *Suppose that* $K \in \mathcal{K}_0(T^g)$ *for some* $g \in G$. *Then* $K \in \mathcal{K}_L(T)$, *and in particular*

$$K \trianglelefteq\trianglelefteq L.$$

Proof. If $K \in \mathcal{K}_L(T)$ then $K \trianglelefteq\trianglelefteq L$ by 12.2.5, and if $J(T) \leq N_L(K)$ then $K \in \mathcal{K}_L(T)$ by 12.2.7. Thus, we may assume now that

$$(*) \qquad J(T) \nleq N_L(K) \overset{12.2.7}{=} N_G(W_K).$$

We will show that this leads to contradiction.

We fix some notation:

$$L_0 := \langle K, J(T) \rangle, \ L^* := L_0 O_2(L), \ L_1 := O_2(L_0)K,$$

and $S \in \mathrm{Syl}_2 G$ such that $J(T) \leq S \cap L^* =: T^* \in \mathrm{Syl}_2 L^*$. Further

$$Z := \Omega(Z(S)), \ V^* := \langle \Omega(Z(T^*))^{L^*} \rangle.$$

(1) $O_2(L_0) \leq N_G(K)$.

By our hypothesis $K \trianglelefteq\trianglelefteq L_1$, so

$$K = O^2(K) \leq O^2(L_1) \leq K.$$

This implies $K = O^2(L_1)$, and (1) follows.

(2) $Z \leq Z(T^*)$ and $V^* \leq Z(O_2(L_0))$.

Note that

$$C_S(O_2(L)) \leq O_2(L) \leq O_2(L^*) \leq T^*$$

and

$$\Omega(Z(T^*)) \leq J(T^*) = J(T) \leq L_0.$$

This implies $Z \leq Z(T^*)$ and $\Omega(Z(T^*)) \leq Z(O_2(L^*)) \cap L_0$. Hence $V^* \leq Z(O_2(L_0))$ follows since $O_2(L_0) \leq O_2(L^*)$.

(3) $W_K = [Z, K] = [V^*, K]$.

By (1) and (2) $[V^*, K] \leq Z(O_2(K))$ since $O_2(K) \leq O_2(L_0)$. The coprime action of K on $Z(O_2(L_0))$ gives $[V^*, K] \leq W_K$. As $|W_K| = 4$ and $Z \leq V^*$ by (2), it suffices to show that $[Z, K] \neq 1$.

Assume that $[Z, K] = 1$. Then

$$K \leq C_L(Z) \overset{\mathcal{Z}}{\leq} N_L(S \cap L) \leq N_L(J(T)),$$

and thus

$$J(T) \leq O_2(L_0) \overset{(1)}{\leq} N_G(K),$$

which contradicts (∗). Hence (3) is proved.

We now derive a final contradiction. As $J(T) = J(T^*)$ and $O_2(L) \leq O_2(L^*)$ we get

$$C_G(O_2(L^*)) \leq C_G(O_2(L)) \leq O_2(L) \leq O_2(L^*),$$

so $L^* \in \mathcal{L}(T^*)$. We apply 12.2.3 to L^* and V^* (in place of L and V).

Let E_i be one of the subnormal subgroups $E_1, ..., E_r$ given there, and set $W_i := [V^*, E_i]$. According to (3) both subgroups E_i and K normalize $W_K W_i$.

On the other hand, by 12.2.3 (d) there are at most 2 L^*-conjugates of W_i in $W_K W_i$ since $|W_K W_i| \leq 2^4$. Hence $K = O^2(K)$ implies that $K \leq N_{L^*}(W_i)$ and either $W_K = W_i$ or $[W_i, K] = 1$.

The first case contradicts (∗). Hence, we have that $[W_i, K] = 1$ for $i = 1, ..., r$. It follows with 12.2.3 d) that

$$[J(T), V^*, K] = 1 = [V^*, K, J(T)],$$

and then with the Three-Subgroups Lemma $[K, J(T)] \leq C^* := C_{L^*}(V^*)$. By 12.2.2 $C^* J(T)$ is 2-closed, so $K \leq N_G(J(T))$. But this contradicts (3) and (∗). □

The next result about the structure of $J(T)$-components will be used at the end of the next section. It is independent of the other results of this section.

12.2.9 *Let $K \in \mathcal{K}(T)$ and $Z_0 := \Omega(Z(J(T)))$. Then*

$$Z_0 = (Z_0 \cap Z(KJ(T)))(Z_0 \cap W_K) \quad and \quad |Z_0 \cap W_K| = 2.$$

Proof. Set $L := KJ(T)$. By $\mathcal{K}_2$ we may assume that $T \in \mathrm{Syl}_2 L$, and $\mathcal{K}_1$ yields

(1) $L/O_2(L) \cong S_3$.

As $J(T) \unlhd T$ and $|L : T| = 3$, there are exactly three conjugates of $J(T)$ in L and thus

(2) $L = \langle J(T), J(T)^d \rangle$ for every $d \in L \setminus T$.

Set

$$V := \langle \Omega(Z(T))^L \rangle \quad (\leq Z(O_2(L))).$$

Then $W_K \cap V \neq 1$ since $W_K \unlhd L$, and $W_K \leq V$ follows. The coprime action of K on V shows that

(3) $W_K = [V, K, K] = [V, K]$ and $C_L(V) = O_2(L)$.

Let $A \in \mathcal{A}(T)$ such that $A \not\leq O_2(L)$. Then (1) and (3) imply that

$$C_A(V) = A \cap O_2(L) \text{ and } |A/C_A(V)| = 2.$$

In particular $|A| \leq |VC_A(V)|$, and the maximality of A yields

$$|A| = |C_A(V)V| = |(A \cap O_2(L))V|;$$

so

(4) $\mathcal{A}(O_2(L)) \subseteq \mathcal{A}(T)$ and $Z_0 \leq Z(J(O_2(L)))$.

Now 9.3.9 on page 247 shows that $[Z_0, K] \leq V$ and thus by (3) $[Z_0, K] \leq W_K$. Set $V_0 := Z_0 W_K$ and $X := Z_0 \cap Z_0^d$, d as in (2). Since $W_K \cap Z(T) \neq 1$ and $|W_K| = 4$, we get that

$$|V_0 : Z_0| = |W_K : (W_K \cap Z_0)| \leq 2 \text{ and } |V_0 : X| \leq 4.$$

On the other hand, by (2) $X \leq Z(L)$ and $Z(L) \cap W_K = 1$, so

$$Z_0 = X \times (Z(T) \cap W_K),$$

and the claim follows. $\square$

12.3 N-Groups of Local Characteristic 2

In this section G is an ZN-group of local characteristic 2 with $O_2(G) = 1$. Moreover, $S \in \mathrm{Syl}_2 G$, $Z = \Omega(Z(S))$, and M is a 2-local subgroup of G containing $N_G(J(S))$.

As $J(S)$ is characteristic in S, we have

$$C_G(Z) \overset{Z}{\leq} N_G(S) \leq N_G(J(S)) \leq M,$$

and the Frattini argument shows that $N_G(M) = M$. Thus

$$M \neq M^x \text{ for all } x \in G \setminus M.$$

If M is not strongly embedded in G, then there exists $x \in G \setminus M$ such that $M \cap M^x$ has even order. This leads to the following notation:

$\mathcal{T}(M)$ is the set of nontrivial 2-subgroups $T \leq M$ satisfying:

(+)   There exists a 2-local subgroup $L \leq G$ such that $T \leq L$ and $L \nleq M$.

12.3.1 M *is strongly embedded in G if and only if $\mathcal{T}(M)$ is empty.*

Proof. By the definition of $\mathcal{T}(M)$ it is evident that M is strongly embedded in G if $\mathcal{T}(M) = \varnothing$.

Assume now that M is strongly embedded in G but $\mathcal{T}(M) \neq \varnothing$. Then

$(')$ $N_G(T) \leq M$ for every $T \in \mathcal{T}(M)$.

Choose $T \in \mathcal{T}(M)$ such that $|T|$ is maximal, and let $T \leq L$ be as in (+); so $L \nleq M$.
By $(')$
$$N_{T_0}(T) \leq M \cap L \text{ for } T \leq T_0 \in \mathrm{Syl}_2 L,$$
and the maximality of T gives $T = N_{T_0}(T) = T_0$. In particular
$$O_2(L) \leq T_0 \leq M \cap L.$$
Hence also $O_2(L) \in \mathcal{T}(M)$, and $(')$ yields
$$L \leq N_G(O_2(L)) \leq M,$$
a contradiction. $\square$

In view of $J(T)$-components introduced in the previous section we now investigate subgroups $T \in \mathcal{T}(M)$ that are maximal with respect to their Thompson subgroups. More precisely, let

$$a(T) := |A|, \quad A \in \mathcal{A}(T).$$

By $\mathcal{T}^*(M)$ we denote the set of $T_0 \in \mathcal{T}(M)$ that are maximal in the following sense:

- If $T \in \mathcal{T}(M)$ then $a(T) \leq a(T_0)$.

- If $T \in \mathcal{T}(M)$ with $a(T) = a(T_0)$ then $|J(T)| \leq |J(T_0)|$.

- If $T \in \mathcal{T}(M)$ with $a(T) = a(T_0)$ and $|J(T)| = |J(T_0)|$ then $|T| \leq |T_0|$.

12.3.2 $N_G(J(T)) \leq M$ *for every* $T \in \mathcal{T}^*(M)$.

Proof. After conjugation in M we may assume that $T \leq S$. If $T = S$ then the choice of M gives yields $N_G(J(T)) \leq M$. If $T < S$ then

$$T < N_S(J(T)) \leq M \cap N_G(J(T)),$$

and the maximality of T yields $N_S(J(T)) \notin \mathcal{T}(M)$; so $N_G(J(T)) \leq M$.   □

12.3.3   *Let* $T \in \mathcal{T}^*(M)$ *and* L *be a 2-local subgroup of* G *such that* $J(T) \leq T_0 \in \mathrm{Syl}_2 L$ *and* $L \nleq M$. *Then*

$$J(T) = J(T_0), \quad T_0 \in \mathcal{T}(M), \quad \text{and} \quad L \in \mathcal{L}(T_0).$$

Proof. By 12.3.2
$$T_1 := N_{T_0}(J(T)) \leq M \cap L,$$
so the maximality of T yields $J(T) = J(T_1)$ and thus $J(T) = J(T_0)$. In particular $T_0 \in \mathcal{T}(M)$. Moreover, again by 12.3.2, $J(T_0) \nleq O_2(L)$; and $L \in \mathcal{L}(T_0)$ follows from 12.2.1.   □

Because of 12.3.3 we are now able to to use the results of Section 12.2.

12.3.4   *Let* L *be a 2-local subgroup of* G *and* $T \in \mathcal{T}^*(M)$. *Suppose that* $T \leq L \nleq M$. *Then* $L \in \mathcal{L}(T)$, *and in particular* $T \in \mathrm{Syl}_2 L$ *and* $\mathcal{K}_L(T) \neq \varnothing$.

Proof. Let $T \leq T_0 \in \mathrm{Syl}_2 L$. Then 12.3.3 shows that $T_0 \in \mathcal{T}(M)$ and $L \in \mathcal{L}(T_0)$. The maximality of T implies $T_0 = T$. The remaining claim follows from 12.2.5.   □

We now use the set $\mathcal{K}_0(T)$ introduced on page 349. For $T \in \mathcal{T}^*(M)$ let $\mathcal{K}_{G \setminus M}(T)$ be the set of $K \in \mathcal{K}_0(T)$ such that

$$K \nleq M \quad \text{and} \quad O_2(\langle K, T \rangle) \neq 1.$$

12.3.5 $\mathcal{K}_{G\backslash M}(T) \neq \varnothing$ *for all* $T \in \mathcal{T}^*(M)$.

Proof. Let $T \in \mathcal{T}^*(M)$. By 12.3.4 there exists $L \in \mathcal{L}(T)$ such that $L \not\leq M$. The Frattini argument gives

$$L = N_L(J(T))J(L),$$

so $J(L) \not\leq M$ by 12.3.2. Hence, by 12.2.3 and 12.2.5 there exists $K \in \mathcal{K}_L(T)$ such that $K \not\leq M$.

It remains to prove that $K \in \mathcal{K}_0(T)$. Let $\widehat{L} := N_G(W_K)$ and $J(T) \leq T_0 \leq$ Syl$_2 \widehat{L}$. Then $\widehat{L} \not\leq M$ since $K \leq L$ but $K \not\leq M$. Hence 12.3.3 implies that $J(T) = J(T_0)$ and thus $K \in \mathcal{K}_0(T)$. $\qquad\square$

12.3.6 Uniqueness Theorem. *Let* $T \in \mathcal{T}^*(M)$ *and* $K \in \mathcal{K}_{G\backslash M}(T)$. *Then* $KJ(T)$ *is contained in a unique maximal 2-local subgroup* L *of* G. *Moreover* $K \trianglelefteq\trianglelefteq L$ *and* $T \in$ Syl$_2 L$.

Proof. Let $\mathcal{L}$ be the set of 2-local subgroups of G that contain $KJ(T)$. Then $\mathcal{L}$ is nonempty since $W_K \neq 1$ and $KJ(T) \leq N_G(W_K)$. Note that the subgroups in $\mathcal{L}$ are not contained in M since $K \not\leq M$.

Let $L \in \mathcal{L}$ and $J(T) \leq T_0 \in \text{Syl}_2 L$. It follows from 12.3.3 that

(1) $\qquad\qquad K \in \mathcal{K}(T_0) \quad \text{and} \quad L \in \mathcal{L}(T_0),$

so 12.2.5 implies

(2) $\qquad\qquad K \trianglelefteq\trianglelefteq L$ for all $L \in \mathcal{L}$.

We will show that $\mathcal{L}$ and K, in place of $\mathcal{U}$ and A, satisfy the hypotheses (1), (2), and (3) of 6.7.3 on page 158. As in 6.7.3 we set

$$\Sigma_L := \{K^g \mid g \in G, \ K^g \trianglelefteq\trianglelefteq L\} \quad (L \in \mathcal{L}).$$

Then (2) implies hypothesis 6.7.3 (1).

Let $\widetilde{L} \in \mathcal{L}$, $K^g \in \Sigma_{\widetilde{L}}$ and $K^g \leq L$. Since $J(T) \leq \widetilde{L}$ we have $K^g \trianglelefteq\trianglelefteq \langle K^g, J(T)\rangle$, and 12.2.8 implies that $K^g \trianglelefteq\trianglelefteq L$. This is hypothesis 6.7.3 (2).

For the verification of hypothesis 6.7.3 (3) let

$$\Sigma := \Sigma_L \cap \Sigma_{\widetilde{L}} \quad \text{and} \quad X := \langle \Sigma \rangle.$$

Then $K \in \Sigma$ and $K \trianglelefteq\trianglelefteq X$ since $X \leq L$. In particular

$$1 \neq O_2(K) \leq O_2(X),$$

and $N_G(O_2(X))$ is a 2-local subgroup of G. As $J(T)$ acts by conjugation on Σ_L and $\Sigma_{\widetilde{L}}$ we get that

$$J(T) \leq N_G(X) \leq N_G(O_2(X)),$$

so $N_G(O_2(X)) \in \mathcal{L}$. This shows hypothesis 6.7.3 (3).

Now 6.7.3 shows that $\mathcal{L}$ contains a unique maximal element L. As the definition of $\mathcal{K}_{G\backslash M}(T)$ yields

$$O_2(\langle K, T \rangle) \neq 1,$$

we also get $T \leq L$. Thus, $T \in \mathrm{Syl}_2 L$ follows from 12.3.4. $\qquad \square$

12.3.7 *Let L, K and T be as in 12.3.6 and $T \leq S$, and let*

$$Z_0 := \Omega(Z(J(T))) \text{ and } Z := \Omega(Z(S)).$$

Then $Z_0 \cap W_K \neq 1$, and one of the following holds:

(a) $Z = Z_0 \cong C_2$.

(b) $Z = Z_0 \cong C_2 \times C_2$ and $T = S$.

(c) $\Omega(Z(T)) = Z_0 \cong C_2 \times C_2$ and $|N_G(Z_0) : C_L(Z_0)| = 2$.

Proof. Since $T \in \mathrm{Syl}_2 L$ and L is a 2-local subgroup we get $Z \leq Z(T)$, i.e.,

(1) $Z \leq Z_0$.

Let $Z_K := C_{Z_0}(K)$. The 12.2.9 implies

(2) $|Z_0 : Z_K| = 2$ and $Z \neq Z_K$;

the latter since by our hypothesis $C_G(Z)$ is 2-closed but not $KJ(T)$. If $Z_0 \cong C_2$ then (a) holds. Thus, we may assume:

(3) $|Z_0| \geq 4$.

We first treat the case

$$N_G(J(T)) \leq L.$$

Then $N_S(J(T)) = T$ and thus $S = T$. Hence, L has the same property as M. In particular $T \in \mathcal{T}^*(L)$ since $M \neq L$.

According to 12.3.5 (with the roles of L and M interchanged) there exists $F \in \mathcal{K}_{G \setminus L}(T)$. Let $Z_F := C_{Z_0}(F)$. As for Z_K we get

(4) $|Z_0 : Z_F| = 2$ and $Z \neq Z_F$.

If $Z_K \cap Z_F \neq 1$ then the uniqueness of L yields

$$\langle F, K, J(S) \rangle \leq N_G(Z_K \cap Z_F) \leq L,$$

which contradicts $F \not\leq L$. Thus, we have

$$Z_K \cap Z_F = 1,$$

and (2), (3), and (4) imply

$$Z_0 \cong C_2 \times C_2 \quad \text{and} \quad Z_K \cong Z_F \cong C_2.$$

If (b) does not hold, then (1), (2), and (4) show that Z has order 2. Moreover, Z is neither conjugate to Z_K nor Z_F since $C_G(Z)$ is 2-closed. As $Z < Z_0 \trianglelefteq S$ this implies that $Z_K^x = Z_F$ for some $x \in S$ and

$$\langle F, K^x, J(S) \rangle \leq N_G(Z_F).$$

On the other hand $x \in S \leq L$; and the uniqueness of L shows that

$$\langle F, K^x, J(S) \rangle \leq L^x = L.$$

This contradicts $F \not\leq L$.

We are left with the case

$$N_G(J(T)) \not\leq L.$$

Let $g \in N_G(J(T)) \setminus L$. If $Z_K \cap Z_K^g \neq 1$, then the uniqueness of L yields

$$KJ(T) \leq N_G(Z_K \cap Z_K^g) \leq L,$$

and similarly, with (K^g, M^g, L^g) in place of (K, M, L),

$$N_G(Z_K \cap Z_K^g) \leq L^g.$$

This shows that $KJ(T) \leq L^g$, so $L^g = L$. Since L is a maximal 2-local subgroup we get $g \in L$, a contradiction.

Hence $Z_K \cap Z_K^g = 1$; and, as above, by (2)

$$Z_0 \cong C_2 \times C_2.$$

Next we show:

(5) $Z_0 \leq Z(T)$.

Otherwise there exists $t \in T$ such that $[Z_0, t] \neq 1$, and in particular $Z < Z_0$. As above Z_K, Z_K^g, and Z are the three subgroups of order 2 in Z_0; and Z is neither conjugate to Z_K nor to Z_K^g. It follows that $Z_K^t = Z_K^g$ and

$$tg^{-1} \in N_G(Z_K) \leq L,$$

which contradicts $g \notin L$. Hence, (5) is proved.

From $\Omega(Z(T)) \leq Z_0$ we get

$$Z_0 = \Omega(Z(T)),$$

and the uniqueness of L gives

$$C_G(Z_0) \leq C_G(Z_K) \leq L.$$

If $Z = Z_0$, then $S = T$ and (b) follow.

Assume that $Z \neq Z_0$. Then $T < S$; so $T < N_S(Z_0)$ and

$$|N_S(Z_0)/C_S(Z_0)| = 2 = |N_G(Z_0)/C_G(Z_0)|,$$

since Z is neither conjugate to Z_K nor to Z_K^g in $N_G(Z_0)$. This is (c). $\square$

12.3.8   *Let* $T \in \mathcal{T}^*(M)$ *and* $K \in \mathcal{K}_{G \backslash M}(T)$. *Then* K *is normal in* $\langle K, T \rangle$.

Proof. Possibly after conjugation in M, we may assume that $T \leq S$. The Uniqueness Theorem (12.3.6) shows that $\langle K, T \rangle$ is contained in a unique maximal 2-local subgroup L and $T \in \operatorname{Syl}_2 L$. Moreover, 12.3.7 implies that

$$Z(T) \cap W_K \neq 1.$$

Now the claim follows from 12.2.6. □

12.3.9 *Let $T \in \mathcal{T}^*(M)$. Then there exist two different $J(T)$-components $K_1, K_2 \in \mathcal{K}(T)$ such that for $P_i := \langle K_i, T \rangle$, $i = 1, 2$, the following hold:*

(a) $C_G(O_2(P_i)) \leq O_2(P_i)$.

(b) $T \in \operatorname{Syl}_2 P_i$.

(c) $P_i/O_2(P_i) \cong S_3$.

(d) $[\Omega(Z(T)), P_i] \neq 1$.

(e) $O_2(\langle P_1, P_2 \rangle) = 1$.

Moreover, P_i is contained in a unique maximal 2-local subgroup L_i of G and $T \in \operatorname{Syl}_2 L_i$.

Proof. By 12.3.5 there exists $K_1 \in \mathcal{K}_{G \setminus M}(T)$, and according to the Uniqueness Theorem (12.3.6) P_1 is contained in a unique maximal 2-local subgroup L of G. In addition $T \in \operatorname{Syl}_2 L$, and thus also $T \in \operatorname{Syl}_2 P_1$. It follows that

$$C_G(O_2(P_1)) \leq C_G(O_2(L)) \leq O_2(L) \leq O_2(P_1).$$

From 12.3.7 we get that $Z(T) \cap W_{K_1} \neq 1$. This shows that $[\Omega(Z(T)), K_1] \neq 1$. Moreover, by 12.3.8 K_1 is normal in P_1, i.e., $P_1 = K_1 T$. Now (a)–(d) follow for $i = 1$.

Assume first that $N_G(T) \not\leq L$. Let $g \in N_G(T) \setminus L$, and set $K_2 := K_1^g$. Then P_2 and P_1 are conjugate, so P_2 also has the properties (a)–(d). In addition L^g is the unique maximal 2-local subgroup of G containing P_2. Hence, (e) follows since otherwise $L = L^g$ and $g \in L$.

Assume now that $N_G(T) \leq L$. Then $T \in \operatorname{Syl}_2 G$, and after conjugation in M we may assume that

$$T = S.$$

From 12.3.7 we get $Z = Z_0$, Z_0 as in 12.3.7; so

$$N_G(J(S)) \leq N_G(Z_0) \overset{\mathcal{Z}}{\leq} N_G(S) \leq L.^3$$

Hence, L has the same property as M. As in the proof of 12.3.7 there exists $K_2 \in \mathcal{K}_{G \setminus L}(T)$; and as above, with L in place of M, the properties (a)–(d) follow for P_2.

For the proof of (e) assume that $O_2(\langle P_1, P_2 \rangle) \neq 1$. Then the Uniqueness Theorem for K_1 shows that $P_2 \leq L$. This contradicts $K_2 \nleq L$.   □

We now use the results of Section 10.3 provided by the amalgam method.

12.3.10   *Let* $T \in \mathcal{T}^*(M)$. *Then there exist two different maximal 2-local subgroups* P_1 *and* P_2 *of* G *such that* $T \in \mathrm{Syl}_2\, P_i$, $i = 1, 2$, *and either*

$$P_1 \cong P_2 \cong S_4 \quad or \quad \mathrm{P}_1 \cong \mathrm{P}_2 \cong S_4 \times C_2.$$

Proof. Let $P_1 \leq L_1$ and $P_2 \leq L_2$ be as in 12.3.9, so in particular

$$T \in \mathrm{Syl}_2\, L_1 \cap \mathrm{Syl}_2\, L_2.$$

From 10.3.11 we get the desired structure for P_i. It remains to prove that $P_i = L_i$ for $i = 1, 2$. We fix i and use the notation

$$(L, K, P) \quad \text{in place of} \quad (L_i, K_i, P_i).$$

Then

$$Z(T) \leq O_2(L) \leq O_2(P) \leq T \in \mathrm{Syl}_2\, L,$$

and in addition $\langle Z(T)^P \rangle = O_2(P)$. It follows that

$$O_2(L) = O_2(P).$$

By 12.2.3 and 12.3.6 K is the unique $J(T)$-component of L since $|O_2(L)| \leq 8$. Hence, again by 12.2.3 and 12.2.5, W_K is normal in L and $|O_2(L)/W_K| \leq 2$. Now 8.2.2 on page 184 shows that $C_L(W_K)$ is a 2-group, so $C_L(W_K) = O_2(L)$. Thus, $L/C_L(W_K) \cong S_3$ implies $L = P$.   □

---

[3]Use the condition equivalent to $\mathcal{Z}$ given on page 335.

We are now able to prove Theorem 1 and Theorem 3 stated in the introduction of this chapter.[4]

Proof of Theorem 3:   Let $H := O^2(G)$ and M be as introduced in the beginning of this section. We assume that M is not strongly embedded in G. Then 12.3.1 implies that $\mathcal{T}(M) \neq \varnothing$. Thus, also $\mathcal{T}^*(M) \neq \varnothing$, and we are allowed to apply 12.3.10. Let P_1, P_2, T be as described there and

$$M_i := H \cap P_i \quad (i = 1, 2).$$

Furthermore, for $i = 1, 2$, set

$Z_i := Z(P_i)$,

$Q_i := O_2(O^2(P_i)) \quad (\leq H \cap T)$, and

$T_0 := \langle Q_1, Q_2 \rangle$.

Then Q_1 and Q_2 are two elementary Abelian subgroups of order 4, and they intersect in a subgroup of order 2 since $Q_1 \neq Q_2$. Hence, T_0 is a dihedral group of order 8. As T_0 is in M_i, we get for $i = 1, 2$

$$M_i \cong S_4 \quad \text{or} \quad M_i = P_i.$$

Assume first that $M_i \cong S_4$, and let L be a maximal 2-local subgroup of H containing M_i. Then

$$N_{O_2(L)}(Q_i) \leq O_2(L) \cap P_i \cap H = O_2(L) \cap M_i = Q_i$$

since $N_G(Q_i) = P_i$. It follows that $Q_i = O_2(L)$ and thus $L = M_i$.

Hence, to prove Theorem 3 it suffices to show that $M_1 \cong S_4 \cong M_2$. We will show that the other case

$$M_i = P_i \cong C_2 \times S_4 \quad (i = 1, 2)$$

leads to a contradiction. Thus, from now on we assume that

$$T \leq H.$$

Set $\langle z \rangle := Z(T_0)$ and $\langle z_i \rangle := Z_i$. Then

---

[4]Theorem 2 has already been proved in Section 12.2.

(1) $O_2(P_i) = \langle z_i \rangle \times Q_i$, $Z(T) = \langle z \rangle \times \langle z_i \rangle$ and $\langle z \rangle = Q_1 \cap Q_2$.

(2) $\mathcal{A}(T) = \{O_2(P_1), O_2(P_2)\}$.

(3) $O_2(P_1) \cup O_2(P_2) = \{x \in T \mid x^2 = 1\}$.

Moreover, z_i is not a square in P_i.[5] The maximality of P_i yields $C_G(z_i) = P_i$. Hence, z_i is also not a square in G. On the other hand, z is a square in T_0, so all involutions of T_0 are squares since they are conjugate to z (in $\langle P_1, P_2 \rangle$). It follows:

(4) $z_i^G \cap T_0 = \varnothing$ for $i = 1, 2$.

Let $S \in \mathrm{Syl}_2 \, G$ such that
$$T \leq S \cap H.$$

If $T = S \cap H$, then $T \in \mathrm{Syl}_2 \, H$ and Thompson's Transfer Lemma (applied to H) shows that z_i is conjugate to an involution of T_0, which contradicts (4). Thus, we have
$$T < S \cap H,$$

in particular $T < N_S(T)$. The maximality of P_i together with $P_1 \neq P_2$ implies that $N_S(T)$ acts transitively (by conjugation) on each of the sets

$$\{Z_1, Z_2\}, \ \{Q_1, Q_2\}, \ \text{and} \ \mathcal{A}(T),$$

and in each case T is the kernel of that action. This gives $|N_S(T)/T| = 2$ and

(5) $\Omega(C_T(x)) = \langle z \rangle$ for every $x \in N_S(T) \setminus T$.

In particular
$$\langle z \rangle = \Omega(Z(S)) = Z.$$

As every element of $\mathcal{A}(N_S(T))$ has order at least 8, (2) and (5) imply

$$T = J(T) = J(N_S(T));$$

and thus $T = J(S)$. In particular $T \trianglelefteq S$. Hence, we have

(6) $|S : T| = 2$, $S \in \mathrm{Syl}_2 \, H$ and $J(S) = J(T) = T$.

---

[5]That is, there is no $x \in P_i$ such that $z_i = x^2$.

Assume first that there exists an involution $t \in S \setminus T$. Let $g \in G$ such that

$$C_S(t) \leq C_{S^g}(t) \in \mathrm{Syl}_2\, C_G(t).$$

If $C_T(t) = Z$, then $C_S(t) = Z\langle t \rangle \cong C_2 \times C_2$; and by 5.3.10 S is a dihedral or semidihedral group. This contradicts the existence of an elementary Abelian subgroup of order 8 in S.

We have shown that $Z < C_T(t)$. By (5) z is a square in $C_T(t)$ and thus in S^g. Since $|S^g : T^g| = 2$ we get

$$z \in T^g.$$

Now (3), applied to T^g, gives a subgroup $A \in \mathcal{A}(T^g)$ such that $|A| = 8$ and $z \in A$. Hence, $\mathcal{Z}$ implies that $A \leq C_G(Z) \leq N_G(S)$, so by (6)

$$A \in \mathcal{A}(T).$$

On the other hand, Thompson's Transfer Lemma shows that t is conjugate to an involution in T. Thus, by (3) there exists an elementary Abelian subgroup B of order 8 in $C_G(t)$, and we may assume that $B \leq S^g$. Again from (6) we get that $B \in \mathcal{A}(T^g)$. Hence, (2) implies that $|B \cap A| \geq 4$ and

$$B \cap A \leq C_T(t),$$

which contradicts (5).

We have shown that there are no involutions in $S \setminus T$. Now we determine the focal subgroup $S \cap H'$. Since $H = O^2(H)$ we get from 7.1.3 on page 166 and (6)

(7) $S = S \cap H' = \langle y^{-1} y^g \mid y \in S,\ y^g \in S,\ g \in G \rangle.$

Let $y, y^g \in S$, $g \in G$. If y and y^g are both in T, then also $y^{-1} y^g \in T$. Assume that $y \notin T$. Then $o(y) > 2$, and (5) shows that $\Omega(\langle y \rangle) = Z$. Either also $y^g \notin T$ and with the same argument $\Omega(\langle y \rangle^g) = Z$, or $y^g \in T$ and again $\Omega(\langle y \rangle^g) = Z$ since z is the only square in T.

We have shown that

$$\Omega(\langle y \rangle) = Z = \Omega(\langle y^g \rangle) \ \text{if}\ o(y) > 2.$$

Thus in this case $Z = Z^g$ and $g \in N_G(S)$ by $\mathcal{Z}$. As $|S : T| = 2$, we again get $y^{-1} y^g \in T$. Hence, (7) implies that that $S = S \cap H' \leq T$, a contradiction.   □

Proof of Theorem 1: As Theorem 2 was already proved at the end of Section 12.1 we may assume that HZ has local characteristic 2. It is evident that HZ is also a ZN-group. Thus, we are allowed to apply Theorem 3 to HZ.

If HZ possesses a strongly embedded subgroup, then (a) of Theorem 1 holds. In the other case H contains a maximal 2-local subgroup P isomorphic to S_4. We want to show that then case (b) of Theorem 1 holds.

We have
$$O_2(P) \cong C_2 \times C_2 \quad \text{and} \quad P = N_H(O_2(P)).$$

Let D be a Sylow 2-subgroup of P, so D is a dihedral group of order 8. After conjugation in H we may assume that

$$D \leq S \cap H =: R.$$

Let
$$Z^* := \Omega(Z(R)).$$

Then $Z^* \leq O_2(P)$, and there exist $t \in O_2(P)$ such that

$$O_2(P) = Z^* \times \langle t \rangle \cong C_2 \times C_2.$$

It follows that
$$C_R(t) = C_R(O_2(P)) = O_2(P),$$

and 5.3.10 on page 117 implies that R is a dihedral or semidihedral group. Hence, (b) of Theorem 1 holds. $\square$

Appendix

In this Appendix we give the results mentioned in Chapters 10 and 12.

Let G be a group. By $Z^*(G)$ we denote the inverse image of $Z(G/O_{2'}(G))$ in G.

Brauer, Suzuki [32]: *Suppose that the Sylow 2-subgroups of G are quaternion groups. Then $Z^*(G)/O_{2'}(G) \cong C_2$.*[1]

Glauberman (Z^*-Theorem) [49]: *Let S be a Sylow 2-subgroup of G. Then*

$$x \in Z^*(G) \iff x^G \cap C_S(x) = \{x\}.$$

Gorenstein, Walter [59]: *Suppose that $O_{2'}(G) = 1$ and that the Sylow 2-subgroups of G are dihedral groups. Then $F^*(G)$ is isomorphic to*

$$\mathrm{PSL}_2(q),\ q \equiv 1\,(\mathrm{mod}\,2),\ \ or\ A_7.$$

Alperin, Brauer, Gorenstein [22]: *Suppose that G is simple and that the Sylow 2-subgroups of G are semidihedral groups. Then G is isomorphic to*

$$\mathrm{PSL}_3(q),\ q \equiv -1\,(\mathrm{mod}\,4),\ \mathrm{PSU}_3(q),\ q \equiv 1\,(\mathrm{mod}\,4),\ \ or\ M_{11}.$$

Bender [29]: *Suppose that G possesses a strongly embedded subgroup.*[2] *Then one of the following holds:*

---

[1]Hence, the Sylow 2-subgroups of $G/Z^*(G)$ are dihedral groups, so the structure of $G/Z^*(G)$ is given by the Theorem of Gorenstein-Walter, below.

[2]A strongly 2-embedded subgroup in the notation of Chapter 10.

(i) *The Sylow 2-subgroups of G are cyclic or quaternion groups.*

(ii) *G possesses a normal series $1 \leq M \leq L \leq G$ such that M and G/L have odd order, and L/M is isomorphic to*

$$\mathrm{PSL}_2(2^n),\ \mathrm{Sz}(2^{2n-1}),\ \text{or}\ \mathrm{PSU}_3(2^n)\,(n \geq 2).$$

Goldschmidt [57]: *Let S be a Sylow 2-subgroup of G and A an Abelian subgroup of S such that*

$$a \in A,\ a^g \in S\ (g \in G) \quad \Rightarrow \quad a^g \in A.^3$$

Suppose that that $G = \langle A^G \rangle$ and $O_{2'}(G) = 1$. Then

$$G = F^*(G), \quad A = O_2(G)\Omega(T),$$

and for every component K of G the factor group $K/Z(K)$ is isomorphic to:

$$\mathrm{PSL}_2(2^n),\ \mathrm{Sz}(2^{2n-1}),\ \mathrm{PSU}_3(2^n)\,(n \geq 2),\ \mathrm{PSL}_2(q),\ q \equiv 3,5\,(\mathrm{mod}\,8),$$
$$\mathrm{R}(3^{2n+1})\,(n \geq 1),\ \text{or}\ \mathrm{J}_1.$$

Thompson [94]: *Let G be a nonsolvable group all of whose p-local subgroups are solvable for every $p \in \pi(G)$. Then $F^*(G)$ is isomorphic to*

$$\mathrm{PSL}_2(q)\,(q > 3),\ \mathrm{Sz}(2^{2n-1})\,(n \geq 2),\ A_7,\ M_{11},\ \mathrm{PSL}_3(3),\ \mathrm{PSU}_3(3),\ \text{or}\ {}^2F_4(2)'.$$

Gorenstein, Lyons [61], **Janko** [73], **Smith** [83]: *Let G be a nonsolvable group all of whose 2-local subgroups are solvable. Then $F^*(G)$ is isomorphic to*

$$\mathrm{PSL}_2(q),\ q > 3,\ \mathrm{Sz}(2^{2n-1}),\ \mathrm{PSU}_3(2^n)\,(n \geq 2),\ A_7,\ M_{11},\ \mathrm{PSL}_3(3),$$
$$\mathrm{PSU}_3(3),\ \text{or}\ {}^2F_4(2)'.$$

Classification Theorem.[4] *Every finite simple group is isomorphic to one of the following groups:*

---

[3] A is **strongly closed** in S with respect to G.
[4] See [10] and the survey article [84].

1. *A cyclic group of prime order,*

2. *An alternating group A_n for $n \geq 5$,*

3. *A classical linear group:* [5]

   $\mathrm{PSL}_n(q)$, $\mathrm{PSU}_n(q)$, $\mathrm{PSp}_{2n}(q)$, *or* $\mathrm{PO}_n^\varepsilon(q)$,

4. *An exceptional group of Lie type:*[6]

   ${}^3D_4(q)$, $E_6(q)$, ${}^2E_6(q)$, $E_7(q)$, $E_8(q)$, $F_4(q)$, ${}^2F_4(2^n)$, $G_2(q)$, ${}^2G_2(3^n)$, *or* ${}^2B_2(2^n)$,

5. *A sporadic simple group:*

   M_{11}, M_{12}, M_{22}, M_{23}, M_{24} (Mathieu-groups); [7]

   J_1, J_2, J_3, J_4 (Janko-groups); [8]

   Co_1, Co_2, Co_3 (Conway-groups);

   HS, Mc, Suz;

   Fi_{22}, Fi_{23}, Fi_{24}' (Fischer-groups);

   F_1 (the Monster[9]), F_2, F_3, F_5; He, Ru, Ly, ON.

---

[5] A description of these groups can be found in [13], and as groups of Lie type in [5].

[6] See [5].

[7] These groups have been found by Mathieu [77] around 1860. The first transparent construction was given by Witt [100] in 1938.

[8] J_1 was found in 1965 [72], after the Mathieu-groups this was first other sporadic group.

[9] F_1 is the largest sporadic group, its order is

$$2^{46} \cdot 3^{20} \cdot 5^9 \cdot 7^6 \cdot 11^2 \cdot 13^3 \cdot 17 \cdot 19 \cdot 23 \cdot 29 \cdot 31 \cdot 41 \cdot 47 \cdot 59 \cdot 71.$$

Bibliography

Books and Monographs

[1]   Aschbacher, M.: *Finite Group Theory.* Cambridge Univ. Press, Cambridge 1986.

[2]   Aschbacher, M.: *Sporadic Groups.* Cambridge Univ. Press, Cambridge 1994.

[3]   Bender, H., Glauberman, G.: *Local Analysis for the Odd Order Theorem,* Cambridge Univ. Press, Cambridge 1994.

[4]   Burnside, W.: *Theory of Groups of Finite Order,* 2nd edn., Cambridge 1911; Dover Publications, New York 1955.

[5]   Carter, R.W.: *Simple Groups of Lie Type.* J. Wiley & Sons, New York 1972.

[6]   Dickson, L. E.: *Linear Groups with an Exposition of the Galois Field Theory,* Leipzig 1901; Dover Publications, New York, 1958.

[7]   Delgado, A., Goldschmidt, D., Stellmacher, B.: *Groups and Graphs: New Results and Methods.* DMV-Seminar, Bd. 6. Birkhäuser Verlag, Basel 1985.

[8]   Doerk, K., Hawkes, T.: *Finite Soluble Groups.* deGruyter, Berlin 1992.

[9]   Feit, W.: *Characters of Finite Groups.* Benjamin, New York 1972.

[10] Gorenstein, D.: *Finite Simple Groups.* Plenum Press, New York 1982.

[11] Gorenstein, D.: *The Classification of Finite Simple Groups.* Plenum Press, New York 1983.

[12] Gorenstein, D.: *Finite Groups.* Harper & Row, New York 1968.

[13] Huppert, B.: *Endliche Gruppen I.* Springer-Verlag, Berlin 1967.

[14] Huppert, B., Blackburn, N.: *Finite Groups* II, III. Springer-Verlag, Berlin 1982.

[15] Jordan, C.: *Traité des substitutions et dés equationes álgébriques.* Paris 1870.

[16] Schmidt, R.: *Subgroup Lattices of Groups.* de Gruyter, Berlin 1994.

[17] Suzuki, M.: *Group Theory* I, II. Springer-Verlag, Berlin 1982, 1986.

[18] Wielandt, H.: *Finite Permutation Groups.* Academic Press, New York 1964.

[19] Zassenhaus, H.: *Lehrbuch der Gruppentheorie.* Leipzig 1937.

Journal Articles

[20] Alperin, J. L.: Sylow intersections and fusion, *J. Algebra* **6** (1967), 222–41.

[21] Alperin, J. L., Gorenstein, D.: Transfer and fusion in finite groups, *J. Algebra* **6** (1967), 242–55.

[22] Alperin, J. L., Brauer, R., Gorenstein, D.: Finite groups with quasi-dihedral and whreathed Sylow 2-subgroups, *Trans. Amer. Math. Soc.* **151** (1970), 1–261.

[23] Baer, R.: Groups with Abelian central quotient groups, *Trans. Amer. Math. Soc.* **44** (1938), 357–86.

[24] Baer, R.: Engelsche Elemente Noetherscher Gruppen, *Math. Ann.* **133** (1957), 256–76.

[25] Baumann, B.: Überdeckung von Konjugiertenklassen endlicher Gruppen, *Geometriae Dedicata* **5** (1976), 295–305.

[26]   Baumann, B.: Über endliche Gruppen mit einer zu $L_2(2^n)$ isomorphen Faktorgruppe, *Proc. AMS* **74** (1979), 215–22.

[27]   Bender, H.: Über den größten p'-Normalteiler in p-auflösbaren Gruppen, *Archiv* **18** (1967), 15–16.

[28]   Bender, H.: On groups with Abelian Sylow 2-subgroups, *Math. Z.* **117** (1970), 164–76.

[29]   Bender, H.: Transitive Gruppen gerader Ordnung, in denen jede Involution genau einen Punkt festläßt, *J. Alg.* **17** (1971), 527–54.

[30]   Bender, H.: A group theoretic proof of Burnside's $p^a q^b$-theorem, *Math. Z.* **126** (1972), 327–38.

[31]   Bender, H.: Goldschmidt's 2-signalizer functor-theorem, *Israel J. Math.* **22** (1975), 208–13.

[32]   Brauer, R., Suzuki, M.: On finite groups of even order whose 2-Sylowsubgroup is a quaternion group, *Proc. Nat. Acad. Sci.* **45** (1959), 175–9.

[33]   Brodkey, J.S.: A note on finite groups with an Abelian Sylow group, *Proc. AMS* **14** (1963), 132–3.

[34]   Burnside, W.: On groups of order $p^a q^b$, *Proc. London Math. Soc.* (2) **1** (1904), 388–92.

[35]   Cameron, P.J.: Finite permutation groups and finite simple groups, *Bull. London Math. Soc.* **13** (1981), 1–22.

[36]   Carter, R.W.: Nilpotent self-normalizing subgroups of soluble groups, *Math. Z.* **75** (1960/61), 136–9.

[37]   Cauchy, A.: Memoire sur le nombre des valeurs on' une function peut acquerir, *ŒUVRES II*, **1**, 64–90.

[38]   Chermak, A., Delgado, A.L.: A measuring argument for finite groups, *Proc. AMS* **107** (1989), 907–14.

[39]   Clifford, H.: Representations induced in an invariant subgroup, *Ann. of Math.* **38** (1937), 533–50.

[40]   Dress, A.W.M., Siebeneicher, Ch., Yoshida, T.: An application of Burnside rings in elementary finite group theory, *Advances in Math.* **91** (1992), 27–44.

[41] Dress, A.W.M.: Still another proof of the existence of Sylow p-subgroups, *Beiträge zur Geometrie und Algebra* **35** (1994), 147–8.

[42] Euler, L.: *Opera Omnia* I 2. Teubner 1915.

[43] Feit, W., Thompson, J. G.: Solvability of groups of odd order, *Pacific J. Math.* **13** (1963), 775–1029.

[44] Fitting, H.: Beiträge zur Theorie der endlichen Gruppen, *Jahresbericht DMV* **48** (1938), 77–141.

[45] Frattini, G.: Intorno alla generazione dei gruppi di operanzoni, *Rend. Atti. Acad. Lincei* **1** (1885), 281–5, 455–77.

[46] Frobenius, G.: Über auflösbare Gruppen IV, *Sitzungsberichte der königl. Preuß. Akad. d. Wiss. zu Berlin* (1901), 1223–25; or in *Gesammelte Abhandlungen*, Bd. III, Springer (1968), 189–209.

[47] Frobenius, G.: Über auflösbare Gruppen V, *Sitzungsberichte der königl. Preuß. Akad. d. Wiss. zu Berlin* (1901), 1324–29; or in *Gesammelte Abhandlungen*, Bd. III, Springer (1968), 204.

[48] Gaschütz, W.: Zur Erweiterungstheorie der endlichen Gruppen, *J. reine angew. Math.* **190** (1952), 93–107.

[49] Glauberman, G.: Central elements in core-free groups, *J. Alg.* **4** (1966), 403–20.

[50] Glauberman, G.: A characteristic subgroup of a p-stable group, *Canadian J. Math.* **20** (1968), 1101–35.

[51] Glauberman, G.: Failure of factorization in p-solvable groups, *Quart. J. Math. Oxford* II. Ser. **24** (1973), 71–7.

[52] Glauberman, G.: On solvable signalizer-functors in finite gorups, *Proc. London Math. Soc.* III. Ser. **33** (1976), 1–27.

[53] Glauberman, G.: *Factorizations in local subgroups of finite groups.* Regional Conference in Mathematics, Vol. 33 (1977).

[54] Goldschmidt, D. M.: A group-theoretic proof of the $p^a q^b$ theorem for odd primes, *Math. Z.* **113** (1970), 373–5.

[55] Goldschmidt, D. M.: 2-signalizer functors on finite groups, *J. Alg.* **21** (1972), 321–40.

[56] Goldschmidt, D. M.: Solvable functors on finite groups, *J. Alg.* **21** (1972), 341–51.

[57] Goldschmidt, D. M.: Strongly closed 2-subgroups of finite groups, *Ann. of Math.* **99** (1974), 70–117.

[58] Goldschmidt, D. M.: Automorphisms of trivalent graphs, *Ann. of Math.* **111** (1980), 377–404.

[59] Gorenstein, D., Walter, J.: The characterization of finite groups with dihedral Sylow 2-subgroups, *J. Alg.* **2** (1964), 354–93.

[60] Gorenstein, D.: On centralizers of involutions in finite groups, *J. Alg.* **11** (1969), 243–77.

[61] Gorenstein, D., Lyon, R.: Nonsolvable groups with solvable 2-local subgroups, *J. Alg.* **38** (1976), 453–522.

[62] Grün, O.: Beiträge zur Gruppentheorie I, *J. reine angew. Math.* **174** (1935), 1–14.

[63] Hall, P.: A note on soluble groups, *J. London Math. Soc.* **3** (1928), 98–105.

[64] Hall, P.: A contribution to the theory of groups of prime-power order, *Proc. London Math. Soc.* (2) **36** (1934), 29–95.

[65] Hall, P.: A characteristic property of soluble groups, *J. London Math. Soc.* **12** (1937), 188–200.

[66] Hall, P.: On the Sylow systems of a soluble group, *Proc. London Math. Soc.* (2) **43** (1937), 198–200.

[67] Hall, P., Higman, P.: The p-length of a p-soluble group and reduction theorems for Burnside's problem, *Proc. London Math. Soc.* **7** (1956), 1–42.

[68] Hölder, O.: Zurückführung einer beliebigen algebraischen Gleichung auf eine Kette von Gleichungen, *Math. Ann.* **34** (1889), 26–56.

[69] Hölder, O.: Die Gruppen der Ordnung p^3, pq^2, pqr, p^4, *Math. Ann.* **43** (1893), 301–412.

[70] Ito, N.: Über das Produkt von zwei abelschen Gruppen, *Math. Z.* **62** (1955), 400–1.

[71]   Iwasawa, K.: Über die Struktur der endlichen Gruppen, deren echte
       Untergruppen sämtlich nilpotent sind, *Proc. Phys. Math. Soc. Japan*
       23 (1941), 1–4.

[72]   Janko, Z.: A new finite simple group with Abelian 2-subgroups,
       Proc. Nat. Acad. Sci. **53** (1965), 657–8.

[73]   Janko, Z.: Nonsolvable groups all of whose 2-local subgroups are
       solvable, *J. Alg.* **21** (1972), 458–511.

[74]   Knoche, H.G.: Über den Frobeniusschen Klassenbegriff in nilpoten-
       ten Gruppen, *Math. Z.* **55** (1951), 71–83.

[75]   Lagrange, J. L.: Reflexions sur la résulotion algébriques de equa-
       tions, *Œuvres* t. **3**, Gauthier-Villans (1938), 205–421.

[76]   Maschke, H.: Über den arithmetischen Charakter der Coeffizien-
       ten der Substitutionen endlicher linearer Gruppen, *Math. Ann.* **50**
       (1898), 482–98.

[77]   Mathieu, E.: Memoire sur le nombre de valeurs que peut acquerir
       une fonction quand on y permut ses variables de toutes les manières
       possibles, *Crelle J.* **5** (1860), 9–42.

[78]   Mathieu, E.: Memoire sur l'étude des fonctions de plusieures quan-
       tités, sur la manière de les formes et sur les substitutions qui les
       laissent invariables, *Crelle J.* **6** (1861), 241–323.

[79]   Mathieu, E.: Sur la fonction cinq fois transitive des 24 quantités,
       Crelle J. **18** (1873), 25–46.

[80]   Matsuyama, H.: Solvability of groups of order $2^a p^b$, *Osaka J. Math.*
       10 (1973), 375–8.

[81]   O'Nan, M.E.: Normal structure of the one-point stabilizer of a
       doubly-transitive permutation group, *Trans. Am. Soc.* **217** (1975),
       1–74.

[82]   Schur, J.: Untersuchungen über die Darstellungen der endlichen
       Gruppen durch gebrochen lineare Substitutionen, *J. reine u. angew.
       Math.* **132** (1907), 85–137.

[83]   Smith, F.: Finite simple groups all of whose 2-local subgroups are
       solvable, *J. Alg.* **34** (1975), 481–520.

[84]   Solomon, R.: On finite simple groups and their classification, *Notices AMS* **42** (1995), 231-9.

[85]   Stellmacher, B.: An analogue to Glauberman's ZJ-theorem, *Proc. Am. Math. Soc.* **109** No. 4 (1990), 925-9.

[86]   Stellmacher, B.: On Alperin's fusion theorem, *Beitr. Alg. Geom.* **35** (1994), 95-9.

[87]   Stellmacher, B.: An application of the amalgam method: The 2-local structure of N-groups of characteristic 2 type, *J. Alg.* **190** (1997), 11-67.

[88]   Stellmacher, B.: A characteristic subgroup for Σ_4-free groups, *Israel J. Math.* **94** (1996), 367-79.

[89]   Sylow, L.: Théorèmes sur les groupes de substitutions, *Math. Ann.* **5** (1872), 584-94.

[90]   Thompson, J.G.: Finite groups with fixed-point-free automorphisms of prime order, *Proc. Nat. Acad. Sci. U.S.A.* **45** (1959), 578-81.

[91]   Thompson, J.G.: Normal p-complements for finite groups, *Math. Z.* **72** (1960), 332-354.

[92]   Thompson, J.G.: Normal p-complements for finite groups, *J. Algebra* **1** (1964), 43-6.

[93]   Thompson, J.G.: Fixed points of p-groups acting on p-groups, *Math. Z.* **80** (1964), 12-13.

[94]   Thompson, J. G.: Nonsolvable finite groups all of whose local subgroups are solvable I–VI, *Bull. AMS* **74** (1968), 383-437; *Pacific J. Math.* **33** (1970), 451-536; **39** (1971), 483-534; **48** (1973), 511-92; **50** (1974), 215-97; **51** (1974), 573-630.

[95]   Timmesfeld, F.: A remark on Thompson's replacement theorem and a consequence, *Arch. Math.* **38** (1982), 491-9.

[96]   Wielandt, H.: Eine Verallgemeinerung der invarianten Untergruppen, *Math. Z.* **45** (1939), 209-44.

[97]   Wielandt, H.: Ein Beweis für die Existenz der Sylowgruppen, *Arch. Math.* **10** (1959), 401-2.

Bibliography

80] Wielandt, H.: Kriterium für Subnormalität in endlichen Gruppen, *Math. Z.* **138** (1974), 199–203.

[99] Wielandt, H.: *Mathematische Werke*, Bd. 1, de Gruyter 1994.

[100] Witt, E.: Treue Darstellung Liescher Ringe, *J. reine angew. Math.* **177** (1938), 152–60.

[101] Witt, E.: Die 5-fach transitiven Gruppen von Mathieu, *Abh. Math. Sem. Univ. Hamburg* **12** (1937), 256–64.

[102] Zassenhaus, H.: Über endliche Fastkörper, *Abh. Math. Sem. Univ. Hamburg* **11** (1935), 187–220.

Index

Universitext *(continued)*